Encyclopedia of Computational Neuroscience

Volume I

Encyclopedia of Computational Neuroscience
Volume I

Edited by **Sophia Nelson**

CLANRYE
INTERNATIONAL

New Jersey

Published by Clanrye International,
55 Van Reypen Street,
Jersey City, NJ 07306, USA
www.clanryeinternational.com

Encyclopedia of Computational Neuroscience: Volume I
Edited by Sophia Nelson

International Standard Book Number: 978-1-63240-179-3 (Hardback)

Printed in the United States of America.

Contents

Preface   VII

Chapter 1   **Composite Match Index with Application of Interior Deformation Field
             Measurement from Magnetic Resonance Volumetric Images of Human
             Tissues**   1
             Penglin Zhang, Xubing Zhang and Jiangping Chen

Chapter 2   **Strategic Cognitive Sequencing: A Computational Cognitive
             Neuroscience Approach**   10
             Seth A. Herd, Kai A. Krueger, Trenton E. Kriete, Tsung-Ren Huang,
             Thomas E. Hazy and Randall C. O'Reilly

Chapter 3   **Spike-Timing-Dependent Plasticity and Short-Term Plasticity Jointly
             Control the Excitation of Hebbian Plasticity Without Weight
             Constraints in Neural Networks**   28
             Subha Fernando and Koichi Yamada

Chapter 4   **Random Bin for Analyzing Neuron Spike Trains**   43
             Shinichi Tamura, Tomomitsu Miyoshi, Hajime Sawai and Yuko
             Mizuno-Matsumoto

Chapter 5   **Emergent Central Pattern Generator Behavior in Gap-Junction-Coupled
             Hodgkin-Huxley Style Neuron Model**   54
             Kyle G. Horn, Heraldo Memelli and Irene C. Solomon

Chapter 6   **Why People Play: Artificial Lives Acquiring Play Instinct to
             Stabilize Productivity**   64
             Shinichi Tamura, Shoji Inabayashi, Waichi Hayakawa, Takahiro Yokouchi,
             Hiroshi Mitsumoto and Hisashi Taketani

Chapter 7   **CUDAICA: GPU Optimization of Infomax-ICA EEG Analysis**   72
             Federico Raimondo, Juan E. Kamienkowski, Mariano Sigman and
             Diego Fernandez Slezak

Chapter 8   **Channel Identification Machines**   80
             Aurel A. Lazar and Yevgeniy B. Slutskiy

Chapter 9   **Color Image Quantization Algorithm Based on Self-Adaptive
             Differential Evolution**   100
             Qinghua Su and Zhongbo Hu

Chapter 10   **Quantitative Tools for Examining the Vocalizations
             of Juvenile Songbirds**   108
             Cameron D. Wellock and George N. Reeke

Chapter 11   **A Comparative Study of Human Thermal Face Recognition Based
on Haar Wavelet Transform and Local Binary Pattern**   121
Debotosh Bhattacharjee, Ayan Seal, Suranjan Ganguly, Mita Nasipuri
and Dipak Kumar Basu

Chapter 12   **Noise-Assisted Instantaneous Coherence Analysis of Brain Connectivity**   133
Meng Hu and Hualou Liang

Chapter 13   **Hippocampal Anatomy Supports the Use of Context in Object Recognition:
A Computational Model**   145
Patrick Greene, Mike Howard, Rajan Bhattacharyya and Jean-Marc Fellous

Chapter 14   **Augmenting Weak Semantic Cognitive Maps with an
"Abstractness" Dimension**   164
Alexei V. Samsonovich and Giorgio A. Ascoli

Chapter 15   **Modeling Spike-Train Processing in the Cerebellum Granular Layer and
Changes in Plasticity Reveal Single Neuron Effects in Neural Ensembles**   174
Chaitanya Medini, Bipin Nair, Egidio D'Angelo, Giovanni Naldi and
Shyam Diwakar

Chapter 16   **Multiobjective Optimization of Evacuation Routes in Stadium Using
Superposed Potential Field Network Based ACO**   191
Jialiang Kou, Shengwu Xiong, Zhixiang Fang, Xinlu Zong and Zhong Chen

Permissions

List of Contributors

Preface

Neuroscience is the study and research in areas such as cell biology, biochemistry, and computer technology in understanding anatomical and physiological studies. Neuroscience is one of the most complex fields of science because of the integration of this field with neuro-scientific studies, psychological studies, psychiatric studies, neurological and neuro-economic studies. The neuro-scientific study focuses on molecular and cellular aspects of a singular nerve cell, along with the imaging of sensory in the brain. Neuroscience provides information about nervous systems of organisms, specifically their structure, development, cellular function, evolution, computation and medical aspects.

This book contains researches that discuss different concepts related to neuroscience. Some important approaches in the field of neurosciences are computational cognitive neuroscience approach, bee colony algorithm and unsupervised approach data analysis on fuzzy possibility of clustering.

The objective of this book is to advance our scientific knowledge and understanding of some of the many unexplored aspects of neuroscience. We bring together an international group of experts, to fill in the gaps in our knowledge of neuroscience, biological adaptation, tools and techniques used for neuro-scientific advancement. The chapters in this book represent a diverse, international set of perspectives on neuroscience studies. The contributors come from different parts of the world and different scientific disciplines. They employ diverse theoretical approaches and scientific methodologies to provide on-the-ground accounts of neuroscientific researches going around the globe.

I would like to thank them for their contributions. I would also like to thank my family for their endless support.

Editor

Composite Match Index with Application of Interior Deformation Field Measurement from Magnetic Resonance Volumetric Images of Human Tissues

Penglin Zhang,[1] Xubing Zhang,[2] and Jiangping Chen[1]

[1] *School of Remote Sensing and Information Engineering, Wuhan University, Wuhan 430079, China*
[2] *College of Mathematics and Computer Science, Wuhan Textile University, Wuhan 430073, China*

Correspondence should be addressed to Xubing Zhang, xubingnational@gmail.com

Academic Editor: Yen-Wei Chen

Whereas a variety of different feature-point matching approaches have been reported in computer vision, few feature-point matching approaches employed in images from nonrigid, nonuniform human tissues have been reported. The present work is concerned with interior deformation field measurement of complex human tissues from three-dimensional magnetic resonance (MR) volumetric images. To improve the reliability of matching results, this paper proposes composite match index (CMI) as the foundation of multimethod fusion methods to increase the reliability of these various methods. Thereinto, we discuss the definition, components, and weight determination of CMI. To test the validity of the proposed approach, it is applied to actual MR volumetric images obtained from a volunteer's calf. The main result is consistent with the actual condition.

1. Introduction

The physical property is the base of the biological simulation, computer-assisted medical applications, such as clinical diagnosis, and surgical simulation, surgical planning. And estimation of internal deformation field or deformation motion for the biological tissues plays a very significant role in physical parameters estimation. Thus, measuring the internal deformation field of biological tissues is becoming the focus research. Magnetic resonance (MR) imaging (MRI) provides superb anatomic images with excellent spatial resolution and contrasts among soft tissues; thus, it is widely used in computer-assisted medical applications, such as clinical diagnosis, surgery simulation, operation planning, and evaluation of physical characteristics of biological tissues. Increasing number of researchers in medical simulation and medical virtual reality focus on the interior deformation field or motion measurement of biological tissues from MR volumetric images, and it has become one of the significant branches of medical image analysis. Generally, approaches for estimating the deformation of MR volumetric images can be classified into two typical types: elastic deformation model-based and feature matching-based methods.

The elastic deformation model-based method can be classified into either parametric or geometric active models [1]. To obtain the deformation information of an object, the parametric active contours, also called snakes, try to minimize a defined cost function so that the function deforms a given initial contour toward the boundary of the object. This method was first introduced by Kass et al. in 1987 [2] and subsequently developed and used by Lang et al. [3], Cho and Benkeser [4], and Matuszewski et al. [5] to estimate deformation motion of nonrigid objects. In the geometric active model [1, 6–8], the curve and the surface of an object are first detected. Then, the deformation propagation of the curve and the surface is used to track the motion. However, irrespective of what elastic deformation models are employed, disadvantages exist in the deformation estimation; for example, the parametric active model cannot handle changes in the topology of the evolving contours when deformation is performed directly, and often, heuristic topology handling procedures are used [8]. In the geometric active model, when contrast is poor and boundaries are not clear or are continuous in the images, the contours tend to leak through the boundary [9]. The tagged images must have a regular grid pattern in the imaging plane because if the number of tagged points is low, the measurement accuracy would be poor. More important than the former two aspects, regardless of what elastic deformation models are used, they can only handle the deformation at the boundary of nonrigid objects and not the interior deformation.

In recent years, researchers have been increasingly concerned on approaches for matching of nonrigid feature points. Typically, thin-plate spline-robust point matching (RPM) is a famous algorithm for matching non-rigid feature points, which can estimate the joint correspondence and non-rigid transformations between two differently sized point sets. However, optimal processing of the energy function utilized in Chui's method may be trapped in bad local minima [10]. Zheng proposed the RPM-local neighborhood structure (LNS) method of matching non-rigid feature points, based on the supposition that relative distances and orientations among feature points in a neighborhood would be preserved [11]. Lee improved the LNS and presented the topology preserving relaxation labeling (TPRL) algorithm. In the TPRL method, log distance and polar angle bins are utilized to capture the coarse location information of the feature points in a neighborhood. Using shape contexts, Belongie proposed a non-rigid point matching method. In this method, every feature point is represented by a histogram descriptor of the distance and orientation between this feature point and its neighbor feature points [12]. In addition, some other useful methods were also proposed for feature-point matching, such as the coherent point drift matching method of non-rigid points [13, 14] and the preservation of local geometrical characteristics [15]. In these methods, a novel objective function is defined to preserve local image-to-image affine transformations across correspondence. In general, some unsolved problems are involved in the aforementioned matching methods of non-rigid points; for example, the optimal processing of the energy function could be trapped in bad local minima, the topology of the neighboring feature points is not always preserved well, and so on. Most importantly, in these methods, useful information of the feature point is considered singly and lacks a comprehensive approach, which can mix up with the useful and significant information in the point matching of deformation measurement.

Therefore, to improve further the proposed feature-matching-based approach and improve the robustness of the matching result, this paper proposes a composite match index (CMI). In Section 2, we introduce some previous work, in Section 3, we describe the concept and definition of CMI, and Section 4 introduces the CMI application on feature matching of image pairs from non-rigid objects. In Section 5, examples and preliminary experimental results are given, and discussion and conclusions are presented in the final section.

2. Previous Work

Feature matching plays a significant role in human visual perception, recognition, and computer vision. In medical imaging, most existing feature matching-based research has focused on non-rigid registration and internal deformation field measurement. The general idea of these works is first to extract enough feature points or markers from medical images acquired from non-rigid objects on natural and deformed states, respectively. Next, the feature matching algorithm is applied on extracted feature-point sets to establish robust corresponding pairs. Finally, corresponding pairs are used as control points in non-rigid registration and are used to calculate sparse deformation fields in internal deformation field measurements. Therefore, finding robust corresponding pairs is a vital problem in the present work. We surveyed existing works on feature-point matching in computer vision. Relaxation is a valid technique to disambiguate matches and improve the robustness of matches. Finding a globally optimal or reasonably good suboptimal solution in relaxation is a difficult task, and such matching techniques in non-rigid medical image processing have been rarely addressed. However, a potential advantage is that harder matching problems can be solved using global optimization techniques.

Papademetris et al. [16] presented a method for the integration of feature and intensity information for non-rigid registration. In this case, a distance-based robust point matching framework was proposed to estimate feature-point correspondences. A disadvantage of the algorithm is that it estimates transformation using weighted least squares, which affects the strength of matching.

Zhang et al. [17] introduced a feature matching-based algorithm and considered the problem of 2D deformation field measurement as an example. Matching strengths are measured using correlation and relative distance between two feature points. Relaxation by the optimization algorithm is deductive of the function of matching strength. In later research [18], after slight revision, the algorithm was extended to a 3D situation because the intensity in a magnetic resonance (MR) image is the information of tissue mapped on an image. Thus, the correlation intensity of regions between two points in matching and relaxation can effectively use the properties of tissue.

The work [19] proposed a local geometric preserving algorithm to find corresponding feature pairs from given feature points set in MR volumes acquired from an object on natural and deformed states, respectively. The main contribution of the algorithm to feature matching is that for a non-rigid tissue, when an outside force is applied on it, the deformation magnitude and orientation are different in different regions. However, for a local region on the object, the difference is actually very slight and can sometimes be ignored.

Problems in image feature-point matching remain as great challenges for medical image processing. Thus, the accuracy of feature matching needs to be further improved. Typically, single factors, such as intensity and distance, are effective in matching algorithm for specific areas. However, total accuracy cannot be improved. The integration of multifactors to form a composite approach can make use of the advantages of each factor to improve total accuracy. The present work proposes a composite framework that can pose multicomponents in a single cost function with associated weights to find corresponding feature pairs.

Composite Match Index with Application of Interior Deformation Field Measurement from Magnetic Resonance
Volumetric Images of Human Tissues

3

TABLE 1: Comparison of the accuracy in different directions.

Approach	Error	Number of landmarks												RMSE
		0#	1#	2#	3#	4#	5#	6#	7#	8#	9#	10#	11#	
CMI	x-error	-12	-8	-7	-13	0	-4	-6	0	0	-7	0	0	6.626965
	y-error	8	-6	7	-7	0	1	6	0	0	5	0	0	4.654747
	z-error	0	-2	2	2	-1	0	-1	3	-1	2	3	2	1.848423
RPFM	x-error	-21	-23	-9	-12	0	-11	0	0	-1	6	0	0	10.61838
	y-error	20	-8	26	-10	0	-13	0	0	-7	4	0	0	11.08302
	z-error	2	2	2	-2	1	3	0	0	3	4	3	1	2.254625

RMSE: root mean square error; x-error: error in the x-direction; y-error: error in the y-direction; z-error: z-error in the z-direction.

3. CMI

CMI-based feature-point matching approach was proposed to address the fusion of different operator types and to improve the reliability of results from single operators. Here, CMI is a scalar quantity that describes the matching possibility of point pairs. Let

$$\mathbf{c}_i = [c_{i,1}, c_{i,2}, \ldots, c_{i,k}]^T,$$
$$\mathbf{w} = [w_1, w_2, \ldots, w_k]^T \tag{1}$$

be the vector of component value and its corresponding weight, respectively. Then, according to the linear weighting method, CMI is defined as

$$\xi_i^{t+1} = \sum_{k=1}^{K} w_k^{t+1} \cdot c_{i,k} \tag{2}$$

subject to

$$\sum_{k=1}^{K} w_k = 1, \tag{3}$$

where ξ_i represents the CMI of the i-th pair, c_k represents the value of the kth component consisting of the CMI, w_k represents the weight of the kth component, and K is the number of components in the CMI. Here, component is a factor that can be used to evaluate feature-point pair similarities. Weight w_k is used to measure the significance of a component for CMI. Various weighting methods have been reported for different research fields. In this case, to consider the independence of each component, the correlation weighted method is used to determine the weight of each component. Let $\mathbf{r} = [r_{1,1}, r_{1,2}, \ldots, r_{1,k}]^T$ be the correlation vector consisting of correlation score of the component 0 and k. Then, the weight of the $t + 1$ time w_k^{t+1} is defined by

$$w_k^{t+1} = \frac{\left| r_{1,k}^t \right|}{\sum_{k=1}^{K} \left| r_{1,k}^t \right|} \tag{4}$$

with

$$r_{1,k} = \frac{\sum_{i=1}^{N} (c_{i,1} - \bar{c}_1)(c_{i,k} - \bar{c}_k)}{\sqrt{\sum_{i=1}^{N} (c_{i,1} - \bar{c}_1)^2} \sqrt{\sum_{i=1}^{N} (c_{i,k} - \bar{c}_k)^2}}, \quad \bar{c}_k = \frac{1}{N} \sum_{i=1}^{N} c_{i,k}, \tag{5}$$

where $c_{i,k}$ is the value of the i-th feature-point pair and N is the total number of match pairs in the potential matching set obtained at time t.

Since feature-point pairs within the potential matching set obtained at time t are used as samples to compute the weight w_k^{t+1} of the kth component in $t + 1$ times iteration, the pairs in potential matching set are different at each time. Thus, w_k values are also different at different times, keeping iterations in the matching process.

CMI is an effective way to fuse multifeature matching algorithm. CMI takes full advantage of all the considered factors to generate a more robust feature matching approach and obtain more accurate matching results. Thus, the feature-point matching algorithm, which decides the strength of matching via a similarity judge function, can theoretically be integrated as a CMI component. In this case, the local geometric persistence (LGP), local intensity similarity (LIS), and local correlation score (LCS) between regions around participants are selected as the components to compute the CMI of a match pair $(\mathbf{p}_{u,i}, \mathbf{p}_{v,i})$ and demonstrate the validity of CMI. The following section will discuss how to compute LCS, LGP, and LIS.

For convenient descriptions, several definitions are first clarified as follows

(1) Initial feature set $\mathbf{p}_u$, feature-point set extracted from the MR volume acquired from the object at a natural state.

(2) Deformed feature set $\mathbf{p}_v$, feature-point set extracted from the MR volume acquired from the object at a deformed state.

(3) PMS, a potential feature match set composed of a match pair $(\mathbf{p}_{u,i}, \mathbf{p}_{v,i})$ if and only if the best match of $\mathbf{p}_{u,i}$ is $\mathbf{p}_{v,i}$ and conversely $\mathbf{p}_{u,i}$ is also the best match of $\mathbf{p}_{v,i}$.

(4) $\mathbf{p}_{u,i}$ represents the feature point i in the initial feature set, and $\mathbf{p}_{v,i}$ represents the feature point i in the deformed feature set.

3.1. LGP.
Let $\mathbf{c}_u$ and $\mathbf{c}_v$ be the moment center computed using the initial feature set and its mapping in the deformed feature set, respectively, let $\mathbf{p}_{u,i}$ be the ith point in the initial feature set, and let the mapping in the deformed feature set be $\mathbf{p}_{v,i}$. Based on the consistent deformation in a local region, the distance ratio of a potential match pair in a local region

far from their moment center is equivalent and thus yields $\mu_{i,j}$:

$$\mu_{i,j} = \frac{d(\mathbf{p}_{u,i,j}, \mathbf{c}_u)}{d(\mathbf{p}_{v,i,j}, \mathbf{c}_v)} \approx \frac{1}{J} \sum_{j=1}^{J} \frac{d(\mathbf{p}_{u,i,j}, \mathbf{c}_u)}{d(\mathbf{p}_{v,i,j}, \mathbf{c}_v)}, \tag{6}$$

where $\mu_{i,j}$ is the distance ratio of the jth potential match pair $(\mathbf{p}_{u,i,j}, \mathbf{p}_{v,i,j})$ in the local region around pair $(\mathbf{p}_{u,i}, \mathbf{p}_{v,i})$, $d(\mathbf{p}_{u,i,j}, \mathbf{c}_u)$ is the Euclidian distance between $\mathbf{p}_{u,i,j}$ and $\mathbf{c}_u$, $d(\mathbf{p}_{v,i,j}, \mathbf{c}_v)$ is the Euclidian distance between $\mathbf{p}_{v,i,j}$ and $\mathbf{c}_v$, and J is the number of potential match pairs in the local region. Ideally, $\mu_{i,j}$ should be a constant in the local region.

Moreover, $\mathbf{d}_{u,i} = [d(\mathbf{p}_{u,i,1}, \mathbf{c}_u), d(\mathbf{p}_{u,i,2}, \mathbf{c}_u), \ldots, d(\mathbf{p}_{u,i,J}, \mathbf{c}_u)]^T$ and $\mathbf{d}_{v,i} = [d(\mathbf{p}_{v,i,1}, \mathbf{c}_v), d(\mathbf{p}_{v,i,2}, \mathbf{c}_v), \ldots, d(\mathbf{p}_{v,i,J}, \mathbf{c}_v)]^T$ are the distance sets of the potential pairs within a local region around pair $(\mathbf{p}_{u,i}, \mathbf{p}_{v,i})$, respectively. Based on the definition of mathematical expectation, we yield

$$E(\mathbf{d}_{u,i}) = \frac{1}{J} \sum_{j=1}^{J} d(\mathbf{p}_{u,i,j}, \mathbf{c}_u),$$
$$E(\mathbf{d}_{v,i}) = \frac{1}{J} \sum_{j=1}^{J} d(\mathbf{p}_{v,i,j}, \mathbf{c}_v). \tag{7}$$

Thus, if $\mathbf{p}_{v,i}$ in the deformed feature set is the best match of a given feature $\mathbf{p}_{u,i}$ in the initial feature set, then, the geometric deformation of potential match pair $(\mathbf{p}_{u,i,j}, \mathbf{p}_{v,i,j})$ within a local region around pair $(\mathbf{p}_{u,i}, \mathbf{p}_{v,i})$ is defined as

$$g_{i,j} = \left| \mu_{i,j} - \eta_i \right| \tag{8}$$

subject to

$$\eta_i = \frac{E(\mathbf{d}_{u,i})}{E(\mathbf{d}_{v,i})}. \tag{9}$$

In a small local region, all the $g_{i,j}(j = 1, 2, \ldots, J)$ should be approximately identical and go to zero; the smaller the value of $g_{i,j}$, the better the geometric persistence of a potential match pair $(\mathbf{p}_{u,i}, \mathbf{p}_{v,i})$. This is called geometric persistence in this case. Thus, the impact factor of the j-th feature pair for the LGP within a small local region is

$$\lambda_{i,j} = \frac{1.0}{1.0 + g_{i,j}}. \tag{10}$$

The geometric property within a local region is approximately consistent in the initial and deformed states. If a pair is the best match for each other, then the correlation of potential matches within a local region around the pair must be a strong one. The correlated score $gc(\mathbf{p}_{u,i}, \mathbf{p}_{v,i})$ of the geometric persistence of PMS in a small local region around $(\mathbf{p}_{u,i}, \mathbf{p}_{v,i})$ can represent the LGP of feature pair $(\mathbf{p}_{u,i}, \mathbf{p}_{v,i})$, specifically:

$$gc(\mathbf{p}_{u,i}, \mathbf{p}_{v,i}) = \frac{\sum_{j=1}^{J} \lambda_{i,j}(d(\mathbf{p}_{u,i,j}, \mathbf{c}_u) - E(\mathbf{d}_{u,i}))}{\sqrt{\sum_{j=1}^{J}(d(\mathbf{p}_{u,i,j}, \mathbf{c}_u) - E(\mathbf{d}_{u,i}))^2}}$$
$$\times \frac{(d(\mathbf{p}_{v,i,j}, \mathbf{c}_v) - E(\mathbf{d}_{v,i}))}{\sqrt{\sum_{j=1}^{J}(d(\mathbf{p}_{v,i,j}, \mathbf{c}_v) - E(\mathbf{d}_{v,i}))^2}}, \tag{11}$$

where J is the number of potential matches within a local region. In (11), if $\lambda_{i,j}$ is large, the pair $(\mathbf{p}_{u,i,j}, \mathbf{p}_{v,i,j})$ may be a strong match pair; thus, its weight must also be large. In addition, the value range of $gc(\mathbf{p}_{u,i}, \mathbf{p}_{v,i})$ should be $[-1, 1]$. Normalizing $gc(\mathbf{p}_{u,i}, \mathbf{p}_{v,i})$ yields normalized LGP as

$$N_{gc}(\mathbf{p}_{u,i}, \mathbf{p}_{v,i}) = \frac{1 + gc(\mathbf{p}_{u,i}, \mathbf{p}_{v,i})}{2}. \tag{12}$$

3.2. LIS. LIS is used to describe the intensity difference between regions around a feature-point pair in the initial and deformed volumes. As mentioned earlier, the tissue within a local region is the same in the initial and deformed states. Thus, based on the MRI principle, the intensity difference is small. The inner product between two regions has the same properties with the invariance of rotation, zoom in, and zoom out. The normalized inner product between regions around $(\mathbf{p}_{u,i}, \mathbf{p}_{v,i})$ is adopted to define the similarity of two regions. Thus,

$$\text{lis}(\mathbf{p}_{u,i}, \mathbf{p}_{v,i}) = \frac{X_{u,i}^\top X_{v,i}}{||X_{u,i}|| \cdot ||X_{v,i}||}, \tag{13}$$

where $X_{u,i}$ is the region in the initial volume centered at feature $\mathbf{p}_{u,i}$ and $X_{v,i}$ is the mapping region of $X_{u,i}$ centered at feature $\mathbf{p}_{v,i}$.

3.3. LCS. Let $I(\mathbf{p}_{u,i,m})$ and $I(\mathbf{p}_{v,i,m})$ be the intensity of the m-th voxel within the region centered at $\mathbf{p}_{u,i}$ and $\mathbf{p}_{v,i}$ in the initial and deformed MR volumes, respectively. Let $\mathcal{O}$ be the local cubic region with a size of $w \times h \times l$. The local correlation score between local cubic regions around feature $\mathbf{p}_{u,i}$ in the initial MR volume and its candidate match feature $\mathbf{p}_{v,i}$ in the deformed MR volume is defined as

$$\text{lcs}(\mathbf{p}_{u,i,m}, \mathbf{p}_{v,i,m}) = \frac{\sum_{m=1}^{M}(I(\mathbf{p}_{u,i,m}) - \bar{a}_{u,i})(I(\mathbf{p}_{v,i,m}) - \bar{a}_{v,i})}{M\sqrt{\sigma^2(I(\mathbf{p}_{u,i,m})) \cdot \sigma^2(I(\mathbf{p}_{v,i,m}))}}, \tag{14}$$

where

$$M = w \times h \times l,$$
$$\bar{a}_{u,i} = \frac{1}{M} \sum_{m=1}^{M} I(\mathbf{p}_{u,i,m}), \qquad \bar{a}_{v,i} = \frac{1}{M} \sum_{m=1}^{M} I(\mathbf{p}_{v,i,m}). \tag{15}$$

Here, $\sigma^2(I(\mathbf{p}_{u,i,m}))$ and $\sigma^2(I(\mathbf{p}_{v,i,m}))$ are the standard derivation of the local region $\mathcal{O}$ around feature $\mathbf{p}_{u,i}$ and $\mathbf{p}_{v,i}$, respectively. They are given by

$$\sigma^2(I(\mathbf{p}_{u,i,m})) = \frac{\sum_{m=1}^{M}(I(\mathbf{p}_{u,i,m}) - \bar{a}_{u,i})^2}{M},$$
$$\sigma^2(I(\mathbf{p}_{v,i,m})) = \frac{\sum_{m=1}^{M}(I(\mathbf{p}_{v,i,m}) - \bar{a}_{v,i})^2}{M}, \tag{16}$$

where $\bar{a}_{u,i}$ and $\bar{a}_{v,i}$ are the averaged intensity in the neighborhood of feature $\mathbf{p}_{u,i}$ and $\mathbf{p}_{v,i}$, respectively.

Composite Match Index with Application of Interior Deformation Field Measurement from Magnetic Resonance
Volumetric Images of Human Tissues

5

4. Application in Feature Matching

This section describes the measurement of internal deformation fields using CMI. First, the cost function is given to obtain optimal feature pairs iteratively. Then, the actual feature matching algorithm is described. Finally, the internal deformation fields are measured using optimal feature pairs.

4.1. Cost Function. CMI is an index that measures the strength between a given feature and its candidate matches in feature matching. In theory, for a given reference feature, its potential match must have the strongest CMI among all the candidates. Thus, for an optimal potential matching set, its whole CMI will also be the strongest. Based on this idea, we yield

$$S^t = \frac{1}{N}\sum_{i=1}^{N}\xi_i^t, \qquad (17)$$

where S is the cost function in iteration and N represents the total number of match pairs in the PMS obtained at time t.

4.2. Actual Matching Algorithm. The objective of the feature matching algorithm is to obtain an optimal PMS ultimately. The idea of PMS optimization is to maximize the aforementioned cost function S iteratively. In each iterative step, the current PMS strength is evaluated by all candidate matches within PMS using the defined cost function S. The iterative steps will stop until S no longer increases or is subjected to stop conditions. Specifically, the inputs are two feature-point sets obtained from MR volumetric images of an object under natural and deformed states, respectively. The output is an optimal PMS. The specific process of the algorithm is summarized as follows.

(0) Compute LCS and LIS. For each given pair $(\mathbf{p}_{u,i}, \mathbf{p}_{v,i})$ consisting of features in initial and deformed volumes, we use a local region (size of $9 \times 9 \times 3$ in this case) centered at features to compute LCS and LIS according to (14) and (13), respectively.

(1) Form initial PMS. The LCS is used as the initial CMI of each match pair in the step of initial PMS formation. In other words, LCS is the only criterion of this step.

(2) Compute LGP. For each given pair $(\mathbf{p}_{u,i}, \mathbf{p}_{v,i})$, we first search for neighbor potential matches within a small window (size of $17 \times 17 \times 3$ in this case) centered at $\mathbf{p}_{u,i}$. The potential matches contained within the window are participants in the LGP computation using the approach in Section 3.1.

(3) Compute $\mathbf{w}$. Compute the weight for each CMI component using potential matches in current PMS as samples. The specific computing method can be seen in (4).

(4) Update the CMI of each pair. For each given pair $(\mathbf{p}_{u,i}, \mathbf{p}_{v,i})$, its corresponding CMI is updated through the weighting sum of the components LCS, LIS, and LCP, which are computed in (0) and (2).

(5) Form PMS and compute the cost function S. The updated CMI of each pair forms new PMS. The cost function in (17) is then computed using potential matches in the current PMS.

(6) Repeat (2) to (5) until S no longer increases.

(7) Return the current PMS.

Although candidate sets LCS and LIS of each pair are constant, PMS is dynamic because of the varying LGP and $\mathbf{w}$ of the component at $t + 1$ times iteration. Thus, the match strength index of CMI is varied. Dynamic cost function will move potential matches into or out of the PMS. The best candidate of a feature-point may also change.

4.3. Measuring Density Deformation Fields. After obtaining the optimal PMS, the internal density deformation fields of non-rigid objects are then obtained. In this study, the method proposed in our previous work [20] is used to obtain the internal density deformation fields. In summary, the internal density deformation fields are interpolated by sparse deformation fields using a finite element model. In detail, the magnitude of the sparse deformation field is first computed by its corresponding pair in PMS using Euclidian distance. The start and end points of a field direction are defined by the points of the corresponding pair. Next, a non-rigid object is reconstructed using tetrahedra, whose nodes are points in the PMS. The density deformation fields can then be interpolated using the finite element method.

Let $\mathbf{P}$ be an arbitrary volume voxel at $\mathbf{x} = (x, y, z)$ within a tetrahedron $\lozenge \mathbf{P}_i \mathbf{P}_j \mathbf{P}_k \mathbf{P}_l$ consisting of nodal points $\mathbf{P}_i$, $\mathbf{P}_j$, $\mathbf{P}_k$, and $\mathbf{P}_l$. Its displacement may be approximated by weighting the finite element node displacements $\mathbf{u}_{i,j,k,l}(\mathbf{x})$ using their shape function [20]:

$$\begin{aligned} \mathbf{u}(\mathbf{x}) = &\mathbf{u}_i(\mathbf{x})N_{i,j,k,l}(\mathbf{x}) + \mathbf{u}_j(\mathbf{x})N_{j,k,l,i}(\mathbf{x}) \\ &+ \mathbf{u}_k(\mathbf{x})N_{k,l,i,j}(\mathbf{x}) + \mathbf{u}_l(\mathbf{x})N_{l,i,j,k}(\mathbf{x}), \end{aligned} \qquad (18)$$

where $\mathbf{u}_i(\mathbf{x})$ is the displacement of nodal i, and the shape function $N_{i,j,k,l}(\mathbf{x})$ on tetrahedron $\lozenge \mathbf{P}_i \mathbf{P}_j \mathbf{P}_k \mathbf{P}_l$ is given by

$$N_{i,j,k,l}(\mathbf{x}) = \frac{\blacklozenge \mathbf{PP}_j \mathbf{P}_k \mathbf{P}_l}{\blacklozenge \mathbf{P}_i \mathbf{P}_j \mathbf{P}_k \mathbf{P}_l}, \qquad (19)$$

where $\blacklozenge \mathbf{PP}_j \mathbf{P}_k \mathbf{P}_l$ and $\blacklozenge \mathbf{P}_i \mathbf{P}_j \mathbf{P}_k \mathbf{P}_l$ are the volume of tetrahedron $\lozenge \mathbf{PP}_j \mathbf{P}_k \mathbf{P}_l$ and $\lozenge \mathbf{P}_i \mathbf{P}_j \mathbf{P}_k \mathbf{P}_l$, respectively.

5. Experiments and Results

Our approach consists of four steps: feature extraction, affine transformation, feature matching, and deformation field measurement. Extracting sufficient features from the initial and deformed volumes is necessary to find enough homologous feature pairs. In this study, high-curvature 3D points were preextracted as features from MR volumetric images. In this case, the two-dimensional Harris operator [21] was extended to a 3D operator by extracting features from the MR volumetric images [22].

Some experiments were designed to demonstrate the performance of the proposed approach. All experiments

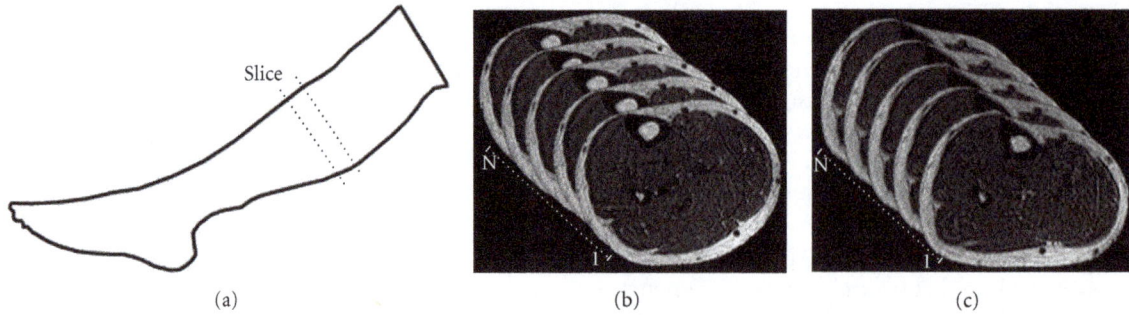

FIGURE 1: Acquired MR volumes. (a) Place of acquired volume; (b) MR slices in volume obtained at natural state; (c) MR slices in volume obtained at deformed state.

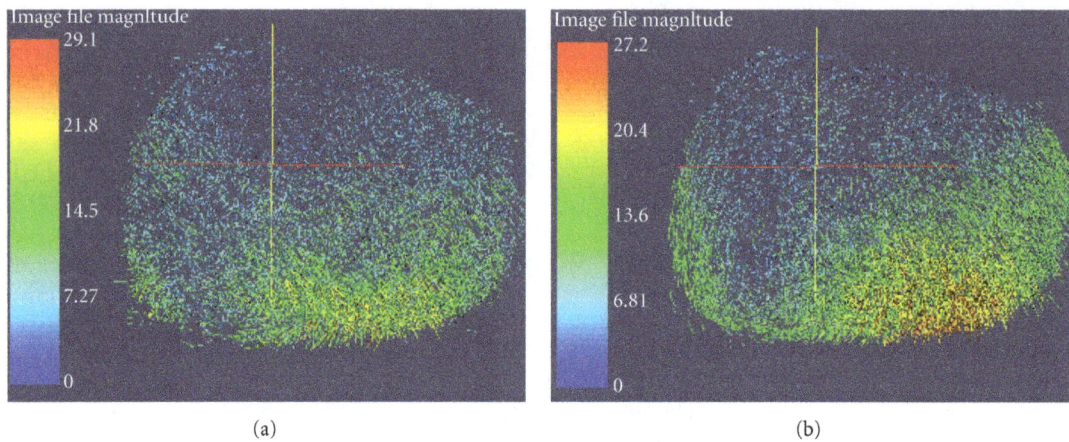

FIGURE 2: Density deformation fields. Deformation fields generated using PMS obtained using the (a) CMI-based feature match algorithm and the (b) RPFM algorithm).

were performed using our own tool developed with Visual C++, which runs on Microsoft Windows XP. All described experimental results were obtained on a Lenovo Portable PC with a 2.20 GHz Intel(R) Core(TM) 2 Duo CPU T6600 and 4 GB of RAM.

In the experiment, the MR images were acquired from a volunteer's calf (Figure 1(a)) using an MRI scanner at natural state and deformed states (initial and under forcing), respectively. In both cases, the FOV was 20×20 cm, and the slice gap was 2 mm. Some slices (Figures 1(b) and 1(c)) placed at the middle section of the calf were selected to form the MR volumes. As a result, initial and deformed volumes with size of $512 \times 512 \times 57$ voxels were generated for the experiment.

First, 500 and 800 features were extracted from the volume acquired on the natural and deformed states, respectively. Next, the proposed CMI-based feature match approach was applied on the two feature-point sets to obtain the optimal PMS. As the result, a PMS with 245 potential match pairs was obtained. The sparse and density deformation fields were computed using the method mentioned in Section 4.3. Figure 2 shows 50000 internal density deformation fields, with large deformation at the bottom of the calf. This result is consistent with the actual situation.

To prove the validity of the proposed CMI-based feature match algorithm, we compared it with a robust point feature matching (RPFM) algorithm proposed by Chen [23]. In the present study, we applied the RPFM algorithm to the same feature-point sets, which resulted in a PMS with 316 potential match pairs.

We selected 12 landmarks in the slice ($z = 40$) of deformed MR volume to test the accuracy of the measured internal deformation fields, as shown in the middle picture of Figures 3 and 4. Then, the landmarks were subjected to reverse moving using the internal deformation fields measured through the CMI-based algorithm and RPFM algorithm. The results on the MR volume acquired at natural state were projected to check the accuracy of the deformation fields. Figures 3 and 4 show the reverse moving results of the landmarks.

In Figures 3 and 4, the center of each red rectangle in the middle picture ($z = 40$) gives the landmark position. Slices that lie on the left and right sides (the middle layer) give the reverse moving result of the landmarks and the z value of different slices, respectively. The outer layer is the zoom in for the reverse moving result of each landmark. In the middle and outer layers, the red rectangles represent the reverse moving position of the landmarks, the green rectangles are actual position of landmarks, and the yellow rectangles represent the reverse moving positions and actual position consistency. From Figures 3 and 4, we note the

Composite Match Index with Application of Interior Deformation Field Measurement from Magnetic Resonance
Volumetric Images of Human Tissues

7

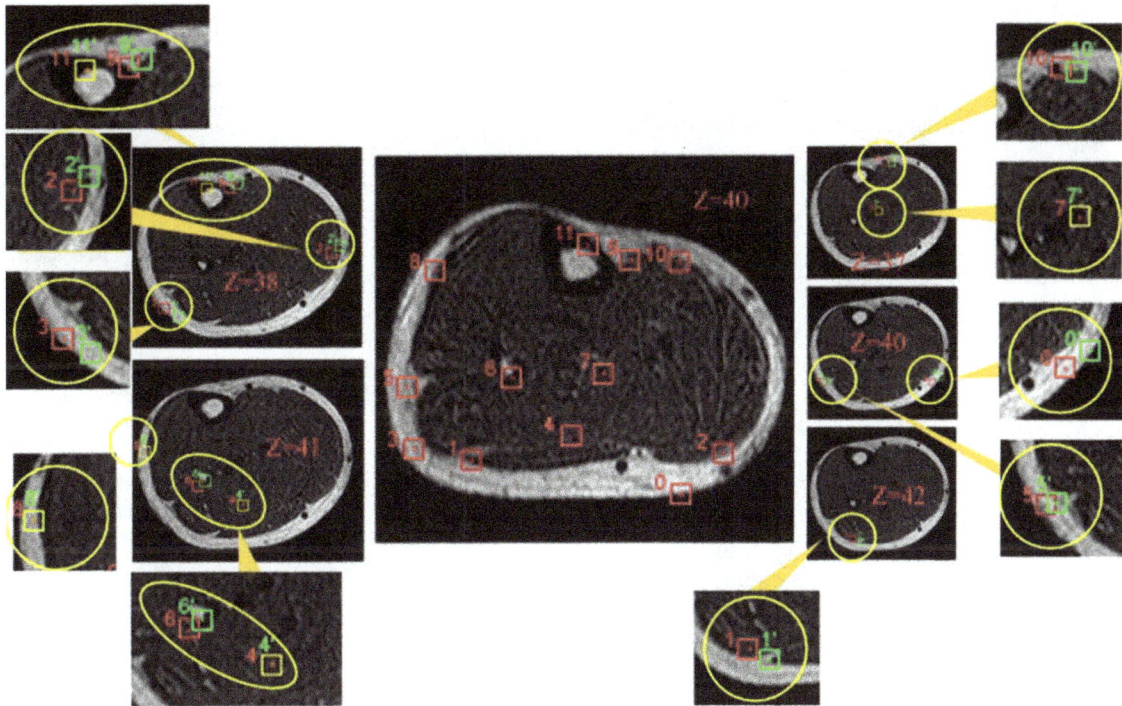

FIGURE 3: Reverse moving result of the landmarks using deformation fields measured through the CMI-based approach.

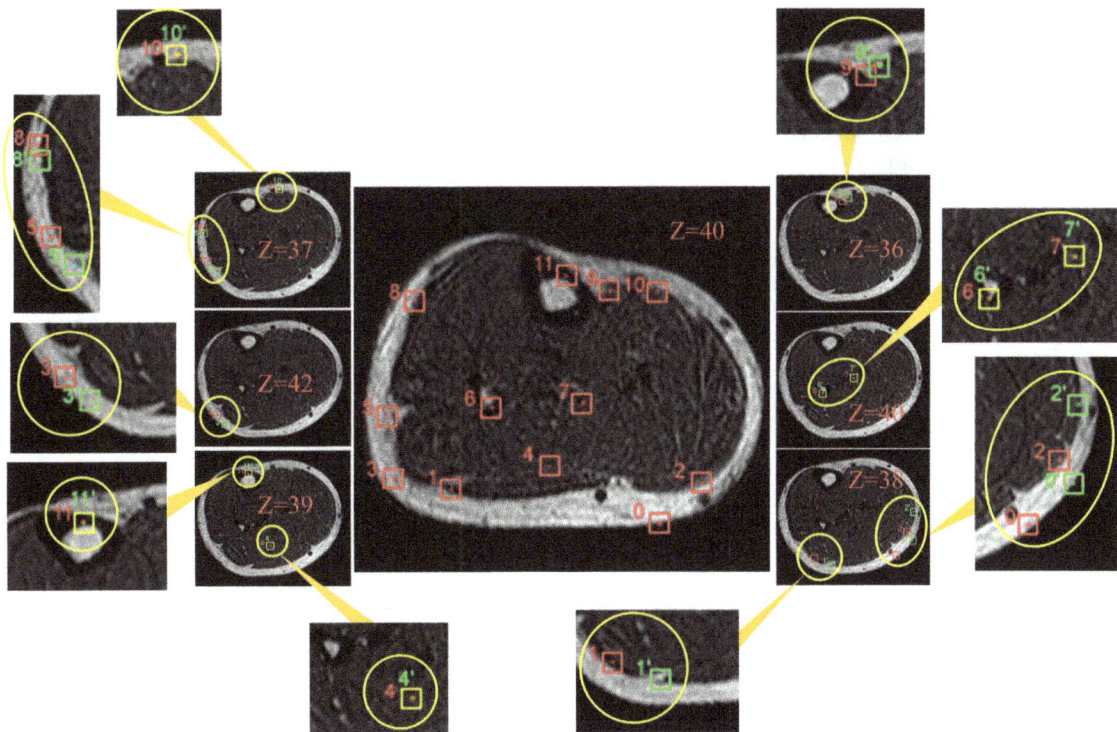

FIGURE 4: The reverse moving result of the landmarks using deformation fields measured through the RPFM approach.

accuracy of the reverse moving position of landmarks using deformation fields calculated by PMS obtained using CMI-based approach obviously is higher than that of RPFM, that is, the reverse moving position of landmarks 0, 1, 2, 3, and 5. Table 1 shows the quantitative accuracy of the reverse moving results of the landmarks using internal deformation fields obtained by PMS via CMI and RPFM.

As shown in Table 1, regardless of the direction (i.e., x-, y-, and z-directions), the accuracy of the deformation fields measured through PMS obtained using the CMI-based approach is better than that using the RPFM algorithm.

The number of potential matches in optimal PMS obtained using the CMI-based feature matching algorithm is fewer than that of RPFM because the CMI-based approach is combined with the multifeatures in feature matching, whereas RPFM is a single-feature approach. In other words, the match requirements of CMI are stricter compared with those of RPFM. The reliability of optimal PMS obtained using the CMI-based algorithm is higher because it has more accurate deformation fields than the RPFM algorithm. This conclusion is supported by the reverse moving results of the landmarks.

6. Conclusions

In this work, a new method called CMI is presented for the integration of feature-based internal deformation field measurements. In general, feature match algorithms using a single property are highly accurate in specific aspects. However, the overall accuracy is limited because the full advantages of different properties in feature-point matching are not fully used. Fusion multialgorithms offer the use of advantages in algorithms to improve accuracy. Such a fusion is necessary for feature matching in non-rigid objects, where the improvement will be more obvious. In addition, the most advantage of the proposed approach is to provide a feasible option to integrate various feature matching algorithms. Each feature matching algorithm can act as the component of the CMI, and if the appropriate weight can be assigned to the component, then, one can obtain more reliable potential matches. Obviously, the effect of the component weight should also be considered. Thus, (1) investigating an approach to determine the appropriate weights should be the focus of future research; (2) the imaging mechanism of MRI should be further considered in component of the CMI to remove the aberrance of machine to improve the accuracy of feature-point matching as possible.

Acknowledgments

The authors greatly appreciate the support of Professor Shinichi Hirai from Ritsumeikan University for this work. The authors would also like to thank the reviewers for their helpful comments and suggestions, which have improved the presentation of this paper. This work was supported by the National Key Technologies R&D Program of China for the 12th Five-year Plan (2012BAJ15B04).

References

[1] Y. Chenoune, E. Deléchelle, E. Petit, T. Goissen, J. Garot, and A. Rahmouni, "Segmentation of cardiac cine-MR images and myocardial deformation assessment using level set methods," *Computerized Medical Imaging and Graphics*, vol. 29, no. 8, pp. 607–616, 2005.

[2] M. Kass, A. Witkin, and D. Terzopoulos, "Snakes: active contour models," *International Journal of Computer Vision*, vol. 1, no. 4, pp. 321–331, 1988.

[3] J. Lang, D. K. Pai, and R. J. Woodham, "Robotic acquisition of deformable models," in *Proceedings of the IEEE International Conference on Robotics and Automation (ICRA '02)*, vol. 1, pp. 933–938, May 2002.

[4] J. Cho and P. J. Benkeser, "Elastically deformable model-based motion-tracking of left ventricle," in *Proceedings of the 26th Annual International Conference of the IEEE Engineering in Medicine and Biology Society (EMBC '04)*, vol. 1, pp. 1925–1928, September 2004.

[5] B. J. Matuszewski, J. K. Shen, L. K. Shark, and C. J. Moore, "Estimation of internal body deformations using an elastic registration technique," in *Proceedings of the International Conference on Medical Information Visualisation-BioMedical Visualisation (MediVis '06)*, pp. 15–20, July 2006.

[6] C. Vicent, C. Francine, C. Tomeu, and D. Francoise, "A geometric model for active contours in image processing," *Numerische Mathematik*, vol. 66, no. 1, pp. 1–31, 1993.

[7] R. Malladi, J. A. Sethian, and B. C. Vemuri, "Shape modeling with front propagation: a level set approach," *IEEE Transactions on Pattern Analysis and Machine Intelligence*, vol. 17, no. 2, pp. 158–175, 1995.

[8] C. Vicent, "Geometric models for active contours," in *Proceedings of the IEEE International Conference on Image Processing*, vol. 3, pp. 9–12, October 1995.

[9] F. Huang and J. Su, "Face contour detection using geometric active contours," in *Proceedings of the 4th World Congress on Intelligent Control and Automation*, vol. 3, pp. 2090–2093, Shanghai, China, June 2002.

[10] H. Chui and A. Rangarajan, "New algorithm for non-rigid point matching," in *proceedings of the IEEE Conference on Computer Vision and Pattern Recognition (CVPR '00)*, pp. 44–51, June 2000.

[11] Y. Zheng and D. Doermann, "Robust point matching for nonrigid shapes by preserving local neighborhood structures," *IEEE Transactions on Pattern Analysis and Machine Intelligence*, vol. 28, no. 4, pp. 643–649, 2006.

[12] S. Belongie, J. Malik, and J. Puzicha, "Shape matching and object recognition using shape contexts," *IEEE Transactions on Pattern Analysis and Machine Intelligence*, vol. 24, no. 4, pp. 509–522, 2002.

[13] A. Myronenko, X. Song, and M. A. Carreira-Perpinan, "Non-rigid point set registration: coherent point drift," *Advances in Neural Information Processing Systems*, vol. 19, pp. 1009–1016, 2007.

[14] A. Myronenko and X. Song, "Point set registration: coherent point drifts," *IEEE Transactions on Pattern Analysis and Machine Intelligence*, vol. 32, no. 12, pp. 2262–2275, 2010.

[15] O. Choi and I. S. Kweon, "Robust feature point matching by preserving local geometric consistency," *Computer Vision and Image Understanding*, vol. 113, no. 6, pp. 726–742, 2009.

[16] X. Papademetris, A. P. Jackowski, R. T. Schultz, L. H. Staib, and J. S. Duncan, "Integrated intensity and point-feature nonrigid registration," in *Proceedings of the 7th International*

Composite Match Index with Application of Interior Deformation Field Measurement from Magnetic Resonance
Volumetric Images of Human Tissues

9

Conference of Medical Image Computing and Computer-Assisted Intervention (MICCAI '04), vol. 3216, pp. 763–770, September 2004.

[17] P. Zhang, S. Hirai, and K. Endo, "A feature matching-based approach to deformation fields measurement from MR images of non-rigid object," *International Journal of Innovative Computing, Information and Control*, vol. 4, no. 7, pp. 1607–1615, 2008.

[18] P. Zhang, S. Hirai, K. Endo, and S. Morikawa, "Local deformation measurement of biological tissues based on feature tracking of 3D MR volumetric images," in *Proceedings of the IEEE/ICME International Conference on Complex Medical Engineering (CME '07)*, pp. 707–712, Beijin, China, May 2007.

[19] P. L. Zhang and S. Hirai, "A local geometric preserving approach for interior deformation fields measurement from MR volumetric images of human tissues," in *Proceeding of the IEEE International Conference on Robotics and Biomimetics*, pp. 437–441, 2010.

[20] P. L. Zhang, S. Hirai, and K. Endo, "A method for non-rigid 3D deformation fields measurement: application to human calf MR volumetric images," in *Proceedings of the Workshop at IEEE/RSJ International Conference on Intelligent Robots and Systems*, pp. 8–13, 2007.

[21] C. Harris and M. J. Stephens, "A combined corner and edge detector," in *Proceedings of the 4th Alvey Vision Conference*, pp. 147–151, 1988.

[22] P. L. Zhang, S. Hirai, and K. Endo, "A feature tracking-based approach for local deformation fields measurement of biological tissue from MR volumes," in *Proceedings of the 3rd Joint Workshop on Machine Perception and Robotics*, December 2007.

[23] G. Q. Chen, "Robust point feature matching in projective space," in *Proceedings of the Robust Point Feature Matching in Projective Space*, pp. 717–722, December 2001.

Strategic Cognitive Sequencing: A Computational Cognitive Neuroscience Approach

**Seth A. Herd, Kai A. Krueger, Trenton E. Kriete, Tsung-Ren Huang,
Thomas E. Hazy, and Randall C. O'Reilly**

Department of Psychology, University of Colorado Boulder, Boulder, CO 80309, USA

Correspondence should be addressed to Seth A. Herd; seth.herd@gmail.com

Academic Editor: Giorgio Ascoli

We address strategic cognitive sequencing, the "outer loop" of human cognition: how the brain decides what cognitive process to apply at a given moment to solve complex, multistep cognitive tasks. We argue that this topic has been neglected relative to its importance for systematic reasons but that recent work on how individual brain systems accomplish their computations has set the stage for productively addressing how brain regions coordinate over time to accomplish our most impressive thinking. We present four preliminary neural network models. The first addresses how the prefrontal cortex (PFC) and basal ganglia (BG) cooperate to perform trial-and-error learning of short sequences; the next, how several areas of PFC learn to make predictions of likely reward, and how this contributes to the BG making decisions at the level of strategies. The third models address how PFC, BG, parietal cortex, and hippocampus can work together to memorize sequences of cognitive actions from instruction (or "self-instruction"). The last shows how a constraint satisfaction process can find useful plans. The PFC maintains current and goal states and associates from both of these to find a "bridging" state, an abstract plan. We discuss how these processes could work together to produce strategic cognitive sequencing and discuss future directions in this area.

1. Introduction

Weighing the merits of one scientific theory against another, deciding which plan of action to pursue, or considering whether a bill should become law all require many cognitive acts, in particular sequences [1, 2]. Humans use complex cognitive strategies to solve difficult problems, and understanding exactly how we do this is necessary to understand human intelligence. In these cases, different strategies composed of different sequences of cognitive acts are possible, and the choice of strategy is crucial in determining how we succeed and fail at particular cognitive challenges [3, 4]. Understanding strategic cognitive sequencing has important implications for reducing biases and thereby improving human decision making (e.g., [5, 6]). However, this aspect of cognition has been studied surprisingly little [7, 8] because it is complex. Tasks in which participants tend to use different strategies (and therefore sequences) necessarily produce data that is less clear and interpretable than that from a single process in a simple task [9]. Therefore, cognitive neuroscience tends to avoid such tasks, leaving the neural mechanisms of strategy selection and cognitive sequencing underexplored relative to the large potential practical impacts.

Here, we discuss our group's efforts to form integrative theories of the neural mechanisms involved in selecting and carrying out a series of cognitive operations that successfully solve a complex problem. We dub this process strategic cognitive sequencing (SCS). While every area of the brain is obviously involved in some of the individual steps in some particular cognitive sequences, there is ample evidence that the prefrontal cortex (PFC), basal ganglia (BG), and hippocampus and medial temporal lobe (HC and MTL) are particularly important for tasks involving SCS (e.g., [10–14]). However, exactly how these brain regions allow us to use multistep approaches to problem solving is unknown. The details of this process are clearly crucial to understanding that process well enough to help correct dysfunctions, to better train it, and perhaps to eventually reproduce it in artificial general intelligence (AGI).

We present four different neural network models, each of a computational function that we consider crucial for strategic cognitive sequencing. The first two models address how sequences are learned and selected: how the brain selects which of a small set of known strategic elements to use in a given situation. The first, "model-free learning," is a model of how dopamine-driven reinforcement learning in the PFC and BG can learn short cognitive sequences entirely through trial and error, with reward available only at the end of a successful sequence. The second, "PFC/BG decision making" (PBDM), shows how cortical predictions of reward and effort can drive decision making in the basal ganglia for different task strategies, allowing a system to quickly generalize learning from selecting strategies on old tasks to new tasks with related but different strategies. The last two models apply to selecting *what* plans or actions (from the large set of possibilities in long-term semantic memory) will be considered by the two "which" systems. The third model, "instructed learning," shows how episodic recall can work with the PFC and BG to memorize sequences from instructions, while the last "subgoal selection" model shows how semantic associative processes in posterior cortex can select representations of "bridging states" which also constitute broad plans connecting current and goal states, each of which can theoretically be further elaborated using the same process to produce elaborate plan sequences.

Because these models were developed somewhat separately, they and their descriptions address "actions," "strategies," "subgoals," and "plans." We see all of these as sharing the same types of representations and underlying brain mechanism, so each model actually addresses all of these levels. All of these theories can be applied either to individual actions or whole sequences of actions that have been previously learned as a "chunk" or plan. This hierarchical relationship between sequence is well understood at the lower levels of motor processing (roughly, supplementary motor areas tend to encode sequences of primary motor area representations, while presupplementary motor areas encode sequences of those sequences); we assume that this relationship holds to higher levels, so that sequences of cognitive actions can be triggered by a distributed representation that loosely encodes that whole sequence and those higher-level representations can then unfold as sequences themselves using identically structured brain machinery, possibly in slightly different, but parallel brain areas.

Before elaborating on each model, we clarify the theoretical framework and background that have shaped our thinking. After describing each model, we further tie each model to our overall theory of human strategic cognitive sequencing and describe our planned future directions for modeling work that will tie these individual cognitive functions into a full process that learns and selects sequences of cognitive actions constituting plans and strategies appropriate for novel, complex mental tasks, one of humans' most impressive cognitive abilities.

2. Theoretical Framework

These models synthesize available relevant data and constitute our attempt at curren best-guess theories. We take a computational cognitive neuroscience approach, in which artificial neural network models serve to concretize and specify our theories. The models serve as cognitive aids in a similar way to diagramming and writing about theories but also serve to focus our inquiries on the computational aspects of the problem. These theories are constrained not only by the data we specifically consider here but also by our use of the Leabra modeling framework [15, 16]. That framework serves as a cumulative modeling effort that has been applied to many topic areas and serves to summarize a great deal of data on neural function. This framework serves as a best-guess theory on cortical function, and individual models represent more specific, but still empirically well-supported and constrained theories of PFC, basal ganglia, reward system, and hippocampal function. Here, we extend these well-developed theories to begin to address SCS.

We also take our constraints from purely cognitive theories of cognitive sequencing. Work on production system architectures serves as elaborate theories of how human beings sequence cognitive steps to solve complex problems [17–19]. The numerous steps by which a production system model carries out a complex task such as air traffic control [20] are an excellent example of cognitive sequencing. Our goal here is to elaborate on the specific neural mechanisms involved, and in so doing, we alter those theories somewhat while still accounting for the behavioral data that has guided their creation.

Neural networks constitute the other class of highly specified and cumulative theories of cognition. However, these are rarely applied to the type of tasks we address here, in which information must be aggregated from step to step, but in arbitrary ways (e.g., first figure out center of a set of points, then calculate the distance from that center of points to an another point, and then based on that distance, estimate the likelihood that the point shares properties with the set). This is essentially because neural networks perform information processing in parallel and so offer better explanations of single-step problem solving. Indeed, we view humans' ability to use strategic cognitive sequences as an exaptation of our ancestral brain machinery, one that makes us much smarter by allowing us to access a range of strategies that lower animals largely cannot use [21, 22].

Because of the weaknesses in each approach and the paucity of other mechanistically detailed, cumulative models of cognition, we take inspiration from the well-developed theories from production systems about how cognitive steps are sequenced [17–19, 23] while focusing on artificial neural network-centered theories on the specifics of how individual cognitive actions are performed. This perspective is influenced by prior work on hybrid theories and cognitive architectures based on ACT-R and Leabra networks for a different purpose [24]. ACT-R [18] is the most extensively developed production system architecture and the one which most explicitly addresses physiology, while Leabra is arguably the most extensively developed and cumulative theory of neural function that spans from the neural to cognitive levels.

In ACT-R, the sequence of cognitive actions is determined by which production fires. This in turn is based upon the "fit" between the conditions of each production and the current state of the cognitive system (which also reflects the state of the environment through its sensory systems). This function has been proposed to happen in the basal ganglia (BG) [25, 26], and this has been borne out through matches with human neuroimaging data [25]. While it is possible that the BG is solely responsible for action selection in well-practiced cases [27], we focus on the learning process and so on less well-practiced cases. In our neural network framework, we divide this functionality between cortical and BG areas, with the cortex (usually PFC) generating a set of possible cognitive actions that might be performed next (through associative pattern matching or "constraint satisfaction"), while the basal ganglia decides whether to perform each candidate action, based on its prior relationship to reward signals in similar circumstances.

In modeling this process, we draw upon previous work from our group in modeling the mechanisms and computations by which the PFC and BG learn to maintain useful information in working memory [28–32]. The prefrontal cortex basal ganglia working memory (PBWM) models developed by O'Reilly and colleagues integrate a wealth of electrophysiological, anatomical, and behavioral data, largely from animal work. Working memory also appears to be a large component of executive function, because in many cases a specific task is performed by virtue of maintaining an appropriate task set [33], in effect remembering what to do. Those maintained representations bias other brain processing through constraint satisfaction. Because it explains the deep question of how we learn our executive function (EF), this theory makes progress in dispelling the "homunculus" [30], by explaining how complex cognitive acts are performed by a collection of systems, each of which supplies a small part of the overall intelligence, decision making, and learning.

In essence, the PBWM framework extends the wealth of knowledge on the role of the basal ganglia in motor control to address working memory and executive function. This is possible because there are striking regularities across areas of frontal cortex, so that the anatomy of cortex and basal ganglia that subserves motor function is highly similar to prefrontal and anterior BG areas known to subserve WM and EF [34]. This anatomy is thought to help select potential motor actions by "gating" that information through thalamus back to cortex, amplifying it and so cleanly selecting one of the several possible candidate actions represented in the cortex (e.g., [35]). The core hypothesis of PBWM is that these same circuits help select which representations will be actively maintained in PFC by fostering local reverberant loops in the cortex, and between cortex and thalamus, and by triggering intrinsic maintenance currents that enable self-sustained persistent firing in cortical pyramidal neurons. The reinforcement learning mechanisms by which BG learns which actions are rewarding also apply to learning what to remember and so what to do.

The primary value and learned value (PVLV) model of dopamine release as change in reward prediction [36, 37] is also a key component of PBWM and is in turn based on electrophysiological and behavioral data from a collection of subcortical areas known to be involved (e.g., [38–41]). The known properties of dopamine release indicate that it serves as a reward prediction error signal [42] which

has informational properties that make it useful for driving learning [43, 44]. This system learns to signal when a new set of representations will likely lead to reward in a biologically realistic variant of the function of the better-known temporal difference (TD) algorithm when it is supplemented with "eligibility trace" information (e.g., [45]). This reward prediction function is crucial, because the difficulty in assessing the benefit of an action (whether it be cognitive or behavioral) is that the actual reward achieved by that action very often occurs later in time and so cannot be used directly as a learning signal [46, 47]. Instead, the system learns to perform actions that are predicted to gain reward. This reinforcement learning trains the striatum and works alongside the more powerful associative and error-driven learning within the PFC portion of PBWM that learns the representations (and therefore the associative semantics) of candidate actions to take.

In the remainder of the paper, we present an overview of four models that elaborate on this process in several ways. The first addresses how the learning mechanisms described previously and elaborated upon in works by various workers in our group [36, 37, 48, 49] can learn short sequences of cognitive actions, when they are sequentially dependent and so must be performed in the right order to achieve reward. The second describes how the hippocampus can achieve instructed learning, participating in the constraint satisfaction process of deciding which action to consider performing, as when we perform a novel task based on memorized instructions. The third model considers how slow, cortical associative learning can contribute to that same "which" process by using constraints of the current state and the goal to arrive at a subgoal that can serve as a viable next step in the sequence. Finally, we close with some discussion of the state of this research and the many remaining questions.

3. Model-Free Reinforcement Learning

Model-free reinforcement learning (RL) can be defined at a high level as learning which actions (which we take to include cognitive "actions") produce reward, without any other knowledge about the world [43]. While the learning mechanisms we describe here are purely trial and error, the same learning mechanisms apply to model-driven or "hypothesis-driven" learning as well. For instance, the same learning principles apply when using actions, explicit plans from memorized instructions, or semantic associations as outlined in the final two models we describe later.

In hopes of better understanding how this process could occur in neural tissue, we have leveraged the prefrontal cortex basal ganglia working memory framework, or PBWM [28–32]. Under this account, a basic actor-critic architecture [43, 50] naturally arises between the prefrontal cortex (PFC), the basal ganglia (BG), and the midbrain dopamine system as modeled by our PVLV system described previously. PVLV serves as the critic, evaluating the state of the network and providing dopamine bursts or dips for better than and worse than expected outcomes, respectively. The BG system is naturally situated to perform the functions of the actor based on its known role in selecting motor actions (and by

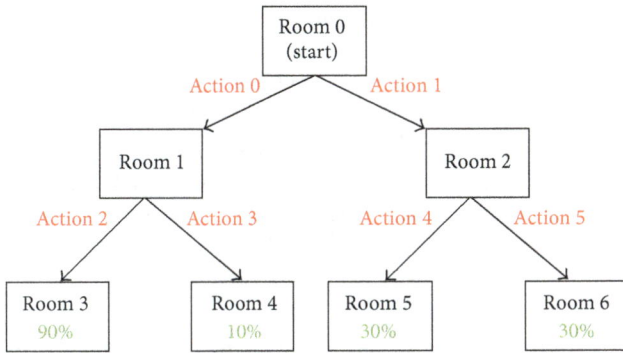

FIGURE 1: Simple state-based room navigation task. The percentages of the last level of rooms at the bottom of the figure represent the probability that the agent will get rewarded if it chooses the path that leads to the respective rooms.

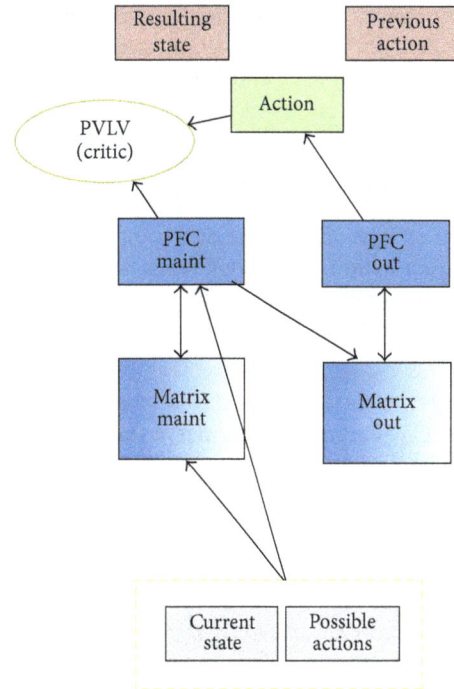

FIGURE 2: Model-free network architecture. Based on both current state and possible actions, the "matrix maint" determines what to maintain in PFC. Based on the stored information in PFC, "matrix out" determines the next chosen action via PFC out. PVLV (consisting of multiple biological systems) evaluates the actions (critic) and helps train the model. See text for in-depth description and functions of the various components of the network. Detailed network architecture is highly similar to the PBDM model discussed later.

hypothesis, selecting cognitive actions with analogous neural mechanisms in more anterior regions of PFC). Using the critic's input, the BG learns from experience a policy of updating segregated portions of the PFC as task contingencies change. The PFC is able to maintain past context and provides a temporally extended biasing influence on the other parts of the system. It is helpful to view this entire process as a "gating" procedure: the BG gating controls that are being actively maintained within the PFC, and therefore subsequently biasing (controlling processing in) other cortical areas. When the gate is closed, however, the contents of the PFC are robustly maintained and relatively protected from competing inputs. Importantly, as task contingencies change and the actor determines that a change is needed, the gate can be opened allowing new, potentially more task appropriate, content into the PFC.

The simple RL-based learning of the PBWM framework allows us to easily and naturally investigate one manner in which the brain may be capable of utilizing model-free RL in order to solve a simple task. In short, the network must learn to maintain the specific actions taken and evaluate this sequence based on either the success or failure of a simulated agent to attain reward. The simple example task we use is a basic state-based navigation task (abstracted at the level of "rooms" as states) in which a simulated agent must navigate a state space with probabilistic rewards as inspired by the work of Fu and Anderson [51] (see Figure 1). The goal of the task is simply to learn an action policy that leads to the highest amount of reward. To achieve this, the agent must make a choice in each room/state it visits to move either to the next room to the left or the next room to right but always moving forward. The only rooms that contain reward are at the final level (as in most tasks). The structure of the reward is probabilistic, so a single room is the most consistent provider of reward (Room 3 in Figure 1), but the others have a lower chance to be rewarding as well. In order for the PBWM framework to ultimately succeed, it must be able to maintain a short history of the actions it took and reward or punish these action choices in the final presence or absence of reward. This is a simple task, but a learning in this way is a valuable tool

when the system must learn basic actions first in order to succeed at more extensive cognitive sequencing tasks.

3.1. Description of the Model. The model-free RL network is depicted in Figure 2. The ultimate goal of the network is to receive reward by determining the best action to take given the reward structure of the simulated environment. There are many models of reinforcement learning in similar domains, and the PBWM and PBDM models have been applied to learning in superficially similar domains. However, some very important differences make the setup of this model unique. Most importantly, the final outcome (what Room the network ends up in based on the action chosen) of the network is not determined in the standard neural network manner of having activation propagate through units and having a competition that determines the winner. Instead, the network chooses an action via the *action* layer, which is the only traditional output layer in the network. The possible actions can be thought of as any atomic action that may result in a change of state, such as "go left" or "go right." After the network chooses an action, a state transition table is used to determine the outcome of the action. More specifically, the network makes a decision about what action to take, and program code determines what the effect is on the environment of the simulated agent. The outcome is reported back to the network via the *resulting state* layer, but for display purposes only (not used in any computation). The example trial stepped through later in this section will help to clarify this process.

3.1.1. Network Layer Descriptions

(i) Action layer: this is the output of the network. The chosen action is used via a state transition table to choose a new room. In the current simulation, the room choice is completely deterministic based on the action.

(ii) CurrentState layer: this is a standard input layer. The CurrentState is the current state (room) that the model is occupying.

(iii) PossibleActions layer: this is the second input layer. The layer is used to specify what "legal" actions are based on the current state that the network is occupying. Importantly, PossibleActions provides the main signal to the simulated basal ganglia to determine the gating policy, as well as the main input to the PFC. This ensures that only legal actions should be chosen (gated) at any given time.

(iv) PreviousAction layer (display only): this is a display only layer. It maintains the last action choice that the network made. This can be useful to understand how the network arrived to its current state.

(v) ResultingState layer (display only): this is a display only layer. The ResultingState is the "room" that the simulated agent will arrive in based on the action that the network produced. The final room is used to determine if the agent should receive reward.

(vi) PVLV layers: the PVLV layer(s) represents various brain systems believed to be involved in the evaluative computations of the critic [36].

(vii) PFC maint and PFC out: simulated prefrontal cortex layers, the maint PFC is used to actively maintain information overextended delay period. The PFC out layer models the process of releasing this information, allowing it to affect downstream cortical areas and drive actual responses.

(viii) Matrix maint and matrix out: these layers are used to model the basal ganglia system and represent the actor portion of the network. They learn to gate portions of the PFC, through experience, using the information provided from the PVLV system.

3.1.2. Task Example

(1) The current state (room) is presented to the network via the CurrentState layer. The inputs correspond to different rooms as shown in Figure 1, where Room 0 corresponds to the first unit in CurrentState layer, Room 1 to the second, Room 2 to the third, and so forth.

(2) Using the CurrentState and the actions maintained within the PFC, the network must decide to go to the room to the left or the room to the right. This decision is reflected by activation in the action layer.

(3) The action that is chosen by the network is used to determine where the simulated agent is in the current state space, and this is accomplished using a standard transition table to look up the next room. The actions are deterministic and move the agent directly to the room based only on the action.

(4) The resulting state of the agent is returned to the network via activation in the CurrentState layer indicating the result of the action. Return to Step 2 unless the agent reaches a terminal room.

(5) If the room reached by the agent is a final room, the reward probabilities for that room are used to determine the likelihood of reward to the agent.

(6) Repeat from Step 1 until task is reliably learned.

3.2. Results. The network is capable of quickly learning the optimal policy of action sequences that optimize its reward on this task. To assess the ability of the network to solve this task, we set up a testing structure which allowed the network 75 "chances" to solve the task per epoch (block). At the end of the epoch, the average rate of reward was recorded for the simulated agent. This was repeated until either the agent received an average reward greater than 85% of the time or for 25 epochs (blocks), whichever came first. Ten simulated agents were ran, and 8 out of the 10 reached criteria of 85% average reward within 25 epochs. On average, it took 4 epochs to achieve this feat. While this may not appear to be a surprising result, the complex nature of the biologically realistic network made this far from a forgone conclusion. Indeed, many insights were gained about the nature of how the actor must balance its exploring of the state space with gaining reward. If the network randomly gets reward in one of the low-reward states, it must still be willing to explore its environment in order to confirm this finding. Conversely, if the network is in the high-reward state and does not receive reward, the (relative) punishment for this nonreward needs to allow a possible return to this same state at some point in the future in order to discover the optimal action policy. The limits of the framework are apparent in the 2 networks that did not reach criteria. In both of these cases, the agent randomly reached the low probability area of state space. In most cases, the agent is able to successfully explore other options and thus find the more rewarding rooms. However, the current PBWM framework will occasionally fail if reward is not present early enough in exploration process. We are investigating biologically inspired mechanisms to bootstrap the learning in more efficient ways. Encouraged by our initial framework, we are actively investigating how a simple model-free approach to learning basic sequences could be utilized by the human brain in order to scaffold up to more complex and interesting sequences. We are hopeful that concentrating on the relevant biological data and learning will provide us with useful insights to help us better understand how people

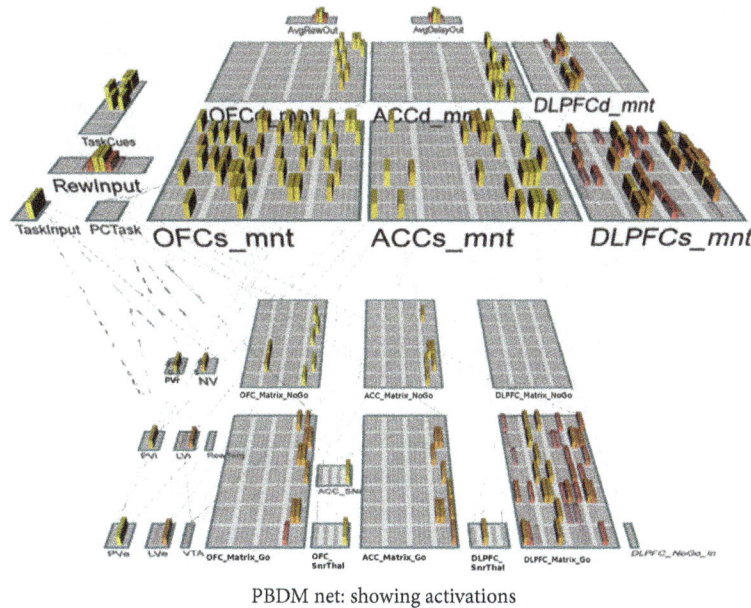

PBDM net: showing activations

FIGURE 3: PBDM decision-making model. This figure shows the PBDM network and the components it models. The bars show the activation strengths of each of the units in the model for a particular point in time.

are capable of such effortless sequencing of extended, diverse, and complex action plans.

We hypothesize that this type of learning aids in cognitive sequencing by allowing humans to discover useful simple sequences of cognitive actions purely by trial and error. While this learning does not likely account for the more impressive feats of human cognition, since these seem to require substantial semantic models of the relevant domain and/or explicit instruction in useful sequences, we feel that understanding what the brain can accomplish without these aids is necessary to understanding how the many relevant mechanisms work together to accomplish useful strategic cognitive sequencing.

4. Prefrontal Cortex Basal Ganglia Decision-Making (PBDM) Model

In the PBDM model, we primarily address decision making at the level of task strategies (task set representations in PFC, primarily dorsolateral PFC (DLPFC)). Decision making is important in many areas, but the selection of strategies for complex tasks is our focus. We believe that the same mechanisms apply to making decisions in many different domains.

The main idea behind PBDM is to computationally model the interactions between basal ganglia and medial prefrontal areas that represent particularly relevant information for making action plan or strategy decisions. Anterior cingulate cortex (ACC) and orbitofrontal cortex (OFC) serve as activation-based monitors of task affective value parameters [52, 53], including action effort in the ACC [54], and probability of reward in the OFC. These then project to the basal ganglia that controls updating in the DLPFC, giving it the necessary information to select choices in favor of lower effort and higher reward strategies. Because the ACC

and OFC are themselves PFC areas with inputs from the same type of basal ganglia/thalamic circuits as motor and working memory areas, they are hypothesized to be able to rapidly update and maintain their value representations and, with a single gating action, change the evaluation to reflect new important information. This confers great flexibility and rapid adaptability to rapidly changing circumstances. Within this framework, several questions remain: what, more precisely, do the ACC and OFC represent? How can these representations drive appropriate gating behavior in the DLPFC BG? How are appropriate representations engaged in novel task contexts?

In the initial version of the PBDM model, described in more detail later and shown in Figure 3, we adopt simple provisional answers to these questions while recognizing that these likely underestimate the complexity of what happens in the real system. In particular, while ACC is often (and in our model) assumed to represent effort, its true role is more complex. The current state of knowledge on these issues is reviewed thoroughly by Kennerley and Walton [55]. The ACC and OFC in our model compute running time-averaged estimates of effort and reward probability, respectively, based on phasic inputs on each trial. If a task is ongoing, the ACC just increases its running average of time-effort by one. When a reward value is received or not (when otherwise expected), the OFC increments its running average estimate of reward probability. We have four different stripes within the ACC and OFC, each of which receives input from and so has a representation determined by one of the task strategies represented in the parietal cortex. These are thought of as very general strategies for dealing with spatial information, and over a lifetime of experience, we build up reasonable estimates of how effortful and rewarding they are on average in similar tasks. In order to in part capture the importance of context, there is also a randomly updated set of task features, which represent specific details about each different task that the model learns to perform. Over time, the model learns to pay attention to the ACC/OFC value representations in selecting a task strategy and pay less attention to these idiosyncratic task cues. Having done so, the model can then generalize to novel task contexts, by paying attention to the underlying spatial task values and ignoring the novel task features. Then, as the novel task progresses, actual experienced reward and effort drive the ACC and OFC representations, providing a more accurate picture for decision making going forward. This is the overall model we think applies to subjects as they engage in novel tasks with multiple possible strategies.

We conceptualize this PBDM process as engaging when people are actively and explicitly considering a new strategy or similar decision. We model an abstract spatial task, in which the strategies consist of individual spatial properties of groups of similar items. Strategies consist of considering one or a combination of these properties. There are 4 different strategies considered (listed by increasing order of both effort and reward probability; the precise values vary by task): Distance Only, Distance + BaseRate, Distance + Radius, and Distance + BaseRate + Radius. These are merely example strategies associated with a hypothetical spatial estimation task and are therefore sometimes also simply referred to as strategies 0 to 3, respectively; the task is not implemented for this model outside of entirely hypothetical probabilities of success (reward) and level of effort (time to implement). The weights for the PBDM component are trained to model a long history of experience with these hypothetical reward and effort values. After this learning (and purely through it), the OFC reward representations primarily bias the Go pathway, while the ACC effort representations bias the NoGo pathway. It is this balance between Go and NoGo that then ultimately determines the strategy selected. In our models, we observe that different random initial weights produce different individual preferences along this tradeoff.

The network performs various tasks (which switch every 10 trials during pretraining, simulating the intermixed variety of spatial tasks a person encounters during their daily life). The probability of reward and the number of trials required are determined by the selected strategy, the task representation that the DLPFC maintains. In reality, the possible strategies and therefore the representational space would be much larger, but we have narrowed it down to just 4 different states in a localist representation, (called Distance, Dist + Base Rate, Dist + Radius, and Dist + BaseRate + Radius; the original relation of these strategies to a particular task is irrelevant since the base task was abstracted to only the strategy component for this model). The inner loop per trial consists of "performing" the task in question, which happens through task-specific areas responding to the DLPFC task representation. We model that process here only at the most abstract level: each strategy takes an amount of time and has a probability of success that varies for each possible task. Thus, the PBDM network only experiences the overall feedback parameters: number of trials and probability of reward at the end of those trials. We do not model the process of carrying out these strategies; each of the models here could also be

applied to understanding how a particular strategy unfolds into an appropriate sequence of cognitive actions.

The overall behavior is thus as follows: select a DLPFC task representation, run a number of blank trials (blank since we assume that the lower-level processes that carry out the strategy have little influence on this level of cortical machinery) according to the "effort" parameter (representing task performance), then receive reward with given probability determined by the PCTask representation that the DLPFC task representation drives, and then repeat. Over time, the BG gating units for the DLPFC are shaped by the effort/delay and reward parameters, to select DLPFC stripes, and associated reps that are associated with greater success and shorter delays.

The BG "Matrix" layer units control gating in DLPFC and so, ultimately, make final decisions on strategy choice. They receive inputs from the ACC and OFC, which learn over time to encode, using dynamic activation-based updating, running time averages of reward and effort, associated with the different strategies on the different tasks. Because we assume that mental effort is equal per unit time across strategies, the effort integration is identical to time integration in this case. Critically, because this is done in activation space, these can update immediately to reflect the current PCTask context. Over time, the BG learns weights that associate each OFC and ACC unit with its corresponding probability of success or effort. Thus, an immediate activation-based update of the ACC and OFC layers will immediately control gating selection of the DLPFC layers, so that the system can quickly change its decision making in response to changing task contexts [52, 56, 57].

Thus, the early part of the network training represents a developmental time period when the ACC and OFC are learning to perform their time-averaging functions, and the DLPFC BG is learning what their units/representations correspond to in terms of actual probability of reward and effort experienced. Then, in the later part, as the DLPFC task representations continue to be challenged with new task cue inputs (different specific versions of this task space), the learned ACC/OFC projections into DLPFC BG enable it to select a good task strategy representation on the first try.

4.1. Details of Network Layer Functions

(i) TaskInput: generalized task control information about the inner loop task being performed projects to DLPFC. We assume that this information comes from abstract semantic representations of the task at hand; this is likely represented in a variety of posterior and prefrontal regions, depending on the type of task. Use the following units/localist representations:

   (a) PERF—performing current task-signals that DLPFC should not update the task representation (see DLPFC NoGo In later); this repeats for the number of trials a given PCTask strategy requires and metes out the delay/effort associated with a given strategy.

(b) DONE—done performing current task-reward feedback will be received in RewInput to OFC and PVe (PVLV); note that there is a "cortical" distributed scalar value representation of reward (RewInput), in addition to the subcortical one that goes directly into the reward learning system (PVe); conceptually these are the same representation, but their implementation differs.

(c) CHOICE—DLPFC should choose a new task representation, based on influences from TaskCues, ACC, and OFC states; the newly gated DLPFC representation will then drive a new PCTask representation, which will then determine how many PERF trials are required and the probability of reward for the next DONE state.

(ii) TaskCues: these are random bit patterns determined by the cur_task_no state, which drives DLPFC (both cortex and BG); they represent all the sensory, contextual, and instructional cues associated with a given specific task.

(iii) PCTask reflects the actual task parameters. In this example, these are Distance, Dist + BaseRate, Dist + Radius, and Dist + BaseRate + Radius, but more generally this would represent a much larger space of task representations that have associated reward and effort parameters for different tasks. This may also reflect a combination of posterior cortical and also more posterior DLPFC representations that provide topdown biasing to these PC task representations and maintain them over shorter durations. The DLPFC in the model is the more anterior "outer loop" DLPFC that maintains higher-level, longer-duration task representations that are "unfolded" into useful sequences by other processes, including but likely not limited to those we address in the models here.

(iv) RewInput: scalar val of reward input level activated during the DONE trial; this also has a −1 state that is activated whenever the network is in PERF task mode, and this is what triggers the incrementing of delay/effort in the ACC layer (i.e., both OFC and ACC feed off of this same basic RewInput layer, pulling out different information). This is overall redundant with signals in PVLV but packages them in a simple way for OFC/ACC to access and for us to manipulate for various experiments.

(v) OFC computes running time average of reward probability/magnitude; only updated when reward occurs (DONE trials), otherwise maintains the current estimate for PERF and CHOICE trials. The network learns coarse-coded distributed representation of this value, not in a scalar value format, through a "decoding" layer (AvgRewOut) that is in scalar value format. But it is the distributed representation that projects to DLPFC to bias its processing. It is not exactly clear what AvgRewOut corresponds to biologically, but the general idea is that there are autonomic level states

in the brainstem, and so forth, that compute low-level time averages based on physiological variables (e.g., longer time average sucrose concentration in the blood), and that is what drives the OFC to learn to compute activation-based running time averages. See (vii) for the way this representation learns to affect DLPFC gating.

(vi) ACC computes running time-average interval between reward trials which constitutes total effort on each task, since we assume roughly equal effort per time. It is updated on each performance trial and maintained during the DONE and CHOICE trials; each time step increases activation. As with OFC, this layer learns coarse-coded distributed representation of this value, not in a scalar value format, through a "decoding" layer (AvgDelayOut), which again reflects longer time-average metabolic cost variables.

(vii) DLPFC encodes current task strategy and learns representations entirely through reinforcement learning stabilization. It receives information about each task from TaskCues; the Matrix layer also receives from ACC and OFC and learns over time to select task representations associated with good values of ACC and OFC (i.e., values of those that have been associated with rewards in the past). DLPFC also projects to PCTask, which in turn projects to ACC and OFC and "conditionalizes" (makes appropriate to the particular task) their representations.

(viii) DLPFC_NoGo_In is our one "hack." It turns on NoGo (strongly) whenever a task is being performed to ensure that the matrix does not update DLPFC midtask. This hard-coded behavior is simply the assumption that the DLPFC task set representation remains active during task performance; that is, people maintain one task set without switching strategies midway through more general learning: when you decide on a strategy, stick with it until you are done (or until it gets "frustrating" by consuming too much time).

4.2. Results

4.2.1. Reward-Only Optimization: OFC Proof of Concept Test. The first proof of concept test sets the probability of reward to .2, .4, .6, and .8 for PCTask units 0–3, respectively (labeled "Distance only," "+BaseRate," "+radius," and "Combined," resp.), with delay set to a constant 1 trial (.2 parameter × 5 trials max delay) for all options. Thus, the best strategy is to select strategy 3, based on OFC inputs. As shown in Figure 4, the network does this through a period of exploration followed by "exploitation" of strategy 3, which is selected automatically and optimally immediately, despite changing TaskCues inputs. All of the batches (10/10) exhibited this same qualitative behavior, with a few stabilizing on strategy 2 instead of 3. This was the second-best strategy, and the fact that the model stabilized on this in some cases shows the stochastic process of sampling success that likely contributes to the selection of nonoptimal strategies in

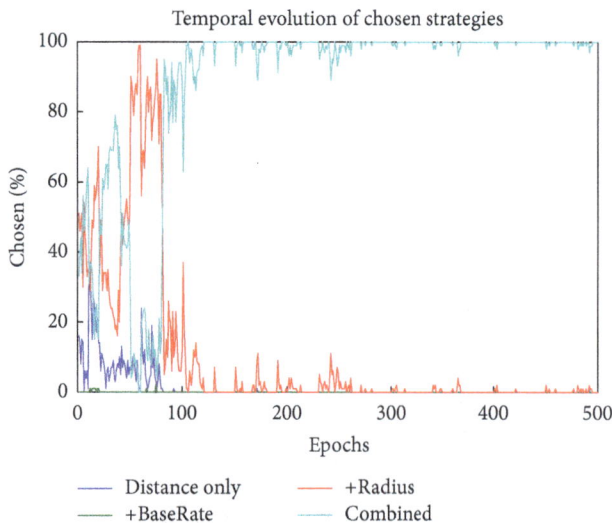

FIGURE 4: Developmental learning trajectory of PCTask selection. Early in learning it explores the different strategies, and later it learns to predominantly select the one (green line, strategy 3 ("Combined")) that produces the best results.

some real-life cases (since after the model stabilizes, it will not learn about potentially better strategies without some sort of external perturbation to force resampling). None stabilized on 0 or 1, since they have substantially lower reward probabilities. As shown in Figure 5, the weights into the Matrix Go stripe that gates DLPFC learned to encode the high-value OFC representations associated with the strategy 3 OFC representation.

4.2.2. Delay-Only Optimization: ACC Proof of Concept Test. Next, we set probability to .6 for all strategies and set the delay factors to 1, 2, 3, and 4 trials of delay, respectively, for strategies 0–3. Without any PVLV feedback at all during the PERF trials, the network does appear to be sensitive to this delay factor, with strategy 0 (1 trial delay) being chosen preferentially. However, this preference is somewhat weak, and to produce stronger, more reliable preferences, we added a direct dopaminergic cost signal associated with delay, as has been shown empirically [58]. This modulation decreased the size of a DA reward burst in proportion to effort/delay (with a small weighting term). In our proof of concept test, this small modulation produced 50% of networks preferring the first (least delay) strategy.

4.2.3. Balanced Reward and Delay (Actual Use Case). To simulate a plausible situation where there is a tradeoff between effort and reward, we set the reward factors to .4, .6, .6, and .8 and the delay factors to .2, .4, .6, and .8. This resulted in a mix of different strategies emerging over training across different random initial weights ("batches") (proportions shown in Figure 6), with some preferring the low-effort, low-reward distance only option, while others going for the full Distance + BaseRate + Radius high-effort, high-reward case, and others falling in between. The particular results are highly

stochastic and a product of our particular choices of reward and effort values; it is easy to push these preferences around by using different weightings of effort versus time.

4.3. Discussion. The PBDM model shows how rapid updating in prefrontal cortex (as captured in the PBWM models and related work on persistent firing in PFC) can aid in decision making by allowing the system to use contextually appropriate representations of predicted reward and effort to drive decisions on task strategy. If the context (e.g., physical environment and task instructions) remains the same, then new learning in the ACC and OFC slowly updates the values of predicted reward and effort through weight-based learning. If, however, the context changes, representations in ACC and OFC will be "gated out," so that a new set of neurons learns about the new context. Detailed predictions about the old context are thus preserved in the synaptic weights to that now silent units (because the learning rule we use, and most others, does not adjust weights to inactive neurons/units).

One way in which this preservation of contextually dependent ACC and OFC representations could be extremely useful is in interaction with episodic memory in the HC. We believe that predictive representations could also be retrieved to ACC and OFC from episodic memory in the hippocampus, a form of PFC-HC interaction similar to but importantly different from that we model in the "instructed learning" model.

This model primarily addresses the "strategic" component of strategic cognitive sequencing, but this type of effortful decision making, bringing the whole predictive power of cortex online to estimate payoff and cost of one possible sequence component, could help bootstrap learning through the mechanisms in either or both of the instructed learning and "model-free" models.

5. Instructed Learning

One source of complex, strategic cognitive sequences is learning them directly from instruction [59–61]. Humans have the remarkable ability to learn from the wisdom of others. We can take advice or follow instruction to perform a particular cognitive sequence. One such example may be observed daily by cognitive scientists who conduct human laboratory experiments. Most normal participants can well implement instructions of an arbitrary novel task with little or no practice. However, in the cognitive neuroscience of learning, reinforcement learning has been the central research topic and instructed learning appears to have been relatively understudied to date. In this section, we contrast reinforcement and instructed learning and outline the dynamics of instruction following in a biologically realistic neural model.

Reinforcement learning adapts behavior based on the consequences of actions, whereas instructed learning adapts behavior in accordance with instructed action rules. As a result, unlike the slow, retrospective process of trial and error in reinforcement learning, instructed learning tends to be fast, proactive, and error-free. In the brain, the neurotransmitter dopamine signals reward prediction errors for the basal ganglia to carry out reinforcement learning of

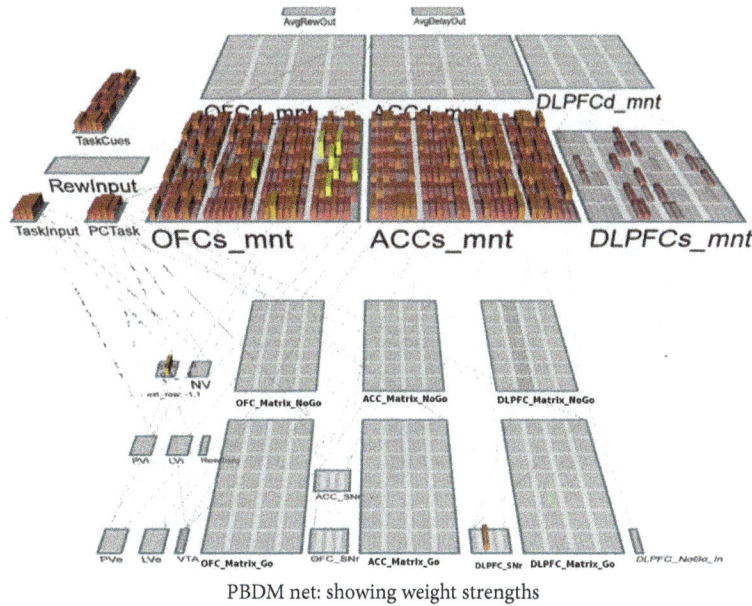

FIGURE 5: PBDM decision-making model. This figure shows the weights from the respective units to a unit in the DLPFC_Matrix_Go layer (green, lower right). It depicts the strength of weights towards the end of learning, at which point there are particularly strong connections from the core OFC distributed representations, which represent strategy's predicted reward value, established through learning.

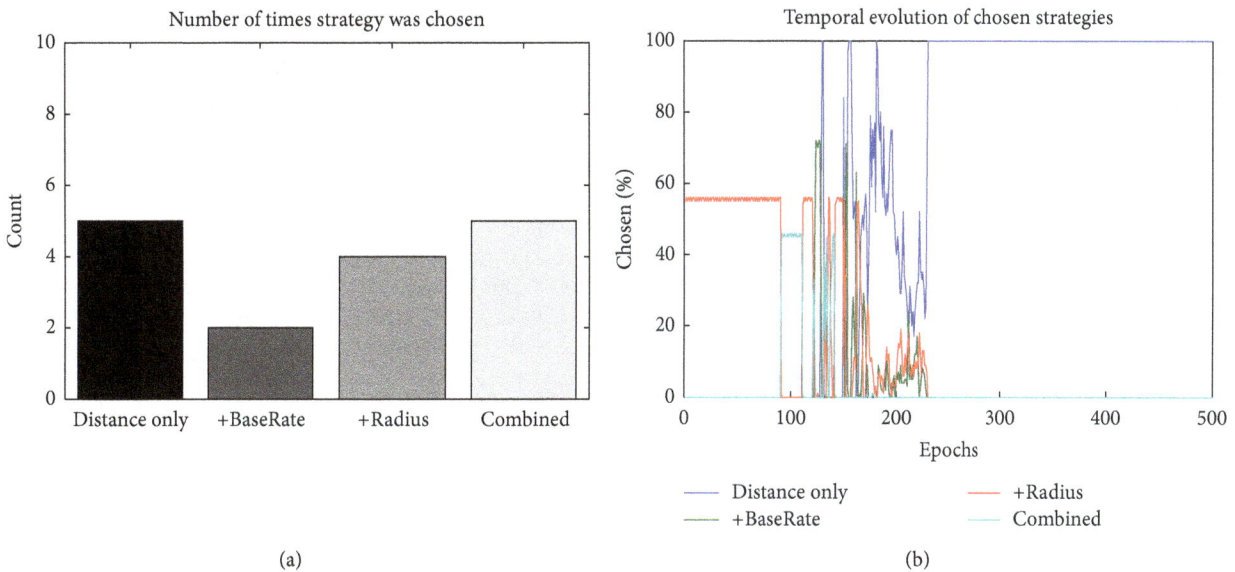

(a)

(b)

FIGURE 6: Balanced reward and delay. The left graph shows the number of times a strategy was chosen over 16 repeats with random initial weights, while the graph on the right shows the temporal evolution of selection for one randomly chosen network. The variability in the equilibrium strategy choice stems from the balance between reward and delay (the higher the reward, the higher the delay) making each strategy approximately equally rational to choose. As discussed in the reward-only case previously, the particular, random history of reward plays a large role in determining the ultimate strategy choice.

reward-linked actions (for a discussion, see [62]). As for instructed learning, the human posterior hippocampus underlies verbal encoding into episodic memory [63] and use of conceptual knowledge in a perceptually novel setting [64].

Compared to reinforcement learning, instructed learning appears effortless. Why is learning so arduous in one mode but effortless in another? How exactly do we perform complex novel tasks on the first attempt? We propose that

instruction offers nothing but a new plan of recombining old tricks that have been acquired through other forms of learning. In other words, instructions quickly assemble rather than slowly modify preexisting elements of perceptual and motor knowledge. For example, we can immediately follow the instruction: "press the left button when seeing a triangle; press the right button when seeing a square," in which the action of button press is a preexisting motor

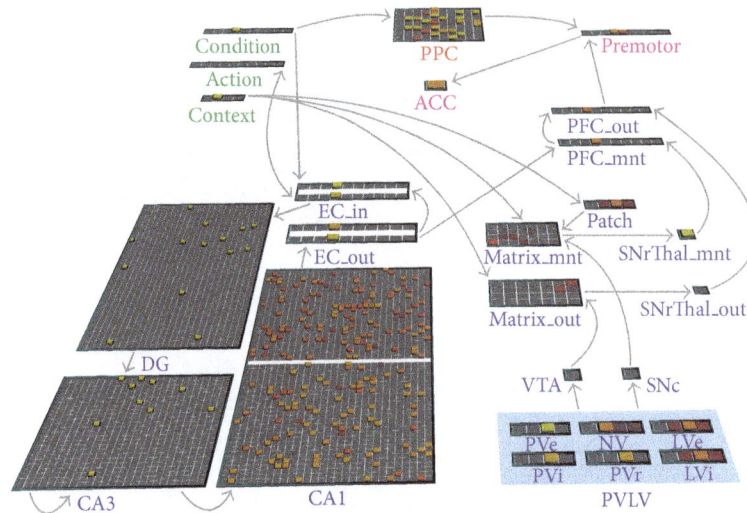

FIGURE 7: The instructed learning model. The model consists of two interactive learning pathways. The hippocampal-prefrontal pathway (i.e., lower part in the diagram) processes newly instructed conditional-action rules, whereas the parietal pathway (i.e., upper part in the diagram) processes habitual actions. The actions suggested by each of these pathways are then gated by the PFC portion.

skill, and visual recognition and categorization of shapes are also an already learned perceptual ability. Note also that understanding the instruction requires a previously learned mapping from language (e.g., the verbal command of "press") to actual behavior (e.g., the motor execution of "press").

To further study how instruction following is carried out from neural to behavioral levels, we constructed a model of instructed learning based upon known neuroanatomical and neurophysiological properties of the hippocampus and the prefrontal-basal ganglia circuits (Figure 7). Specifically, the model basal ganglia (BG) carries out reinforcement learning of motor execution (abstracted in the model to premotor); the model hippocampus rapidly encodes instructions as action episodes that can be contextually retrieved into the prefrontal cortex (PFC) as a goal for guiding subsequent behavior. Unlike a single-purpose neural network that slowly rewires the whole system to learn a new sensorimotor transformation, this general purpose instructable model separates motor from plan representations and restricts plan updating to lie within the fast-learning hippocampus, which is known to rapidly bind information into episodic memories.

As a concrete example, the proposed model is instructed with 10 novel pairs of if-then rules (e.g., if you see A, then do B) and evaluated for its success in performing conditional actions (e.g., do B) when encountering a specific condition (e.g., seeing A). In the model, each of the "Condition," "Action," and "Premotor" layers consists of 10 localist representations of conditions, verbal actions, and (pre-)motor outputs, respectively. The model is pretrained with action-to-motor mappings (i.e., from verbal commands to premotor responses) during the Pretraining stage and then trained with condition-to-action mappings (i.e., if-then rules) during the Instruction stage. Finally, during the Performance stage, it is tested with Condition-to-Motor mappings without any inputs from the "Action" layer. The simulation results are shown in Figure 8. The model quickly learns an if-then rule

in just few trials during the Instruction stage, and without further practice, it makes no error in carrying out these instructions for response during the Performance stage, just as human subjects often do after being presented with clear instructions and a short practice period.

Inside the model, learning occurs in multiple parts of the architecture. During the Pretraining stage, the hippocampus learns to perform identity mapping for relaying information from the "Action" layer to the corresponding motor representations in the PFC layers. Meanwhile, BG learns to open the execution gate for PFC to output a motor decision to the "Premotor" layer. During the Instruction stage, the hippocampus associates inputs from the "Condition" and "Action" layers and learns each condition-action pair as a pattern. During the Performance stage, all the model components work together using mechanisms of pattern completion, and the hippocampus recalls instructions about what action to do based on retrieval cues from the "Condition" layer, and its downstream PFC either maintains a retrieved premotor command in working memory when BG closes the execution gate or further triggers a motor response in the "Premotor" layer when BG opens the execution gate.

Compared to earlier work on instructable networks [65], our model further explicates how different parts of the brain system coordinate to rapidly learn and implement instructions. Albeit simple, our instructed learning mechanisms can support strategic cognitive sequencing in that a cognitive sequence can be constructed from an ordered set of instructed or self-instructed operations. Beside sequential behavior, the model is being extended to also explain the interactions between instructions and experience (e.g., [66–69]) in the context of confirmation bias and hypothesis testing. The modeled ability of the hippocampus to memorize specific contingencies in one shot undoubtedly contributes an important piece of our ability to learn complex goal-oriented sequences of cognitive actions. Beyond simply memorizing

Pretraining stage: learn to map verbal actions to motor responses

Sensory inputs: condition-**action** → Hippocampus: condition-**action** → BG/PFC: motor → Motor outputs: motor

Instruction stage: learn condition-action associations

Sensory inputs: condition-action → Hippocampus: condition-action → BG/PFC: motor → Motor outputs: motor

Performance stage: respond to conditions with instructed actions

Sensory inputs: condition-action → Hippocampus: condition-action → BG/PFC: motor → Motor outputs: motor

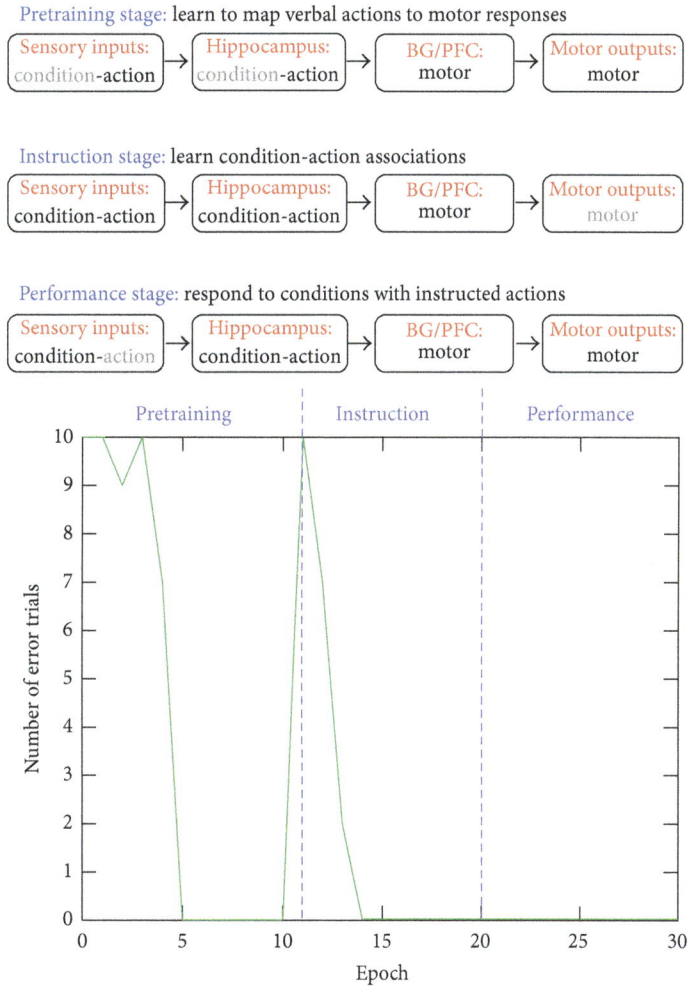

FIGURE 8: Instructed learning stages and simulation results. In the upper panel, black and grey texts denote present and absent representations, respectively. In the bottom panel, each epoch consists of 10 trials. Note that no error is produced during the Performance stage, since the prememorized mappings can be recalled perfectly after four repetitions during the Instruction period.

instructions given by others, it can also aid in "self-instructed" learning by remembering successful steps learned by trial and error or other means for assembly into new sequences.

6. Planning through Associative Discovery of Bridging States

We explore the idea that the active maintenance of long-term goals in the PFC can work in conjunction with a network's semantic knowledge to identify relevant subgoals and then use those individual subgoals in a similar manner to bias action selection in the present. One fundamental question motivates this research. Given some ultimate goal, possibly associated with explicit reward, how does the system identify subgoals that lead to the final goal? Our hypothesis revolves around semantics, that is, knowledge about how the world works. Our model uses this knowledge to perform constraint satisfaction by using active representations of the current state (where I am) and the desired goal (where I want to be) to associatively arrive at a representation of a subgoal that

"bridges" between the two states. This subgoal can serve as the focus for a strategy or plan to achieve the larger goal.

6.1. *Description of the Model.* There is a tension that exists between the temporal sequencing over one or more subgoals versus a multiple constraint-satisfaction approach that does things all in one step. It seems clear that both can be involved and can be important. So, when does the brain do one versus the other? We have adopted the following heuristic as a kind of corollary of Occam's razor. In general, the brain will by default try to do things in a single time step if it can; as an initial hypothesis, we suspect that bridging over a single subgoal is probably about as much as can be done in this way. When no such plan exists, a more complex process of navigating the modeled task-space through stepwise simulations of intervening states can be undertaken; because this process is among the most complex that humans undertake, a model that does this in a biologically realistic way is a goal for future research. Thus, our initial objective here is to try to demonstrate a one-step constraint satisfaction solution to

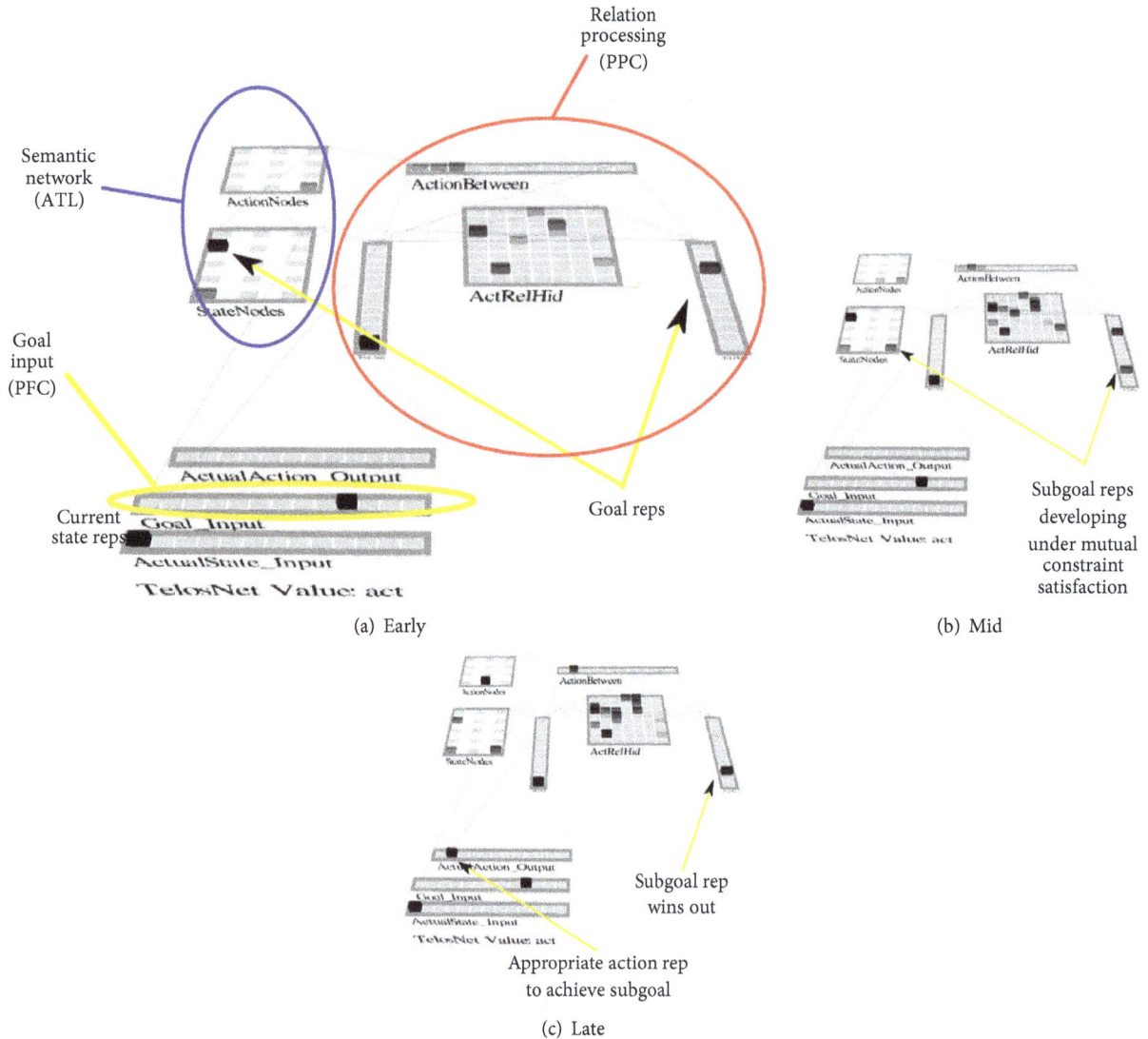

Semantic
network
(ATL)

Relation
processing
(PPC)

Goal
input
(PFC)

Goal reps

Current
state reps

Subgoal reps
developing
under mutual
constraint
satisfaction

(a) Early

(b) Mid

Subgoal rep
wins out

Appropriate action rep
to achieve subgoal

(c) Late

FIGURE 9: Subgoaling through constraint satisfaction. This figure shows settling of activations of the current state and goal state in both the *Semantic Network* (see text) and *Relation Area* (see text). (a) shows activations early in the settling process of a trial. (b) Activations midway into settling for a trial. The activation of two units in the rightmost Goal layer shows the constraint satisfaction process selecting two plausible subgoals. (c) Activations late in settling when they have converged. The network has settled onto the single most relevant subgoal through constraint satisfaction (simultaneous association from the current state and maintained goal state).

a simple three-state problem: current state and end state to subgoal ("bridging") state.

Another major issue is the tension that exists between state representations sometimes having to compete with one another (e.g., "What is the current state?,") versus sometimes needing to coexist as in spreading activation so as to represent a full motor plan or model of state space (e.g., representing all three of the states in the previous three-state problem). The solution we have settled on is a division of labor between a relation processing area, possibly in the posterior parietal cortex (PPC, circled in red in Figure 9), and a semantic association area, possibly in the anterior temporal lobe (ATL, circled in blue). Because many brain areas are involved in semantics, the precise areas can be expected to vary with the semantic domain, but the mechanisms we describe are expected to be general across those variances. Figure 9 later

illustrates these two distinct areas. The PFC (not explicitly modeled) is envisioned to represent the goal state and thus to bias processing in these two areas. The relation processing area is based on the ideas described in "Semantic Cognition" by Rogers and McClelland [70].

Training: the network is trained on the semantics of the State-Action-State triad relation (parietal/anterior temporal cortex) but includes connections to the semantic part of the network. The idea is that the relation area will learn the specific role relations between the states (before, after) and the actions (between states), while the semantic area will learn simple associations between the nodes. The former is dominated by a tight inhibitory competition, while the latter is more free to experience spreading activation. In this way, pre-training on all the individual S-A-S relations enables the bridging over an intermediate subgoal state and biases the correct action in the current state, under the biasing influence of the goal state.

As illustrated in Figure 9(a), which shows a network trained only on pairwise transitions between adjacent states, when a current state and a remote goal state are input, both are activated in both the semantic network and relation engine early in settling. At this very early stage of settling, there are three action units active in the ActionBetween layer (Relation Engine), which are all of the possible actions that can be taken in the current state (S0). Later in settling (Figure 9(b)), a third state unit comes on, which is the intermediate state between the current state and the goal. It becomes the only active unit due to a constraint satisfaction process that includes both bottom-up input from the current state and top-down input from the goal state. This in turn drives the intermediate state unit ON in AfterState layer in the RelationEngine module.

Finally, late in settling (Figure 9(c)), the intermediate state outcompetes the goal unit in the AfterState layer due to the attractor associated with the prior training of contiguous state transitions. This is associated with the third action unit in the ActionBetween and ActionNodes (Semantic Network) layers. This is the correct answer. This model illustrates how constraint satisfaction to find bridging states can work as one component of more complex planning.

6.2. Discussion. Subgoals in this context are conceived as a version of "cold" goals, defined as teleological representations of a desired state of the world that, in and of itself, does not include primary reward. Thus, in a sense, cold goals (here subgoals) are "just a means to an end."

In thinking about the role of subgoals, a number of important issues can be identified. First, as already noted, a fundamental issue concerns how brain mechanisms create useful subgoals, if they are not provided externally. In addition, a second important issue is whether there are one or more biologically plausible mechanisms for rewarding the achievement of subgoals. This in turn has two subcomponents: (1) learning how to achieve subgoals in the first place (e.g., how to grind coffee in support of making coffee in the morning) and (2) learning how/when to exploit already familiar subgoal in the service of achieving a master goal (e.g., learning that having ground coffee is a precursor to enjoying a nice fresh cup of hot coffee for yourself and/or receiving kudos from your significant other). It is interesting to note that these two learning categories exhibit a mutual interdependence. Usually, learning how to achieve subgoals must precede learning to exploit them, although an interesting alternative can sometimes occur: if a learner is allowed to use its what-if imagination. For example, if a learner can do thought experiments like: "IF I had ground coffee, and cold water, and a working coffee maker, THEN I could have hot coffee." Thinking about it over and over could transfer (imagined) value from the hot coffee to the ground coffee, and so forth, *which then* could be used as secondary reinforcement to motivate the learning of instrumental subgoals. This scenario-spinning behavior is not modeled in any realistic cognitive model of which we are aware; achieving this will be

difficult but an important step toward understanding human intelligence.

A third critical issue is how subgoals are actually used by the system (in a mechanistic sense) in the service of pursuing the master goal. Here, the simple idea that serves as a kind of working hypothesis in our work is that the active maintenance of subgoals can serve to bias the behavior that produces their realization in a kind teleological "pull of the future" way. Finally, there then still needs to be some sort of cognitive sequencing control mechanism organizing the overall process, that is, the achievement of each subgoal in turn. Ultimately, in our way of thinking, this whole process can be biased by keeping the master goal in active maintenance throughout the process.

In sum, this model demonstrates a rough draft of one aspect of human high-level planning: abstract state representations allow constraint satisfaction processes based on associative learning to find a bridging state between current and goal states. We hypothesize that this process is iterated at different levels of abstraction to create more detailed plans as they are needed. However, we do not as yet have a model that includes the movement between different levels of plan abstraction. The other models presented here represent some of the mechanisms needed for this process but have yet to be integrated into a coherent, let alone complete, model of human planning.

Explaining how brains perform planning requires understanding the computational demands involved. The more abstract literature on the algorithmic and computational properties of planning in artificial intelligence research has thoroughly explored the problem space of many types of planning (e.g., [71–73]). Consistent with this proposed biological model, algorithmic constraint satisfaction solvers have been an important part of AI planning algorithms (e.g., [74, 75]). Other extensions and combinations of these models are also suggested by AI planning work; search-based algorithms (e.g., [76, 77]) show that sequencing, storing, and retrieval of state (as in the model-free and instructed sequencing model) are essential for flexible planning. We address some such possible combinations and extensions later.

7. General Discussion

The four models here represent an incomplete start at fully modeling human strategic cognitive sequencing. A full model would explain how basic mammalian brain mechanisms can account for the remarkable complexity and flexibility of human cognition. It would address the use of elaborate cognitive sequences which constitute learned "programs" for solving complex problems and how people generalize this ability to new problems by selecting parts of these sequences to construct appropriate strategies for novel tasks in related domains. A complete model is thus a long-term and ambitious project, but one with important implications for understanding human cognition.

The following primarily addresses the limitations in the work described and our plans to extend these models toward a more complete explanation of complex human sequential cognition. Although learning was integral to all presented models, demonstrating the feasibility of bootstrapping such flexible cognitive systems, the learning in these initial models was mostly still domain specific: models were trained within the class of tasks to be performed from a naive state. While the instructed model could generalise to a variety of unseen if-then rules and the constraint satisfaction model generalizes to unseen state-goal pairings, they were both only trained on their respective tasks.

In future work, we plan to extend this to a more sophisticated pre-training or scaffolding of networks that are more general and ecologically valid. Instead of beginning training of specific task structures from a naive network, the idea is to train the networks on a large variety of distinct tasks, progressing from simple to complex. The PBDM model, for instance, was trained in a relatively ecologically valid way but did not learn increasing complexity of tasks as it mastered simple ones as humans do. With increasing number of tasks trained, the network should learn to extract commonality between tasks, abstracting the essence of tasks into distinct representations. While it remains unclear what these task representations might look like on the finer biological scale, either from experimentation or computational modeling, it seems likely that representations for some basic computational building blocks of cognitive sequencing exist.

Such representations must, at an abstract level, include some of those found in any standard computer programming language, such as sequencing, loops, storing, and recalling of state. While the models presented here cannot accomplish any of these functions as they stand, we already have a rough basis for these basic task building blocks. All of the previous "program flow" functions can be seen as subsets of conditional branching (e.g., if you have not yet found the goal object, use a sequence that looks for it). The other models presented here (planning, model-free sequence learning, and decision making) address important aspects of how sequences are learned and used, but the instructed learning model alone is enough to understand one way in which the brain can exhibit such program flow control once a relevant sequence is learned. This behavior requires extending the model to store and use state information. This minor extension would include working memory updates in the potential actions and make action pairs conditional on those working memory representations as well as sensory inputs.

Dayan [78] has already explored this behavior in a more abstract version of PBWM. This model includes storage actions and dependency upon stored information consistent with the role for which PBWM was primarily developed, understanding how mechanisms evolved for gating motor actions control storage in working memory. Making memorized pairings dependent upon state information in working memory is also straightforward, and known basal ganglia connectivity suggests such a convergence of information between prefrontal working memory and posterior sensory cortices for the purpose of gating decisions. Dayan [78] also includes a match detection function to allow nonmatch criteria that do not arise naturally from the associative nature of neural networks, an important consideration for our future development of these models.

The models presented here are also generally consistent with the most well-developed models in this domain, procedural models such as ACT-R [60], from which our approach draws inspiration. While our work is generally compatible, we hope to provide more constraints on these theories by considering the wealth of data on detailed aspects of neural function.

In particular, our learning neural network approach will also allow us to constrain theories of exactly what representations are used to produce cognitive sequences by how they are learned. By studying learning over a large number of tasks, we aim to address the question of how these representations emerge on a developmental time scale from a young infant to the fully developed capability of an adult. This focus addresses *learning to learn*, a phenomenon that has both been extensively studied in psychology as well as in machine learning and robotics [79–81]. In both cases, learning to learn transfers beneficial information from a group of tasks to new ones, speeding up learning of new tasks. While in machine learning, many different algorithms have been proposed to achieve transfer learning or learning to learn, a good proportion is based upon representational transfer [79, 82]; that is, due to the efficient and general representations learned in prior tasks, new tasks can be learned more rapidly or more effectively instructed.

To address these questions, we will draw on our and others' work on learning of abstract categories from sensory data (e.g., [83]). Generalizing from prior learning usefully categorizes novel sensory inputs through neural processing that is now relatively well understood. Such category generalization, when combined with the models presented here, offers one explanation of learning to learn. When strategic cognitive sequencing is performed based upon categorical representations (e.g., substitute "input A" in the instructed learning model for "signal to stop and wait for instructions"), learning will generalize to new sensory inputs that can be correctly categorized. This type of generalized matching bears a resemblance to the variable matching rule in recent versions of ACT-R (e.g., "if the word was (word X, previously stored), press the red button"). Modeling this process in greater neural detail will provide more constraints on what types of generalization and matching can be learned and performed by realistic neural networks.

Perhaps because such high-level cognition inevitably involves interactions between many brain regions, computational modeling and other forms of detailed theory construction have, as yet, made little progress. However, the enormous accumulation of work aimed at understanding the contributions from individual brain areas have rendered this complex but important domain a potentially productive target for detailed modeling and computational-level theory.

Acknowledgments

The authors thank members of the CCN Lab at CU Boulder for helpful comments and discussion. The paper is supported by ONR N00014-10-1-0177, NIH MH079485, and the Intelligence Advanced Research Projects Activity (IARPA) via Department of the Interior (DOI) Contract no. D10PC20021.

The US Government is authorized to reproduce and distribute reprints for governmental purposes notwithstanding any copyright annotation thereon. The views and conclusions contained hereon are those of the authors and should not be interpreted as necessarily representing the official policies or endorsements, either expressed or implied, of IARPA, DOI, or the US Government.

References

[1] A. M. Owen, "Tuning in to the temporal dynamics of brain activation using functional magnetic resonance imaging (fMRI)," *Trends in Cognitive Sciences*, vol. 1, no. 4, pp. 123–125, 1997.

[2] T. Shallice, "Specific impairments of planning," *Philosophical transactions of the Royal Society of London B*, vol. 298, no. 1089, pp. 199–209, 1982.

[3] L. Roy Beach and T. R. Mitchell, "A contingency model for the selection of decision strategies," *The Academy of Management Review*, vol. 3, no. 3, pp. 439–449, 1978.

[4] P. Slovic, B. Fischhoff, and S. Lichtenstein, "Behavioral decision theory," *Annual Review of Psychology*, vol. 28, pp. 1–39, 1977.

[5] M. Chi and K. VanLehn, "Meta-cognitive strategy instruction in intelligent tutoring systems: how, when, and why," *Educational Technology & Society*, vol. 13, no. 1, pp. 25–39, 2010.

[6] J. M. Unterrainer, B. Rahm, R. Leonhart, C. C. Ruff, and U. Halsband, "The tower of London: the impact of instructions, cueing, and learning on planning abilities," *Cognitive Brain Research*, vol. 17, no. 3, pp. 675–683, 2003.

[7] A. Newell, "You can't play 20 questions with nature and win: projective comments on the papers of this symposium," in *Visual Information Processing*, W. G. Chase, Ed., pp. 283–308, Academic Press, New York, NY, USA, 1973.

[8] L. B. Smith, "A model of perceptual classification in children and adults," *Psychological Review*, vol. 96, no. 1, pp. 125–144, 1989.

[9] M. J. Roberts and E. J. Newton, "Understanding strategy selection," *International Journal of Human Computer Studies*, vol. 54, no. 1, pp. 137–154, 2001.

[10] J. Tanji and E. Hoshi, "Role of the lateral prefrontal cortex in executive behavioral control," *Physiological Reviews*, vol. 88, no. 1, pp. 37–57, 2008.

[11] A. Dagher, A. M. Owen, H. Boecker, and D. J. Brooks, "Mapping the network for planning: a correlational PET activation study with the tower of London task," *Brain*, vol. 122, no. 10, pp. 1973–1987, 1999.

[12] O. A. van den Heuvel, H. J. Groenewegen, F. Barkhof, R. H. C. Lazeron, R. van Dyck, and D. J. Veltman, "Frontostriatal system in planning complexity: a parametric functional magnetic resonance version of Tower of London task," *NeuroImage*, vol. 18, no. 2, pp. 367–374, 2003.

[13] A. Dagher, A. M. Owen, H. Boecker, and D. J. Brooks, "The role of the striatum and hippocampus in planning: a PET activation study in Parkinson's disease," *Brain*, vol. 124, no. 5, pp. 1020–1032, 2001.

[14] K. Shima, M. Isoda, H. Mushiake, and J. Tanji, "Categorization of behavioural sequences in the prefrontal cortex," *Nature*, vol. 445, no. 7125, pp. 315–318, 2007.

[15] R. C. O'Reilly and Y. Munakata, *Computational Explorations in Cognitive Neuroscience: Understanding the Mind By Simulating the Brain*, The MIT Press, Cambridge, Mass, USA, 2000.

[16] R. C. O'Reilly, T. E. Hazy, and S. A. Herd, "The leabra cognitive architecture: how to play 20 principles with nature and win!,"

in *The Oxford Handbook of Cognitive Science*, S. Chipman, Ed., Oxford University Press, In press.

[17] A. Newell and H. A. Simon, *Human Problem Solving*, Prentice Hall, Englewood Cliffs, NJ, USA, 1972.

[18] J. R. Anderson, *Rules of the Mind*, Lawrence Erlbaum Associates, Hillsdale, NJ, USA, 1993.

[19] R. Morris and G. Ward, *The Cognitive Psychology of Planning*, Psychology Press, 2005.

[20] C. Lebiere, J. R. Anderson, and D. Bothell, "Multi-tasking and cognitive workload in an act-r model of a simplified air traffic control task," in *Proceedings of the 10th Conference on Computer Generated Forces and Behavioral Representation*, 2001.

[21] T. Suddendorf and M. C. Corballis, "Behavioural evidence for mental time travel in nonhuman animals," *Behavioural Brain Research*, vol. 215, no. 2, pp. 292–298, 2010.

[22] S. J. Shettleworth, "Clever animals and killjoy explanations in comparative psychology," *Trends in Cognitive Sciences*, vol. 14, no. 11, pp. 477–481, 2010.

[23] D. Klahr, P. Langley, and R. Neches, Eds., *Production System Models of Learning and Development*, The MIT Press, Cambridge, Mass, USA, 1987.

[24] D. J. Jilk, C. Lebiere, R. C. O'Reilly, and J. R. Anderson, "SAL: an explicitly pluralistic cognitive architecture," *Journal of Experimental and Theoretical Artificial Intelligence*, vol. 20, no. 3, pp. 197–218, 2008.

[25] J. R. Anderson, D. Bothell, M. D. Byrne, S. Douglass, C. Lebiere, and Y. Qin, "An integrated theory of the mind," *Psychological Review*, vol. 111, no. 4, pp. 1036–1060, 2004.

[26] J. R. Anderson, *How Can the Human Mind Occur in the Physical Universe?* Oxford University Press, New York, NY, USA, 2007.

[27] H. H. Yin and B. J. Knowlton, "The role of the basal ganglia in habit formation," *Nature Reviews Neuroscience*, vol. 7, no. 6, pp. 464–476, 2006.

[28] M. J. Frank, B. Loughry, and R. C. O'Reilly, "Interactions between frontal cortex and basal ganglia in working memory: a computational model," *Cognitive, Affective and Behavioral Neuroscience*, vol. 1, no. 2, pp. 137–160, 2001.

[29] M. J. Frank, L. C. Seeberger, and R. C. O'Reilly, "By carrot or by stick: cognitive reinforcement learning in Parkinsonism," *Science*, vol. 306, no. 5703, pp. 1940–1943, 2004.

[30] T. E. Hazy, M. J. Frank, and R. C. O'Reilly, "Banishing the homunculus: making working memory work," *Neuroscience*, vol. 139, no. 1, pp. 105–118, 2006.

[31] T. E. Hazy, M. J. Frank, and R. C. O'Reilly, "Towards an executive without a homunculus: computational models of the prefrontal cortex/basal ganglia system," *Philosophical Transactions of the Royal Society B*, vol. 362, no. 1485, pp. 1601–1613, 2007.

[32] R. C. O'Reilly and M. J. Frank, "Making working memory work: a computational model of learning in the prefrontal cortex and basal ganglia," *Neural Computation*, vol. 18, no. 2, pp. 283–328, 2006.

[33] K. Sakai, "Task set and prefrontal cortex," *Annual Review of Neuroscience*, vol. 31, pp. 219–245, 2008.

[34] G. E. Alexander, M. R. DeLong, and P. L. Strick, "Parallel organization of functionally segregated circuits linking basal ganglia and cortex," *Annual Review of Neuroscience*, vol. 9, pp. 357–381, 1986.

[35] M. J. Frank, "Dynamic dopamine modulation in the basal ganglia: a neurocomputational account of cognitive deficits in medicated and nonmedicated Parkinsonism," *Journal of Cognitive Neuroscience*, vol. 17, no. 1, pp. 51–72, 2005.

[36] R. C. O'Reilly, M. J. Frank, T. E. Hazy, and B. Watz, "PVLV: the primary value and learned value Pavlovian learning algorithm," *Behavioral Neuroscience*, vol. 121, no. 1, pp. 31–49, 2007.

[37] T. E. Hazy, M. J. Frank, and R. C. O'Reilly, "Neural mechanisms of acquired phasic dopamine responses in learning," *Neuroscience and Biobehavioral Reviews*, vol. 34, no. 5, pp. 701–720, 2010.

[38] J. M. Fuster and A. A. Uyeda, "Reactivity of limbic neurons of the monkey to appetitive and aversive signals," *Electroencephalography and Clinical Neurophysiology*, vol. 30, no. 4, pp. 281–293, 1971.

[39] E. K. Miller, "The prefrontal cortex and cognitive control," *Nature Reviews Neuroscience*, vol. 1, no. 1, pp. 59–65, 2000.

[40] T. Ono, K. Nakamura, H. Nishijo, and M. Fukuda, "Hypothalamic neuron involvement in integration of reward, aversion, and cue signals," *Journal of Neurophysiology*, vol. 56, no. 1, pp. 63–79, 1986.

[41] S. A. Deadwyler, S. Hayashizaki, J. Cheer, and R. E. Hampson, "Reward, memory and substance abuse: functional neuronal circuits in the nucleus accumbens," *Neuroscience and Biobehavioral Reviews*, vol. 27, no. 8, pp. 703–711, 2004.

[42] W. Schultz, P. Dayan, and P. R. Montague, "A neural substrate of prediction and reward," *Science*, vol. 275, no. 5306, pp. 1593–1599, 1997.

[43] R. S. Sutton and A. G. Barto, "Time-derivative models of pavlovian reinforcement," in *Learning and Computational Neuroscience*, J. W. Moore and M. Gabriel, Eds., pp. 497–537, MIT Press, Cambridge, Mass, USA, 1990.

[44] J. P. O'Doherty, P. Dayan, K. Friston, H. Critchley, and R. J. Dolan, "Temporal difference models and reward-related learning in the human brain," *Neuron*, vol. 38, no. 2, pp. 329–337, 2003.

[45] R. S. Sutton and A. G. Barto, *Reinforcement Learning: An Introduction*, MIT Press, Cambridge, Mass, USA, 1998.

[46] R. Stuart Sutton, *Temporal credit assignment in reinforcement learning [Ph.D. thesis]*, University of Massachusetts Amherst, Amherst, Mass, USA, 1984.

[47] P. Dayan and B. W. Balleine, "Reward, motivation, and reinforcement learning," *Neuron*, vol. 36, no. 2, pp. 285–298, 2002.

[48] R. C. O'Reilly, S. A. Herd, and W. M. Pauli, "Computational models of cognitive control," *Current Opinion in Neurobiology*, vol. 20, no. 2, pp. 257–261, 2010.

[49] C. H. Chatham, S. A. Herd, A. M. Brant et al., "From an executive network to executive control: a computational model of the n-back task," *Journal of Cognitive Neuroscience*, vol. 23, no. 11, pp. 3598–3619, 2011.

[50] D. Joel, Y. Niv, and E. Ruppin, "Actor-critic models of the basal ganglia: new anatomical and computational perspectives," *Neural Networks*, vol. 15, no. 4-6, pp. 535–547, 2002.

[51] W.-T. Fu and J. R. Anderson, "Solving the credit assignment problem: explicit and implicit learning of action sequences with probabilistic outcomes," *Psychological Research*, vol. 72, no. 3, pp. 321–330, 2008.

[52] J. D. Wallis, "Orbitofrontal cortex and its contribution to decision-making," *Annual Review of Neuroscience*, vol. 30, pp. 31–56, 2007.

[53] M. P. Noonan, N. Kolling, M. E. Walton, and M. F. S. Rushworth, "Re-evaluating the role of the orbitofrontal cortex in reward and reinforcement," *European Journal of Neuroscience*, vol. 35, no. 7, pp. 997–1010, 2012.

[54] P. L. Croxson, M. E. Walton, J. X. O'Reilly, T. E. J. Behrens, and M. F. S. Rushworth, "Effort-based cost-benefit valuation and the human brain," *The Journal of Neuroscience*, vol. 29, no. 14, pp. 4531–4541, 2009.

[55] S. W. Kennerley and M. E. Walton, "Decision making and reward in frontal cortex: complementary evidence from neurophysiological and neuropsychological studies," *Behavioral Neuroscience*, vol. 125, no. 3, pp. 297–317, 2011.

[56] M. J. Frank and E. D. Claus, "Anatomy of a decision: striato-orbitofrontal interactions in reinforcement learning, decision making, and reversal," *Psychological Review*, vol. 113, no. 2, pp. 300–326, 2006.

[57] J. M. Hyman, L. Ma, E. Balaguer-Ballester, D. Durstewitz, and J. K. Seamans, "Contextual encoding by ensembles of medial prefrontal cortex neurons," *Proceedings of the National Academy of Sciences of the United States of America*, vol. 109, no. 13, pp. 5086–5091, 2012.

[58] J. J. Day, J. L. Jones, R. M. Wightman, and R. M. Carelli, "Phasic nucleus accumbens dopamine release encodes effort- and delay-related costs," *Biological Psychiatry*, vol. 68, no. 3, pp. 306–309, 2010.

[59] J. Duncan, M. Schramm, R. Thompson, and I. Dumontheil, "Task rules, working memory, and fluid intelligence," *Psychonomic Bulletin & Review*, vol. 19, no. 5, pp. 864–8870, 2012.

[60] J. R. Anderson, *The Architecture of Cognition*, Harvard University Press, Cambridge, Mass, USA, 1983.

[61] P. M. Fitts and M. I. Posner, *Human Performance*, Belmont, Mass, USA, 1967.

[62] P. Redgrave and K. Gurney, "The short-latency dopamine signal: a role in discovering novel actions?" *Nature Reviews Neuroscience*, vol. 7, no. 12, pp. 967–975, 2006.

[63] G. Fernández, H. Weyerts, M. Schrader-Bölsche et al., "Successful verbal encoding into episodic memory engages the posterior hippocampus: a parametrically analyzed functional magnetic resonance imaging study," *The Journal of Neuroscience*, vol. 18, no. 5, pp. 1841–1847, 1998.

[64] D. Kumaran, J. J. Summerfield, D. Hassabis, and E. A. Maguire, "Tracking the emergence of conceptual knowledge during human decision making," *Neuron*, vol. 63, no. 6, pp. 889–901, 2009.

[65] D. C. Noelle and G. W. Cottrell, "A connectionist model of instruction following, pages," in *Proceedings of the 17th Annual Conference of the Cognitive Science Society*, J. D. Moore and J. F. Lehman, Eds., pp. 369–374, Lawrence Erlbaum Associates, Mahwah, NJ, USA, January 1995.

[66] G. Biele, J. Rieskamp, and R. Gonzalez, "Computational models for the combination of advice and individual learning," *Cognitive Science*, vol. 33, no. 2, pp. 206–242, 2009.

[67] B. B. Doll, W. J. Jacobs, A. G. Sanfey, and M. J. Frank, "Instructional control of reinforcement learning: a behavioral and neurocomputational investigation," *Brain Research*, vol. 1299, pp. 74–94, 2009.

[68] J. Li, M. R. Delgado, and E. A. Phelps, "How instructed knowledge modulates the neural systems of reward learning," *Proceedings of the National Academy of Sciences of the United States of America*, vol. 108, no. 1, pp. 55–60, 2011.

[69] M. M. Walsh and J. R. Anderson, "Modulation of the feedback-related negativity by instruction and experience," *Proceedings of the National Academy of Sciences of the United States of America*, vol. 108, no. 47, pp. 19048–19053, 2011.

[70] T. T. Rogers and J. L. McClelland, *Semantic Cognition: A Parallel Distributed Processing Approach*, MIT Press, Cambridge, Mass, USA, 2004.

[71] S. Russell and P. Norvig, *Artificial Intelligence: A Modern Approach*, Prentice Hall, 1995.

[72] F. Bacchus, "AIPS '00 planning competition: the fifth international conference on Artificial Intelligence Planning and Scheduling systems," *AI Magazine*, vol. 22, no. 3, pp. 47–56, 2001.

[73] E. D. Sacerdoti, "Planning in a hierarchy of abstraction spaces," *Artificial Intelligence*, vol. 5, no. 2, pp. 115–135, 1974.

[74] M. B. Do and S. Kambhampati, "Planning as constraint satisfaction: solving the planning graph by compiling it into CSP," *Artificial Intelligence*, vol. 132, no. 2, pp. 151–182, 2001.

[75] P. Gregory, D. Long, and M. Fox, "Constraint based planning with composable substate graphs," in *Proceedings of the 19th European Conference on Artificial Intelligence (ECAI '10)*, H. Coelho, R. Studer, and M. Wooldridge, Eds., IOS Press, 2010.

[76] A. L. Blum and M. L. Furst, "Fast planning through planning graph analysis," *Artificial Intelligence*, vol. 90, no. 1-2, pp. 281–300, 1997.

[77] E. Fink and M. M. Veloso, "Formalizing the prodigy planning algorithm," Tech. Rep. 1-1-1996, 1996.

[78] P. Dayan, "Bilinearity, rules, and prefrontal cortex," *Frontiers in Computational Neuroscience*, vol. 1, no. 1, pp. 1–14, 2007.

[79] S. Thrun and L. Pratt, "Learning to learn: introduction and overview," in *Learning To Learn*, S. Thrun and L. Pratt, Eds., Springer, New York, NY, USA, 1998.

[80] J. Baxter, "A bayesian/information theoretic model of learning to learn via multiple task sampling," *Machine Learning*, vol. 28, no. 1, pp. 7–39, 1997.

[81] G. Konidaris and A. Barto, "Building portable options: Skill tran sfer in reinforcement learning," in *Proceedings of the 20th International Joint Conference on Artificial Intelligence*, M. M. Veloso, Ed., pp. 895–900, 2006.

[82] K. Ferguson and S. Mahadevan, "Proto-transfer learning in markov decision processes using spectral methods," in *Proceedings of the Workshop on Structural Knowledge Transfer for Machine Learning (ICML '06)*, 2006.

[83] R. C. O'Reilly, D. Wyatte, S. Herd, B. Mingus, and D. J. Jilk, "Recurrent processing during object recognition," *Frontiers in Psychology*, vol. 4, article 124, 2013.

Spike-Timing-Dependent Plasticity and Short-Term Plasticity Jointly Control the Excitation of Hebbian Plasticity without Weight Constraints in Neural Networks

Subha Fernando[1] and Koichi Yamada[2]

[1] Information Science and Control Engineering, Graduate School of Engineering, Nagaoka University of Technology,
1603-1 Kamitomioka-machi, Nagaoka, Niigata 940-2188, Japan
[2] Management and Information Systems Science, Faculty of Engineering, Nagaoka University of Technology,
1603-1 Kamitomioka-machi, Nagaoka, Niigata 940-2188, Japan

Correspondence should be addressed to Subha Fernando, s085160@stn.nagaokaut.ac.jp

Academic Editor: Vince D. Calhoun

Hebbian plasticity precisely describes how synapses increase their synaptic strengths according to the correlated activities between two neurons; however, it fails to explain how these activities dilute the strength of the same synapses. Recent literature has proposed spike-timing-dependent plasticity and short-term plasticity on multiple dynamic stochastic synapses that can control synaptic excitation and remove many user-defined constraints. Under this hypothesis, a network model was implemented giving more computational power to receptors, and the behavior at a synapse was defined by the collective dynamic activities of stochastic receptors. An experiment was conducted to analyze can spike-timing-dependent plasticity interplay with short-term plasticity to balance the excitation of the Hebbian neurons without weight constraints? If so what underline mechanisms help neurons to maintain such excitation in computational environment? According to our results both plasticity mechanisms work together to balance the excitation of the neural network as our neurons stabilized its weights for Poisson inputs with mean firing rates from 10 Hz to 40 Hz. The behavior generated by the two neurons was similar to the behavior discussed under synaptic redistribution, so that synaptic weights were stabilized while there was a continuous increase of presynaptic probability of release and higher turnover rate of postsynaptic receptors.

1. Introduction

Even though Hebbian synaptic plasticity is a powerful concept which explains how the correlated activity between presynaptic and postsynaptic neurons increases the synaptic strength, its value has been diminished as a learning postulate because it does not provide enough explanation how synaptic weakening occurs. In simple mathematical interpretation of Hebbian learning algorithm, an increase of the synaptic strength between two neurons can be seen if their activity is correlated otherwise it is decreased [1]. This interpretation to Hebbian plasticity allows boundless growth or weakening of synaptic strength between the two neurons [2]. Even though Hebbian plasticity has been supported by the biological experiments on long-term plasticity, it is still not completely understood how Hebbian plasticity can avoid synaptic saturation and bring the competition between synapses to balance the excitation of Hebbian neurons. Normalization of weight [2], BCM theory [3], and spike timing-dependent plasticity (STDP) [4] are the most biologically significant mathematical mechanisms that have been discussed in the literature to address this issue effectively. Weight normalization has been introduced either in additive or multiplicative modes to scale the synaptic weights and to control the continuous growth or weakening of synaptic strength; however, these user-defined weight constraints significantly affect the dynamic behavior of the applied neural network and limit the performance of learning [5].

BCM theory is another significant approach that explains synaptic activity as a temporal competition between input patterns. Synaptic inputs that drive postsynaptic firing to higher rate than a threshold value result in an increase of synaptic strength while inputs that make postsynaptic firing to lower rate than the threshold value result in a decrease of synaptic strength. BCM approach has mainly considered instantaneous postsynaptic firing frequencies for its threshold updating mechanism instead of spike arrival time to the synapses. As per the recent literature, it has recognized STDP [4] as a key mechanism of how information is processing in the brain. STDP is a form of long-term plasticity that merely depends on the relative timing of presynaptic and postsynaptic action potentials [6, 7]. Although the process and the role of STDP in information passing in some area of the human brain in the development stages are still not clear [8, 9], it has been shown that average case versions of the perception convergence theorem hold for STDP in simple models of spike neurons for both uncorrelated and correlated Poisson input spike trains. And further it has shown that not only STDP changes the weight of synapses but also STDP modulates the initial release probability of dynamic synapses [10]. Moreover, STDP has been tested on a variety of computational environments, especially to balance the excitation of Hebbian neurons by introducing synaptic competition [11–13] and to identify the repetitive patterns in a continuous spike trains [14, 15]. These experimental studies on synaptic competition using STDP are conducted in two forms: additive form and multiplicative form. In additive form, for example, as in [11], synapses competed against each other to control the timing of postsynaptic firing but this approach assumed that synaptic strength does not scale synaptic efficacy and hard constraints were used to define the efficacy boundaries. In the multiplicative form synaptic scaling was separately introduced to synaptic weight as a function of postsynaptic activity [12, 13]. However, because of the reduced competition between synapses, for strong spike input correlations all synapses stabilized into similar equilibrium. In sum, many applications based on STDP to control the excitation of Hebbian neuron depend on the user-defined constraints on weight algorithm which ultimately limit the performance of learning. To alleviate this limitation in the learning process of using hard weight constraints, another significant approach has been discussed in the literature to remove the correlation in input spike trains by using recurrent neural networks [16]. Their results claim the possibility of reducing the correlation in the spike inputs by recurrent network dynamics. The experiment was conducted on two types of recurrent neural networks; with purely inhibitory neurons and mixed inhibitory-excitatory neurons. At low firing frequencies, response fluctuations were reduced in recurrent neural network with inhibitory neurons when compared to feed-forward network with inhibitory neurons. Moreover, in the case of homogeneous excitatory and inhibitory subpopulation, negative feedback helps to suppress the population rate in both recurrent neural network and feed-forward network. Because inhibitory feedback effectively suppresses pairwise correlations and population rate fluctuations in recurrent neural networks,

they suggested using inhibitory neurons to de correlate the input spike correlations. Moving one step further by combining the underlying concepts in [17, 18], that is, using nonlinear temporally asymmetric Hebbian plasticity and recent experimental observation of STDP in inhibitory synapses, Luz and Shamir [19] have discussed the stability of Hebbian plasticity in feed-forward networks. Their findings supported the fact that temporally asymmetric Hebbian STDP of inhibitory synapses is responsible for the balance the transient feed-forward excitation and inhibition. Using STDP rules, the stochastic weights on inhibitory synapses were defined to generate the negative feedback and stabilized into a unimodal weight distribution. The approach was tested on two forms of network structure; feed-forward inhibitory synaptic population and feed forward inhibitory and excitatory synaptic population. The former structure converged to a uniform solution for correlation input spikes but later destabilized and excitatory synaptic weights were segregated according to the correlation structure in input spike train. Even though the proposed model in the presence of inhibitory neurons of the learning is more sensitive to the correlation structure, the stability of the network is needed to be validated when the correlation between the excitatory synapses and inhibitory synapses is present.

However, the specifics of a biologically plausible model of plasticity that can account for the observed synaptic patterns have remained elusive. To get biologically plausible model and remove the instability in Hebbian plasticity many mechanisms have been discussed in recent findings. One remarkable suggestion is to combine STDP with multiple dynamic and stochastic synaptic connections which enable the neurons to contact each other simultaneously through multiple synaptic communication pathways that are highly sensitive to the dynamic updates and stochastically adjust their states according to the activity history. Furthermore, strength of these individual connections between neurons is necessarily a function of the number of synaptic contacts, the probability of neurotransmitter release, and postsynaptic depolarization [20]. These synapses are further capable of adjusting their own probability of neurotransmitter release (p_r) according to the history of short-term activity [20, 21] which provides an elegant way of introducing activity-dependent modifications to synapses and to generate the competition between synapses [22]. Based on this hypothesis many approaches have been proposed by modeling the behavior at synapses stochastically [23, 24]; here the model we have proposed differs from others because of the computational power that has been granted to the modelled receptors, so that behavior at a single synapse was determined by collective activities of these dynamic stochastic receptors. Using this model, an experiment was conducted to find the answers to the following two questions: first, can STDP and short-term plasticity control the excitation of Hebbian neurons in neural networks without weight constraints? Second, if the excitation was controlled what parameters help STDP in such a controlling activity?

A fully connected neural network was developed with two neurons in which each neuron consisted of thousands of computational units. These computational units were categorized as transmitters and receptors according to the role they played on the network. A unit was called a transmitter if it transmitted signals to other neurons and a unit was called a receptor if it received the signals into the neuron. The receptors of a given neuron were clustered into receptor groups. According to the excitation and the inhibition of the model neuron these computational units could update their states dynamically from active state to inactive state or vice versa. Only when a computational unit was in active state it could successfully transmit signals between neurons. Transmitters from presynaptic neuron and receptors of the corresponding receptor group of the postsynaptic neuron together simulated the process of a single synapse. Transmitter at presynaptic neuron can be considered as a synaptic vesicle which can release only a single neurotransmitter at a time and the model receptors can be considered as postsynaptic receptors at synaptic cleft. With these features, excitation of a neuron at a particular synapse in our network was determined by the function of the number of active transmitters in the presynaptic neuron, transmitters' release probability, and the number of active receptors at the corresponding receptor group of the postsynaptic neuron. First, in order to analyze how network with two neurons could balance the excitation when Poisson inputs with mean rates 10 Hz and 40 Hz were applied, Only one neuron was fed by Poisson inputs while letting the other neuron to adjust itself according to the presynaptic fluctuations. Neurons stabilized its weight for both Poisson inputs while the weight values of Poisson inputs with mean rate 10 Hz were stabilized into higher range compared to when Poisson inputs with mean rate 40 Hz was applied. The analysis into internal dynamics of neurons shows that neurons have behaved similar to the process discussed in synaptic redistribution when long-term plasticity interacts with short-term depression. Further, neurons have played complementary roles to maintain the network's excitation in an operational level. These compensatory roles have not damaged the network biological plausibility as we could see that neurons worked as integrators that integrate higher synaptic weighted inputs to lower output and vice versa. Finally the network behavior was evaluated for other Poisson inputs with mean rates in the range of 10 Hz to 40 Hz and observed as the mean rate of the Poisson inputs increases, the immediate postsynaptic neuron increases its synaptic weights, while the immediate presynaptic neuron of those inputs was settle, into a complementary state to the immediate postsynaptic neuron.

2. Method

A fully connected network with two neurons was created. Each neuron was attached to thousands of computational units which were either in active state or inactive state according to the excitation and the inhibition of the attached neuron. Units attached to a neuron were classified into two groups based on the role they played on the neuron. A computational unit that transmitted the signal from the attached neurons to other neurons was called a transmitter

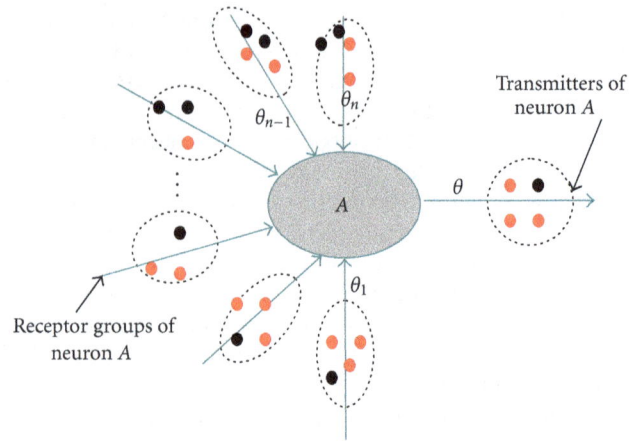

FIGURE 1: Structure of neuron A.

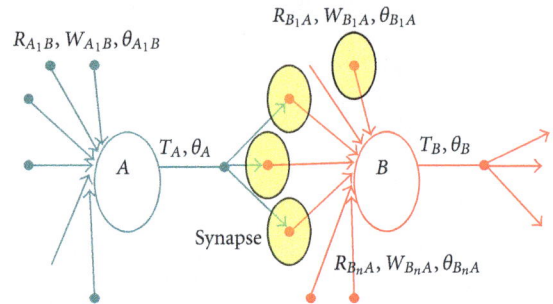

FIGURE 2: Structure of the neural network.

and a computational unit that received the signals to the attached neurons from other neurons was called a receptor. Further, receptors attached to a neuron were clustered into groups so that transmitters from presynaptic neuron could contact the postsynaptic neuron simultaneously through multiple synaptic connections. Figure 1 shows the structure of our modeled neuron A, with n receptor groups and a transmitter set. Moreover, transmitters in our presynaptic neurons were similar to the synaptic vesicles in real neurons with a single neurotransmitter. The states, either active or inactive, of these transmitters and receptors were modeled using two-state stochastic process as explained in the next section. Only when the units were in active states, they were reliable to successfully transmit or receive the signals to or from other neurons.

The transmitters from presynaptic neurons contacted the receptors of a particular receptor group of postsynaptic neurons by forming a synapse between the two neurons; see Figure 2. Through multiple receptor groups of postsynaptic neurons, presynaptic transmitters could make multiple synaptic connections simultaneously forming dynamic and stochastic synapses. As depicted in the Figure 2 each receptor group R of postsynaptic neuron and transmitter set T of presynaptic neuron jointly measured the excitation at

Spike-Timing-Dependent Plasticity and Short-Term Plasticity Jointly Control the Excitation of Hebbian Plasticity without Weight Constraints in Neural Networks

31

the attached synapse w and balanced the excitation using threshold θ as discussed in the next section.

2.1. Process at Dynamic Stochastic Synapses.

When defining the process under dynamic stochastic synapses we have only concerned with the properties and mechanisms of use-dependent plasticity from the few milliseconds to several minutes time scales. Therefore, use-dependent activity to our modeled network was introduced using short-term plasticity; facilitation and depletion [25, 26]. When defining the probability of neurotransmitter release at our modeled transmitters it was assumed that facilitation at biological synapses depends only on the external ca^{+2} ions that influx to the biological synapse after arriving of an action potential and residual ca^{+2} ion concentrations that the synapse already has. And depletion has no influence on ca^{+2} ion concentrations and merely depends on use activity of the synapse. Then signal release probability at a transmitter in a synapse was adopted by the model proposed in [27] which determines the signal release probability p_r as a function of ca^{+2} ions influx to synapse, vesicle depletion, and signal arriving time to the transmitter. Only the influx of ca^{+2} ions after arriving of neurotransmitters into receptors of postsynaptic neuron was considered when determining states of the receptors.

If $Ps(t_i)$ is the probability that signal is released by a transmitter S at time t_i and train $\underline{t} = \{t_1, t_2, \ldots, t_n, \ldots\}$ consists of exact signal releasing times of the S, $S(t)$ consists of the sequences of times where S has successfully released the signals. The map $\underline{t} \rightarrow S(t)$ at S forms a stochastic process with two states, that is, Release (R) for $t_i \in S(t)$ and Failure of Release (F) for $t_i \notin S(t)$. The probability $Ps(t_i)$ in (1) describes a signal release probability at time t_i by S as a function of facilitation $C(t) \geq 0$ in (2) and a depletion $V(t) > 0$ in (4) at time t. C_0 and V_0 are the facilitation and depression constants, respectively. Function $C'(s)$ given in (3) defines the response of $C(t)$ to presynaptic signal that had reached to S at time $t - s$; α is the magnitude of the response. Similarly $V'(s)$ given in (5) models the response of $V(t)$ to the preceding releases of the synapse S at time $t - s \leq t$ and τ_c and τ_v are time decay constants of facilitation and depression. Maass and Zador [27] allowed S to release the received signal at time t, if $Ps(t_i) > 0$. We updated this rule by introducing a new θ threshold so that if $Ps(t_i) > \theta$, a transmitter S is allowed to release the received signal. And we called it as in active state. Receptors in the postsynaptic neuron were modeled using the same model of Maass and Zador except that they were not involved in the process of vesicle depletion. Therefore, the states of the receptors were determined by setting the depletion $V(t)$ in (1) into a unit. According to the recent biological findings of [22], parameters were initialized to $C_0 = 20$, $V_0 = 10$, $\tau_c = 100$ ms, $\tau_v = 800$ ms, and $\alpha = 4$:

$$Ps(t_i) = 1 - \exp(-C(t_i) \cdot V(t_i)), \tag{1}$$

$$C(t) = C_0 + \sum_{t_i < t} C'(t - t_i), \tag{2}$$

$$C'(s) = \alpha \cdot \exp\left(-\frac{s}{\tau_c}\right), \tag{3}$$

$$V(t) = \max\left(0, V_0 - \sum_{t_i < t, t_i \in S(t)} V'(t - t_i)\right), \tag{4}$$

$$V'(s) = \exp\left(-\frac{s}{\tau_v}\right). \tag{5}$$

A modeled neuron maintained threshold values θ for each receptor groups and set of transmitters. Let $R_{J_j I}$ denote the jth receptor group in postsynaptic neuron J that contact transmitters in the presynaptic neuron I, and let $X_{J_j I}(t)$ the output and $\theta_{J_j I}(t)$ be the threshold value of $R_{J_j I}$ at time step t. Similarly let T_I denote the transmitters in neuron I and let $O_I(t)$ be the output and $\theta_I(t)$ the threshold value of T_I at time step t. The threshold value of the receptor group $R_{J_j I}$ was defined as in (6) and it was exponentially increased as the activity of $R_{J_j I}$ to T_I is increasing (or decreased when the activity of $R_{J_j I}$ to T_I is decreasing). Threshold value for transmitters in neuron I, that is, T_I, was defined as a function of total synaptic inputs from all its synaptic connections to the neuron I into the total output of the neuron as in (7). Every 60 time steps threshold values of both neurons were updated:

$$\theta_{J_j I}(t) = f\left(\frac{X_{J_j I}(t)}{O_I(t)}\right) \tag{6}$$

Let g be the number of receptor groups a neuron has, then

$$\theta_I(t) = f\left(O_I(t) \cdot \sum_{i=1}^{g} X_{I_j J}(t)\right) \tag{7}$$

$f(x) = 1/(1 - e^{-x})$, $X_{J_j I}(t) = |R_{J_j I}^{Act}(t)|/|R_{J_j I}|$; $i = 1, 2, \ldots, g$; $O_I(t) = |T_I^{Act}(t)|/|T_I|$. $|G|$ is the number of components in G and $|G^{Act}(t)|$ is the number of active components in G at time step t.

Moreover, according to the following predefined behavioral rule signal was propagated between neurons.

Rule 1. When a receptor receives a signal from the corresponding presynaptic neuron at time step t, the signal is propagated within the network according to the following conditions.

Condition 1. Once a received signal is applied to a receptor if the receptor is updated to inactive state then the received signal is inactivated otherwise the signal is propagated to a randomly selected transmitter of the same neuron.

Condition 2. Once a transmitter of a particular neuron receives a signal at time step t, the signal is transmitted to a randomly selected receptor of the randomly selected receptor group of the postsynaptic neuron if updated state of the transmitter is active otherwise the received signal is inactivated.

The above behavioral rule defines the underlying mechanism of signal transmission between the presynaptic neuron and the postsynaptic neuron; that is, when the related computational units from the two neurons are active only, the signal is successfully transmitted. Therefore, the number of active receptors in a receptor group of the postsynaptic neuron and the number of active transmitters in the presynaptic neuron jointly define the efficacy at a given synapse. In addition to this short-term plasticity and homeostatic synaptic plasticity [28, 29] adjustments (it was shown that under similar conditions, neurons processed similar to Hebbian neurons [30] and the defined threshold mechanism functioned as a homeostatic synaptic plasticity process; see [31]) our dynamic stochastic synapses are subject to long-term plasticity induced by STDP as discussed next.

2.2. Bin the Process at Synapses. The process at synapses where transmitters from the presynaptic neuron contacted the receptors in a particular receptor group of the postsynaptic neuron were binned to analyze the synapse's excitation. Bin is an array of seven columns, $n_b = 7$, which stored data of a given synapse of successive seven time steps. A single cell of a bin contains data at a time step t, namely, the number of active transmitters in the presynaptic neuron, the number of active transmitters in the postsynaptic neuron, the number of active receptors in the corresponding receptor group of the postsynaptic neuron, and the mean release probability of transmitters in presynaptic neuron. Let C_i be ith cell of kth bin, the time gap between two consecutive cells is set to 5 ms as in (8).

This allowed us to define the time represented by each cell in a bin from its first cell as in (9); see Figure 3. This arrangement of bin was necessary in our model to satisfy the condition $(t_{c_1} = 0 \, \text{ms}) < (\tau_+ = \tau_- = 20 \, \text{ms}) < (t_{c_7} = 30 \, \text{ms})$, where τ_+ and τ_- are membrane constants for potentiation and depression (discussed later):

$$\Delta t_{c_{i+1}-c_i} = 5 \, \text{ms}, \quad i = 1, \ldots, 6, \tag{8}$$

$$t_{c_i} = (0, t_{c_i}) = \sum_{i=1}^{7} 5 \cdot (i-1). \tag{9}$$

Let $AT_{\text{pre}} = \{AT_{\text{pre},1}, AT_{\text{pre},2}, \ldots, AT_{\text{pre},n_b}\}$ be random variables of the number of active transmitters in presynaptic neuron at successive seven time steps of a bin; similarly let $AT_{\text{post}} = \{AT_{\text{post},1}, AT_{\text{post},2}, \ldots, AT_{\text{post},n_b}\}$ be random variables of the number of active transmitters in postsynaptic neuron and let $AR_{\text{post},s} = \{AR_{\text{post},s,1}, AR_{\text{post},s,1}, \ldots, AR_{\text{post},s,n_b}\}$ be random variables of the number of active receptors in receptor group s that correspons to synapse s in kth bin (B^k). Since the activity between the presynaptic transmitters and receptors in receptor group s is not independent, we defined mean, $\mu_{B^k,s}$, and variance, $\sigma^2_{B^k,s}$, of the kth bin on synapse s as in (10) and (11):

$$\mu_{B^k,s} = \mu_{AT_{\text{pre}}} + \mu_{AR_{\text{post},s}}, \tag{10}$$

$$\sigma^2_{B^k,s} = \text{Var}\left(AT_{\text{pre}} + AR_{\text{post},s}\right)$$
$$= \sigma^2_{AT_{\text{pre}}} + \sigma^2_{AR_{\text{post},s}} - 2\text{Cov}\left(AT_{\text{pre}}, AR_{\text{post},s}\right), \tag{11}$$

where $\mu_{AT_{\text{pre}}}$ and $\sigma^2_{AT_{\text{pre}}}$ are the mean and variance of AT_{pre}. Similarly, $\mu_{AR_{\text{post},s}}$ and $\sigma^2_{AR_{\text{post},s}}$ are the mean and variance of

$AR_{\text{post},s}$. The mean and variance of both AT_{pre} and $AR_{\text{post},s}$ were estimated using maximum likelihood estimators, so that $\mu_{B^k,s}$ in (10) can be written as in (12) if $\overline{AT_{\text{pre}}} = \sum_{j=1}^{n_b} AT_{\text{pre}}^j/n_b$ and $\overline{AR_{\text{post},s}} = \sum_{j=1}^{n_b} AR_{\text{post},s}^j/n_b$ are the sample means, and $\sigma^2_{B^k,s}$ in (11) can be written as in (13) if $S^2_{AT_{\text{pre}}} = \sum_{j=1}^{n_b} (AT_{\text{pre}}^j - \overline{AT_{\text{pre}}})/(n_b - 1)$, and $S^2_{AR_{\text{post},s}} = \sum_{j=1}^{n_b} (AR_{\text{post},s}^j - \overline{AR_{\text{post},s}})/(n_b - 1)$ are the sample variances of AT_{pre} and $AR_{\text{post},s}$ respectively. The covariance of AT_{pre} and $AR_{\text{post},s}$ is defined in (14):

$$\hat{\mu}_{B^k,s} = \overline{AT_{\text{pre}}} + \overline{AR_{\text{post},s}}, \tag{12}$$

$$\sigma^2_{B^k,s} = S^2_{AT_{\text{pre}}} + S^2_{AR_{\text{post},s}} - 2\text{Cov}\left(AT_{\text{pre}}, AR_{\text{post},s}\right), \tag{13}$$

$$\text{Cov}\left(AT_{\text{pre}}, AR_{\text{post},s}\right)$$
$$= \frac{\sum_{j=1}^{n_b} \left(AT_{\text{pre}}^j - \overline{AT_{\text{pre}}}\right)\left(AR_{\text{post},s}^j - \overline{AR_{\text{post},s}}\right)}{n_b - 1}. \tag{14}$$

The mean release probability of the presynaptic transmitters within a given bin, say B^k, can be defined as in (15), if $\overline{P}_{T_i}$ be the mean release probability of the transmitters in presynaptic neuron at time step $t + i$:

$$M_{B^k_s}^T = \frac{\sum_{i=1}^{n_b} \overline{P}_{T_i}}{n_b}. \tag{15}$$

2.3. Defining Synapse's Activity Using Bins' Activity. STDP is a form of long-term modification to synaptic strength that depends on the action potential arriving timing between presynaptic neuron t_{pre} and postsynaptic neuron t_{post} [4] and can be described by weight window function defined in (16). This weight function defines how strength between the two neurons can be adjusted for a single pair of action potential within the time window $\Delta t = |t_{\text{pre}} - t_{\text{post}}|$. As defined in (16) if presynaptic action potential occurs before the postsynaptic action potential then it strengths the synaptic strength and called long-term potentiation. Conversely if postsynaptic potential occurs before the postsynaptic action potential, it weakens the synaptic strength and called long-term depression:

$$W(\Delta t) = \begin{cases} A_+ \cdot e^{-(t_{\text{post}}-t_{\text{pre}})/\tau_+} & \text{if } t_{\text{pre}} < t_{\text{post}}, \\ -A_- \cdot e^{-(t_{\text{pre}}-t_{\text{post}})/\tau_-} & \text{if } t_{\text{pre}} \geq t_{\text{post}}. \end{cases} \tag{16}$$

Here $A_+, A_- > 0$ and τ_+, τ_- are membrane constants of long-term potentiation and long-term depression. The values for A_+ and A_- need to satisfy the condition $A_+\tau_+ < A_-\tau_-$ as it required the integral of the weight window to be negative to generate stable synaptic strength based on STDP [4]. Furthermore, recent biologically observations [32] have estimated τ_+ and τ_- roughly to 20 ms. Thus, in order to generate stable synaptic strength, it is required to have $A_+ < A_-$. In our model, the weight window function was applied in bin level at each synapse in order to apply long-term modifications to neuron. Let $H^{B^k}_{\text{pre},c_{\text{pre}}}$ be the highest amount of active transmitters recorded from the presynaptic neuron during bin B^k and it was at cell c_{pre} as defined

Spike-Timing-Dependent Plasticity and Short-Term Plasticity Jointly Control the Excitation of Hebbian Plasticity without Weight Constraints in Neural Networks

33

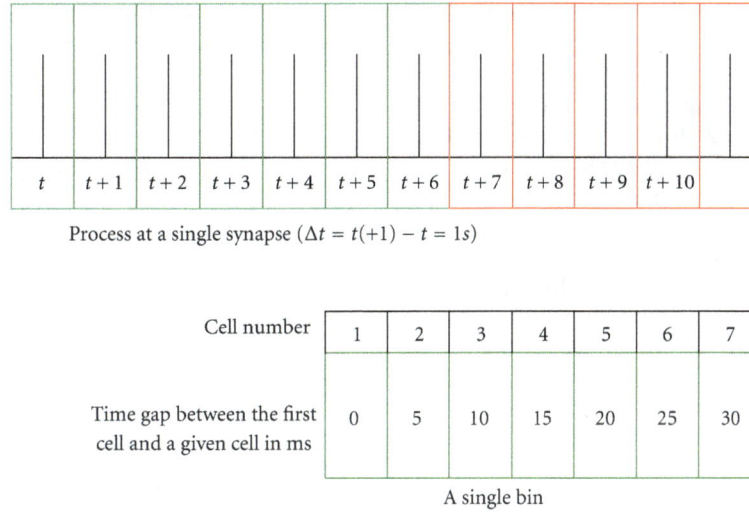

Process at a single synapse ($\Delta t = t(+1) - t = 1s$)

Cell number	1	2	3	4	5	6	7
Time gap between the first cell and a given cell in ms	0	5	10	15	20	25	30

A single bin

FIGURE 3: Bin the process at a single synapse.

in (17). Similarly let $H^{B_k}_{\text{post},c_{\text{post}}}$ be the highest amount of active transmitters recorded from the postsynaptic neuron during bin B^k and it was at cell c_{post} as defined in (18). Then STDP weight window function was applied on bin's level by mapping $H^{B_k}_{\text{pre},c_{\text{pre}}}$ as an action potential occurred in the presynaptic neuron during bin B^k which could significantly update the synaptic strength presynaptically at the corresponding synapse and $H^{B_k}_{\text{post},c_{\text{post}}}$ as the action potential occurred in the postsynaptic neuron during bin B^k which could significantly update the synaptic strength postsynaptically on the same synapse. Here we have assumed that within the duration of a bin only the highest hitter of that bin can significantly update the synaptic strength. Subsequently we mapped t_{pre} to $(c_{\text{pre}} - 1) \times 5$ ms and t_{post} to $(c_{\text{post}} - 1) \times 5$ ms. Therefore, if postsynaptic hitter occurs after the presynaptic hitter, it leads to the potentiation, and if presynaptic hitter is followed by the postsynaptic hitter, it depresses the synapses during the given bin:

$$\forall i \ AT_{\text{pre},i} > AT_{\text{pre},j}, \quad i = 1,\ldots,7, \ j = 1,\ldots,7, \ i \neq j$$

$$H^{B_k}_{\text{pre},c_{\text{pre}}} = \left(AT_{\text{pre},i}, c_{\text{pre}} = i\right) \tag{17}$$

$$\forall i \ AT_{\text{post},i} > AT_{\text{post},j}, \quad i = 1\ldots7, \ j = 1\ldots7, \ i \neq j$$

$$H^{B_k}_{\text{post},c_{\text{post}}} = \left(AR_{\text{post},i}, c_{\text{post}} = i\right) \tag{18}$$

2.4. Mean and Variance of a Synapse. Learning based on STDP was implemented on synapses assuming that bins of a given synapse are mutually independent and the impact that each bin made on the synapse sums linearly. Then mean $\mu_{S^s_k}$ and variance $\sigma^2_{S^s_k}$ of the sth synapse (S^s) can be defined as in (19) and (20) when kth bin (B^k) is interacted with the synapse. The mean $\mu_{S^s_k}$ and the variance $\sigma^2_{S^s_k}$ of the synapse S^s were estimated using maximum likelihood estimators as shown in (21) and (22), respectively. Further, the total mean

release probability at synapse S^s at B^k was defined using bin's mean release probabilities as in (23):

$$\mu_{S^s_k} = \mu_{S^s_{k-1}} + \mu_{B^k,s}; \quad \mu_{S^s_0} = 0, \ k = 1, 2, \ldots, \tag{19}$$

$$\sigma^2_{S^s_k} = \text{Var}\left(S^s_{k-1} + B^k\right) = \sigma^2_{S^s_{k-1}} + \sigma^2_{B^k,s};$$
$$\sigma^2_{S^s_0} = 0, \quad k = 1, 2, \ldots, \tag{20}$$

$$\hat{\mu}_{S^s_k} = \hat{\mu}_{S^s_{k-1}} + \hat{\mu}_{B^k,s}$$
$$\overline{X_{S^s_k}} = \overline{X_{S^s_{k-1}}} + \overline{X_{B^k,s}}; \quad k = 1, 2, \ldots, \tag{21}$$

$$\hat{\sigma}^2_{S^s_k} = \hat{\sigma}^2_{S^s_{k-1}} + \hat{\sigma}^2_{B^k,s}$$
$$S^2_{S^s_k} = S^2_{S^s_{k-1}} + S^2_{B^k,s}; \quad k = 1, 2 \ldots, \tag{22}$$

$$M^T_{S^s_k} = M^T_{S^s_{k-1}} + M_{B^k,s}; \quad M^T_{S^s_0} = 0, \ k = 1, 2, \ldots. \tag{23}$$

In order to generate action potentials real neurons are necessary to be in a nonquiescence state. If a neuron is in a quiescence state, it is not possible for the neuron to generate action potentials that can change the synaptic strength significantly. Therefore, STDP was applied on synapses only if model presynaptic and postsynaptic neurons were not in quiescence states. We defined that a neuron is not in a quiescence state when the average output produced by the neuron during bin B^k is greater than the average output it had produced so far. That is, if the current mean number of active transmitters of a particular neuron is less than the mean number of active transmitters during the kth bin, neuron was recognized as in a nonquiescence state at bin B^k. That is mathematically if $M^{S^s_{k-1}}_{AT_{\text{pre}}} \leq M_{AT^{B^k}_{\text{pre}}}$ and $M^{S^s_{k-1}}_{AT_{\text{post}}} \leq M_{AT^{B^k}_{\text{post}}}$, the weight was updated on the synapse S^s at bin B^k as discussed next.

2.5. Learning Based on STDP and Release Probability. According to the model proposed in [33], the amplitude of the excitatory postsynaptic current A_k of the kth spike in a spike train is proportional to the weight at that synapse and the release probability at the kth spike. In our approach A_k is proportional to the impact that made by transmitters in the presynaptic neuron and receptors in the corresponding receptor group of the postsynaptic neuron during B^k on the synapse S^s. If we applied the model proposed in [33] to our kth bin instead of kth spike, we can express A_k as in (24). Moreover, biological evidence supports the fact that the amount of change on weight is also dependent on the initial synaptic size [34]. Depression is independent of the synaptic strength, whereas strong synapses are less potentiated than weak synapses. By assuming that there is an inverse relationship between the initial synaptic strength and the amount of potentiation, potentiation can be expressed for the kth bin, (B^k) as in (25) if $W^p_{S^s,k}$ is the amount of potentiation during kth bin at synapse S^s:

$$A_{k,S^s} \propto W_{k,S^s} \cdot U_{k,S^s}, \tag{24}$$

$$W_{k,S^s} \propto \frac{1}{W^p_{k,S^s}}. \tag{25}$$

If we combined (16), (24), and (25), the amount of weight updated during kth bin at synapse S^s, $W_{S^s}(\Delta k)$, can be defined as in (26) and the synaptic weight at S^s at the end of bin B^k is determined as in (27):

$$W_{S^s}(\Delta k)$$
$$= \begin{cases} \gamma_p \cdot \dfrac{A_{k,S^s}}{U_{k,S^s}} \cdot \dfrac{1}{W^p_{k,S^s}} \cdot e^{-(t_{\text{post}} - t_{\text{pre}})/\tau_+} & \text{if } t_{\text{pre}} < t_{\text{post}} \\[2em] -\gamma_d \cdot \dfrac{A_{k,S^s}}{U_{k,S^s}} \cdot e^{-(t_{\text{pre}} - t_{\text{post}})/\tau_-} & \text{if } t_{\text{pre}} \geq t_{\text{post}} \\[1.5em] & \qquad k = 1 \ldots, \ s = 1 \ldots \end{cases}$$
$$\tag{26}$$

$$W_{S^s}(k) = W_{S^s}(k-1) + W_{S^s}(\Delta k), \tag{27}$$

where $\gamma_p = 0.005$ and $\gamma_d = 0.00525$ are potentiation and depression learning rates [11]. The amplitude A_k during the kth bin was estimated by the proportion of the deviation that made by the bin compared to its mean, to the deviation that synapse had made so far compared to its overall mean. That is, in statistically amplitude A_k during the kth bin can be expressed as a proportion of the coefficient variation (CV = σ/μ) during the kth bin to the coefficient variation of the synapse S^s has as given in (28). The release probability U_{k,S^s} during the kth bin was determined as a proportion of mean release probability during kth bin to the total mean release probability at synapse S^s has as in (29). Median of the weight distribution at synapse S^s was taken as an estimator for W^p_{k,S^s} as in (30). This is merely because median provides a good approximation about the center of the weight distribution

than mean; that is, the median is not affected by the outliers, whereas the mean is affected by the outliers:

$$A^S_{k,S^s} = \frac{\text{CV}_{B^k,s}}{\text{CV}_{S^s}} = \frac{(\sigma^2_{B^k,s}/\mu_{B^k,s})}{(\sigma^2_{S^s_k}/\mu_{S^s_k})}; \quad k = 1 \ldots, \ s = 1 \ldots \tag{28}$$

$$U_{k,S^s} = \frac{M^T_{B^k,s}}{M^T_{S^s_k}}; \quad k = 1, \ldots, \ s = 1 \ldots \tag{29}$$

$$W^p_{k,S^s} = \text{median}\{W_{S^s}(i) \mid i = 1, \ldots, k-1, \ s = 1 \ldots\}. \tag{30}$$

3. Balancing the Excitation of the Network

An experiment was arranged to find the possibility that can STDP and short-term plasticity together balance the excitation of a network with two Hebbian neurons without defining any constraints on the weight learning algorithm. A fully connected network with two neurons, say neuron A and neuron B, was developed; each neuron had ten receptor groups, making a presynaptic neuron to contact the postsynaptic neuron through ten dynamic stochastic synapses simultaneously. Both neurons, A and B had equal number of transmitters $n_T = 30000$ and receptors $n_R = 30000$; and receptors attached to neurons were uniformly distributed among receptor groups. At the onset one percent of transmitters and one percent of receptors in each receptor-group was set to active state. Poisson inputs with mean firing rates (λ) 10 Hz and 40 Hz were applied to all the receptor groups of neuron A simultaneously while giving enough space for neuron B to adjust itself according to the feedbacks of neuron A, see Figure 4. Each input was applied around two hours continuously to neuron A; and the behavior of the network was analyzed after the synaptic connections have established the effect of the altered activity and network activity has developed. The values were fed to the system according to following rule: the generated Poisson distribution was converted to byte stream by the following rule: if generated value is greater than the median of the Poisson distribution then it was considered to represent value 1 otherwise it was considered to represent value 0. Only when the represented value is equal to one, the signal was generated and fed to neuron A.

Figures 5 and 6 show the distribution of weights of both postsynaptic neurons A and B at Poisson inputs with mean rates 10 Hz and 40 Hz. As shown in these figures, the weight distributions of both neurons at each, synapse have stabilized around 175 bins. After the weights distribution was stabilized, the median of the weight distribution was calculated and these calculated median values are shown in Figure 7. As shown in the figure of Poisson inputs with low mean rate, that is, at 10 Hz, medians of the all the synapses of the postsynaptic neurons reach to higher value compared to when Poisson inputs with higher mean rate, that is, at 40 Hz, were applied. The network balanced its excitation by pushing synaptic weights towards higher values for inputs with low mean rate and for inputs with higher mean rate the network has pulled down the synaptic weights into a lower weight values. This dynamic behavior of both neurons is a necessary

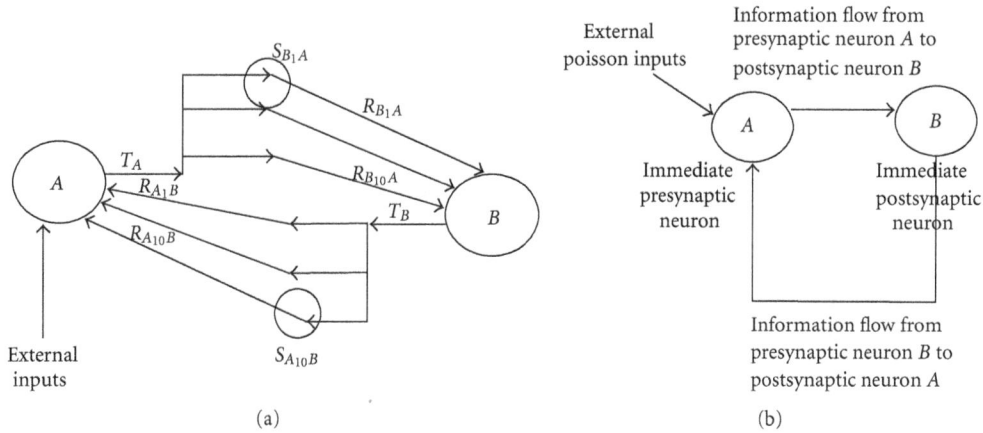

(a)   (b)

FIGURE 4: Network structure with ten synaptic connections. (a) shows the developed fully connected network to test how neurons could balance the excitation after external input was applied to part of it. T_A and T_B are the outputs (the number of active transmitters attached to the neuron at a given time step) of neuron A and neuron B, respectively. Receptor groups R_{AB} and receptor groups R_{BA} symbolized the number of active receptors in the corresponding receptor groups of postsynaptic neuron A and postsynaptic neuron B, respectively. S_{AB} are the synapses where receptor groups of postsynaptic neuron A contact the transmitters of neuron B. Similarly, S_{BA} are the synapses where receptor groups of postsynaptic neuron B contact the transmitters of neuron A. (b) shows how information flows between the two neurons. When signals are passing from A to B, A is called presynaptic neuron and B is called postsynaptic neuron. Similarly, when signals are passing from B to A, B is called presynaptic neuron and A is called postsynaptic neuron. Since external Poisson inputs are applied to neuron A only, neuron A becomes the immediate presynaptic neuron and B becomes the immediate postsynaptic neuron for the external inputs.

adjustment to balance neurons' excitation and subsequently to balance the network excitation while adjusting to external manipulations.

Next we were interested to know what makes the neuron to stabilize its activity without being overexcited or overdepressed in a network which has no controlling constraints. To understand that we analyzed the internal behaviors of neurons A and B in terms of their mean release probability (the mean of the release probabilities of transmitters attached to the neuron) and the coefficient variation which measures the given synapse's excitation (CV = $CV_{B^k,s}$ in (28)) in terms of the number of active transmitters in the presynaptic neuron and the number of active receptors in the corresponding receptor groups of the synapse. Thus, the value of CV shows the extent of variability of the given synapse in relation to the synapse's mean and portraits effectively the synapse's internal dynamics. So that higher CV value implies the higher internal fluctuations and higher deviation of the synaptic mean. Figure 5 shows the mean release probability of the presynaptic neurons at 10 Hz while Figure 8 depicts what is happening inside the synapse in terms of CV at 10 Hz. As shown in these figures, the neuron that has made higher synaptic weight has maintained higher CV compared to the other neuron in the network. Notably, the neuron which had the higher synaptic weight has produced a lower mean release probability. For example, if we analyze the behavior of neuron B, as shown in Figure 7, its synapses S_{BA}, from presynaptic neuron A to postsynaptic neuron B, have scored higher synaptic weights at 10 Hz compared to synaptic weights of postsynaptic neuron A. Further, neuron B has maintained higher CV at all its synapses S_{BA} compared to the values of CV of the synapses of postsynaptic neuron A, that is, S_{AB}. In contrast, the

neuron B has maintained lower mean release probability as a presynaptic neuron at 10 Hz of Poisson inputs at its all synapses compared to neuron A. These opposing and balancing behaviors of the two neurons are consistent in Poisson inputs with mean firing rate 40 Hz as given in Figures 6 and 9.

Moreover, if the difference of the value of CVs between two neurons was considered, it is clearly shown in the Figures 8 and 9 that this difference was reduced to amount 0.0001 after the two neurons have adjusted to the external input and stabilized themselves. However, most importantly, even that we could see the stabilize activity of the two neurons in terms of synaptic weights and CV the mean release probabilities have not reached to any stable position, but instead either continuously positively or negatively increasing. The positive correlation between the synaptic weights and the CV, and the negative correlation between the synaptic weights and the mean release probability at a same neuron have proven that neurons could act as integrators that integrate the excited synaptic weights and controlled the excitation via higher CV fluctuations and produced balanced output that help to balance the network activity. This reduced excitation in the output flow that has allowed the other neuron to play compensatory role and to balance the network activity.

Finally we would like to understand the behavior of the network when applying Poisson inputs in the range of 10 Hz and 40 Hz. The Poisson inputs with mean rates, 15 Hz, 20 Hz, 25 Hz, 30 Hz, and 35 Hz were also presented to neuron A's receptor groups and studied the behavior of the both neurons on the same network. Figure 10 shows the average value of the medians of synapses of each neuron. When magnitude of the Poisson inputs mean rate is greater than the STDP potentiation and depression time constants

$\lambda = 10\,\mathrm{Hz}$

Synapse number = 1   Synapse number = 2   Synapse number = 3

Synapse number = 4   Synapse number = 5   Synapse number = 6

Synapse number = 7   Synapse number = 8   Synapse number = 9

Synapse number = 10

(Each subfigure) $m_B^I = 0.00126$ $m_B^{II} = 0.00149$ W_{BA} R_A R_B W_{AB} $m_A^I = 0.000798$ $m_A^{II} = 0.00118$

Axis labels: Mean release probability; Weight distribution; Bin number

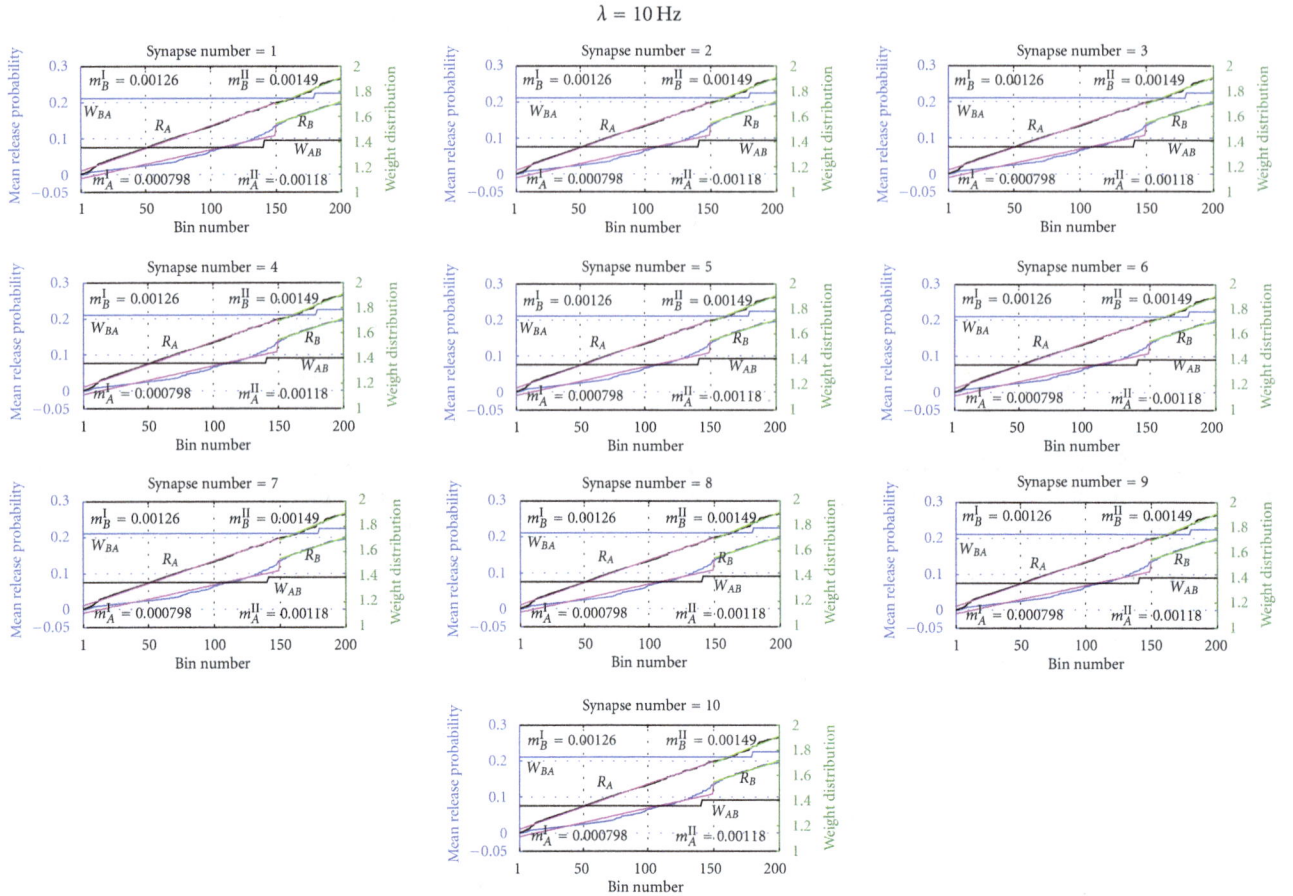

FIGURE 5: Distribution of the weights and release probabilities of neurons at Poisson inputs with mean rate 10 Hz. Each subfigure in the figure depicts the distribution of the weight algorithm and the mean release probability at the given synapse of postsynaptic neuron A and postsynaptic neuron B at Poisson inputs with mean firing rate 10 Hz. For example, the leftmost top subfigure shows the variation of the mean release probability and the weight distribution at the first synapse of the ten synapses. As shown in the figure, the network, both neuron A and neuron B, spent around 150 bins to adjust to the external Poisson inputs and subsequently reach the stability. W_{BA} gives the distribution of the weights of the synaptic connections from presynaptic neuron A to postsynaptic neuron B. Similarly W_{AB} gives the distribution of the weights of the synaptic connections from presynaptic neuron B to postsynaptic neuron A. R_B is the distribution of the mean release probability of transmitters of presynaptic neuron B and R_A is the distribution of mean release probability of transmitters of presynaptic neuron A at postsynaptic connections of neuron B. Moreover, the slopes of the mean release probabilities, m_B^I and m_B^{II}, were determined using linear regression analysis and give the slope of mean release probability of S_{BA} from bin 1 to 150 and bin 150 to 200, respectively. Similarly, m_A^I and m_A^{II} give the mean release probability of S_{AB} from bin 1 to bin 150 and from 150 to 200, respectively.

$25 < \lambda$, the medians of the stabilized synaptic weights of synapses of neuron B as the immediate postsynaptic neurons of external inputs have continuously increased as the mean rate of Poisson inputs is increased. Again the compensatory behavior from neuron A could be seen as it has generally decreased its medians of the stabilized synaptic weights as the mean rate is increasing. These complementary behaviors of two neurons seem to be necessary to stabilize the overall network activity. When $20 > \lambda$, both neurons have worked together to control the overall excitement of the network. Intriguingly, when $\lambda = 20$ and $\lambda = 25$, the excitation of the entire network was equally balanced between the two neurons as their average value of the medians of the stabilized synaptic weights becomes almost equal to each other. This might be because of the effect of $\tau_+ = \tau_- = 20\,\mathrm{ms}$ that we selected for STDP potentiation and depression

time constant. This is important observation which provides the possibility that postsynaptic neurons could excited and stabilized into the same level of the presynaptic neuron if STDP time constants are highly correlated with the mean rate of the Poisson inputs applied. Therefore, STDP with different time constants for potentiation and depression might be a good solution to scale down external inputs effectively into neuronal level.

4. Discussion

As per the literature, a synapse can be strengthened either by increasing the probability of transmitter release presynaptically or by increasing the number of active receptors postsynaptically. This general functionality at the synapses can be varied by the interplay between long-term plasticity

Spike-Timing-Dependent Plasticity and Short-Term Plasticity Jointly Control the Excitation of Hebbian Plasticity
without Weight Constraints in Neural Networks

37

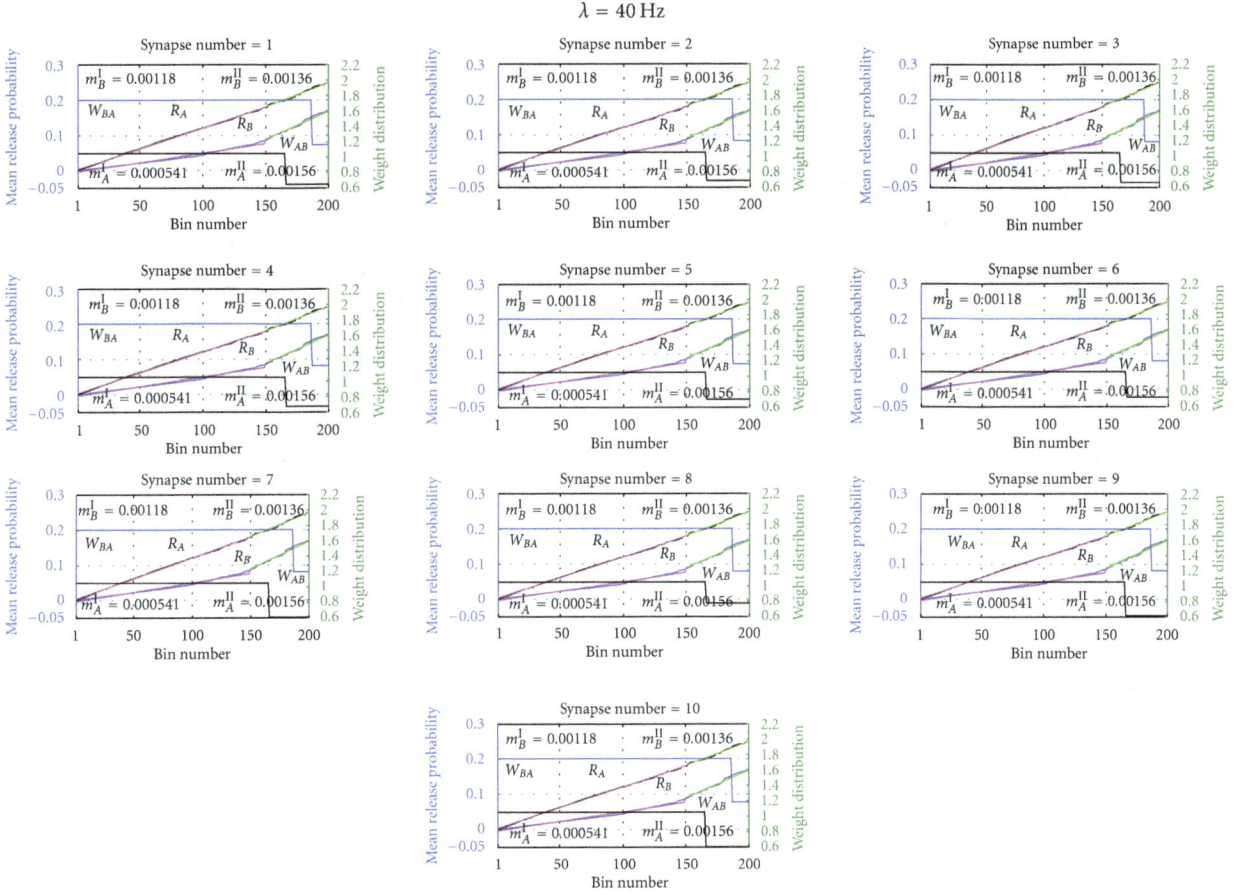

FIGURE 6: Distribution of the weights and release probabilities of neurons at Poisson inputs with mean rate 40 Hz. Each subfigure in the figure depicts the distribution of the weight algorithm and the mean release probability at the given synapse of postsynaptic neuron A and postsynaptic neuron B at Poisson inputs with mean firing rate 40 Hz. For example, the leftmost top subfigure shows the variation of the mean release probability and the weight distribution at the first synapse of the ten synapses. As shown in the figure, the network, both neuron A and neuron B, spent around 150 bins to adjust to the external Poisson inputs and subsequently reach the stability. W_{BA} gives the distribution of the weights of the synaptic connections from presynaptic neuron A to postsynaptic neuron B. Similarly, W_{AB} gives the distribution of the weights of the synaptic connections from presynaptic neuron B to postsynaptic neuron A. R_B is the distribution of the mean release probability of transmitters of presynaptic neuron B and R_A is the distribution of mean release probability of transmitters of presynaptic neuron A at postsynaptic connections of neuron B. Moreover, the slopes of the mean release probabilities, m_B^I and m_B^{II}, were determined using linear regression analysis and give the slope of mean release probability of S_{BA} from bin 1 to 150 and bin 150 to 200, respectively; Similarly, m_A^I and m_A^{II} give the mean release probability of S_{AB} from bin 1 to bin 150 and from 150 to 200, respectively.

and short-term dynamics, especially short-term depression. Short-term depression is mainly based on vesicle depletion which is the use-dependent reduction of neurotransmitter release in the readily releasable pool [4]. When long-term plasticity is interacting with the short-term depression it is called synaptic redistribution [7]. The role of this synaptic redistribution is not yet clearly identified. However, this synaptic redistribution allows the presynaptic neuron to increase the probability of release and thereby increase the signal transmission between the two neurons. In our developed network, the two neurons have simulated a behavior similar to the effect of synaptic redistribution. Neuron B as the immediate postsynaptic neuron of the external inputs has scored the higher synaptic weights compared to neuron A. That is, its synapses, where the transmitters from immediate presynaptic neuron A contacted each receptor group of

the postsynaptic neuron B, have scored higher weights compared to the other neuron. The presynaptic neuron A in this functional process has maintained higher mean release probability. Therefore, first, the synaptic weights of neuron B have been increased presynaptically by increasing the probability of neurotransmitter release. Second, the analysis of CV of these synapses of postsynaptic neuron B shows that it is laying higher range than the function of CV of neuron A, confirming the possibility of increasing the synaptic weights by higher turnover rate of active receptor component of postsynaptic neuron. This behavior of postsynaptic neuron B is also supported at Poisson inputs with mean firing rate 40 Hz. Intriguingly, if the behavior of postsynaptic neuron B at 40 Hz after neuron has adjusted to the external inputs and stabilized, was analyzed, the higher fluctuations of CV and comparatively lesser synaptic weights of neuron

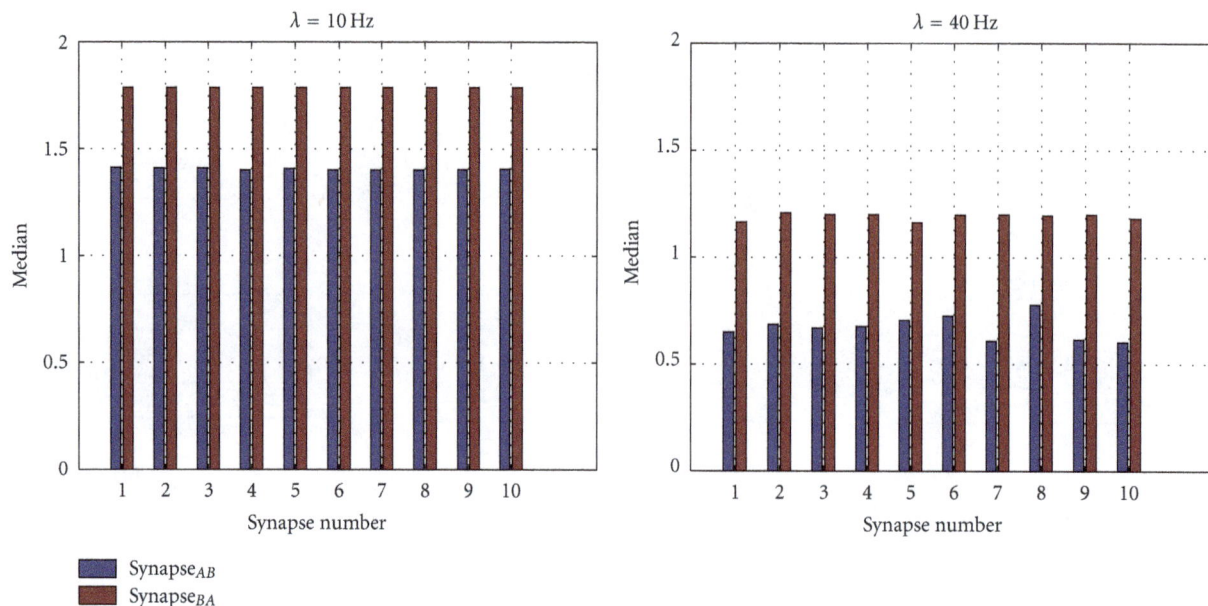

FIGURE 7: Distribution of the medians of synaptic weights at each synapse at Poisson inputs. The figure shows the variation of the median of the weight distribution at each Poisson input; λ is the mean Poisson firing rate. The median in the above figures was determined after synaptic connections have established the effect of external inputs and network stabilized. This stabilization happened after 170 bins; see Figure 5. Medians of postsynaptic neuron B, Synapse$_{BA}$ show the variations of the medians of the weight distributions of synaptic connection from presynaptic neuron A to postsynaptic neuron B. Similarly, Synapse$_{AB}$ show the variations of the median of the weight distributions of synaptic connections from presynaptic neuron B to the postsynaptic neuron A.

B at Poisson inputs with mean rate 40 Hz were observed compared to the postsynaptic neuron B's behavior at Poisson inputs with mean rate 10 Hz, showing the possibility that synaptic redistribution can increase the synaptic weights for Poisson inputs with low mean rate at steady state and not for Poisson inputs for higher mean rate. And for higher mean rates it is only a higher turnover rate of active receptor components. Further, STDP potentiation and depression time constants have made higher impact on the behavior of those two neurons; that is, it has controlled the level of excitation of each neuron equally when the magnitude of the mean rate laid near the magnitude of the STDP time constants. As the mean rate of the Poisson inputs is increasing, the complementary roles have been initiated into the two neurons.

STDP has successfully interplayed with short-term plasticity to control the excitation or inhibition of neural network according to external adjustments. Notably, these adjustments are consistent and are also biologically plausible. The stabilization of synaptic weights in an operational level without controlling constraints seems to be possible if STDP as long-term plasticity interacts with the short-term dynamics. The dynamic behavior of short-term activity is necessary to propagate and balance the excitation of neural network without damaging the synaptic weight distribution; similar to how CV and probability of release have played with STDP to balance the excitation. When compared to Luz and Shamir [19] findings, instead of specifically using inhibitory neurons to generate negative feedback to stabilize

the excitation and inhibition of the network, here we have used ground plasticity mechanisms that observed in biology to alleviate the excitation and inhibition of the network. Even though two approaches have used different derivatives of temporally asymmetric STDP to implement the stochastic response of neurons, both have once again proven the possibility of stabilization of the network excitation due to Hebbian plasticity using STDP. However, our approach differs from their mechanism because of the integration of the sensitivity of the release probability and turnover rate of active components attached to a synapse. Instead of the inhibition made on network by the inhibitory neuron by negative feedback to impinge the excitation generate by excited correlated spikes, our mechanism absorbed the high firing frequency excitation or overcomes the low firing frequency inhibition in terms of appropriate turnover rate of active components attached to a given synapse and adjusting release probabilities of the attached active transmitters. The mechanism under this manipulation of excitation and inhibition is similar to the synaptic redistribution discussed in biology, which drives our approach more towards to the biological plausibility. However, both systems are still needed to be evaluated and examined on larger networks and the approach of Luz and Shamir [19] needs to be tested when correlation between the inhibitory and excitatory synapses is present. Moreover, we compare the result of [11] against our findings on how excitation was balanced. In their model excitation was balanced by introducing synaptic competition in which synapses competed against

Spike-Timing-Dependent Plasticity and Short-Term Plasticity Jointly Control the Excitation of Hebbian Plasticity without Weight Constraints in Neural Networks

39

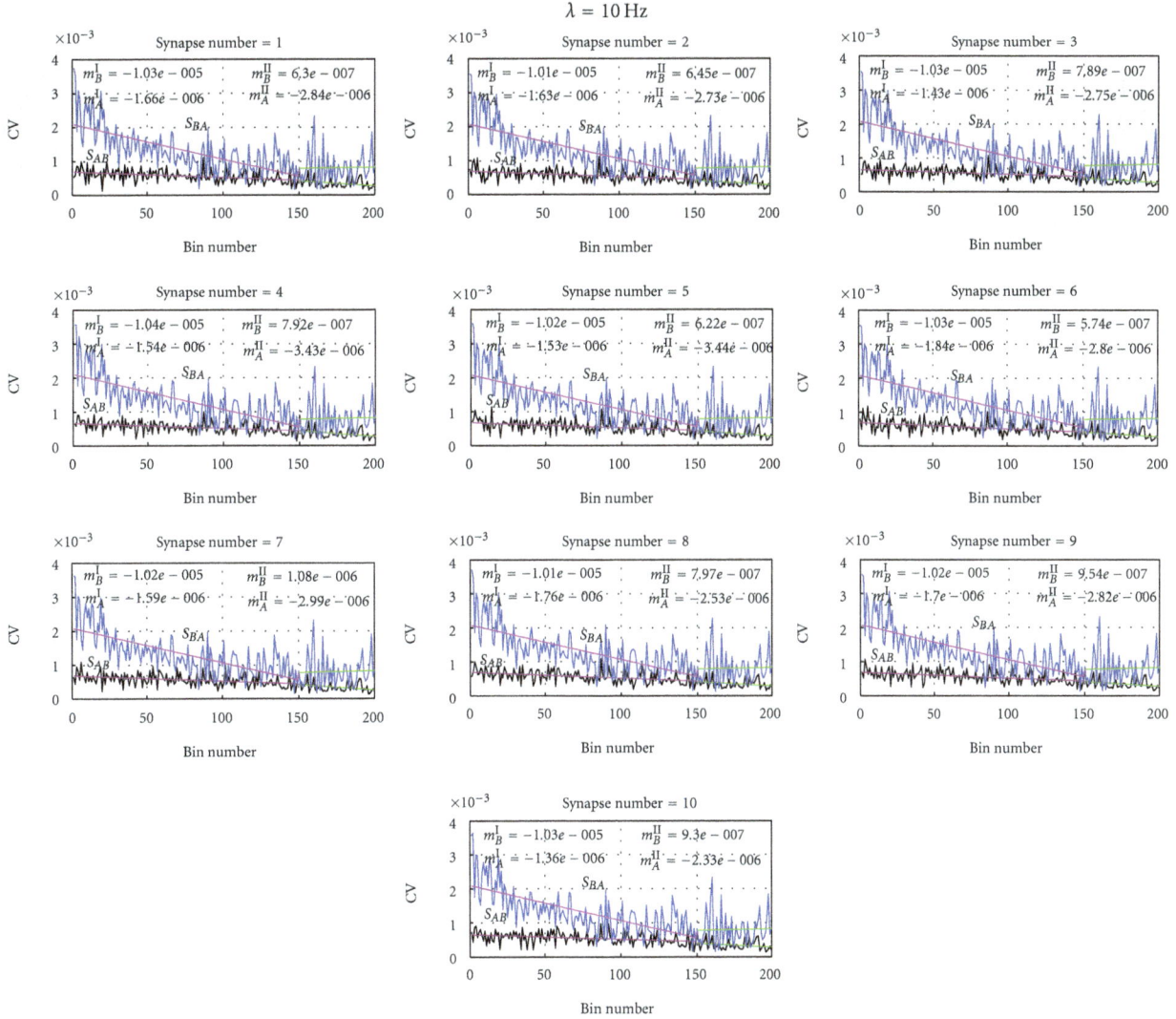

FIGURE 8: Distribution of the coefficient of variation (CV) at each synapse at Poisson inputs with 10 Hz. Each subfigure Figure 8 depicts the distribution of the CV at the given synapse of both neuron A and neuron B at Poisson inputs mean firing rate 10 Hz. For example, As shown in the leftmost subfigure S_{AB} shows the variation of CV of synapses from presynaptic neuron B to postsynaptic neuron A. Similarly, S_{BA} depicts the distribution of CV of synapses from presynaptic neuron A to postsynaptic neuron B. Moreover, the slopes of the mean release probability, m_B^I and m_B^{II}, were determined using linear regression analysis and give the slope of CV of S_{BA} from bin 1 to 150 and bin 150 to 200, respectively. Similarly, m_A^I and m_A^{II} give the CV of S_{AB} from bin 1 to bin 150 and from 150 to 200, respectively. As depicted in all these subfigures, at all the synapses when its around 200 bins, the difference of the CVs of neuron A and neuron B was about 0.001 and time functions of CVs of both neurons were paralleled to each other.

each other to control the postsynaptic firing times; further, this competition was introduced by scaling the synaptic efficacy using hard boundary conditions. This model balanced the excitation of Poisson inputs 10 Hz and 40 Hz, so that for 10 Hz more synapses approach to the higher limit of synaptic efficacy and for higher inputs more synapses remained in lower limits. However, once the system reached stability it was hardly disturbed by the presynaptic firing frequency. Therefore, the stability that the system reached is moderately stronger than in our case. Although our model also exhibited similar characteristics at Poisson inputs 10 Hz and 40 Hz, no boundary conditions were defined

to achieve this stability. Furthermore, compared to their moderately strong equilibrium discussed in terms of synaptic efficacy, internal dynamics of our neurons were continuously fluctuating around the equilibrium allowing neurons to remain dynamically active even at the equilibrium as similar to many natural systems.

The model proposed in this research is a computational model to investigate the internal dynamics of neural networks when STDP, Hebbian plasticity and short-term plasticity, are interacting with each other. The model itself has few drawbacks; mainly the neural network of our model has spent around 150 bins to show the adjustment to external

$$\lambda = 40\,\text{Hz}$$

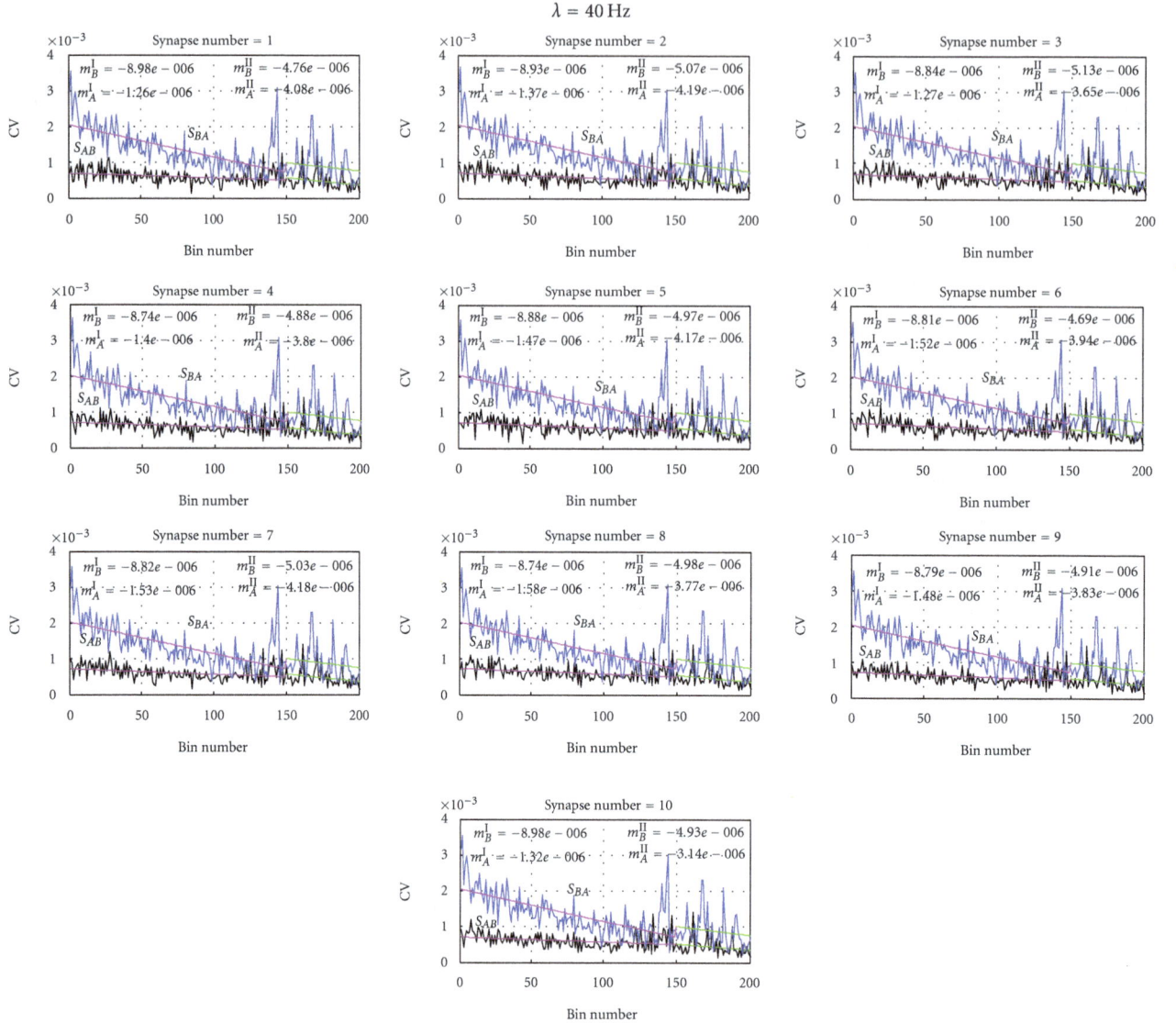

FIGURE 9: Distribution of the CV at each synapse at Poisson inputs with mean firing rate 40 Hz. Each subfigure in the figure depicts the distribution of the CV at the given synapse of both neuron A and neuron B at Poisson inputs mean firing rate 40 Hz. For example, As shown in the leftmost subfigure S_{AB} shows the variation of CV of synapses from presynaptic neuron B to postsynaptic neuron A. Similarly, S_{BA} depicts the distribution of CV of synapses from presynaptic neuron A to postsynaptic neuron B. Moreover, the slopes of the mean release probability, m_B^I and m_B^{II}, were determined using linear regression analysis and give the slope of CV of S_{BA} from bin 1 to 150 and bin 150 to 200, respectively. Similarly, m_A^I and m_A^{II} give the CV of S_{AB} from bin 1 to bin 150 and from 150 to 200, respectively. As depicted in all these subfigures, at all the synapses when its around 200 bins, the difference of the CVs of neuron A and neuron B was about 0.001 and time functions of CVs of both neurons were paralleled to each other.

modifications; this is mainly because we selected the median of the weight distribution as the amount of synaptic potentiated as a response to the pair of presynaptic and postsynaptic spike (in (30)). This static quantifier is less sensitive to the sudden changes that occurred in the tail of the distribution until those changes are visible via many elements of the distribution. On the other hand, this quantifier effectively quantifies the distribution of the network into a range where many elements of the distribution are approximately lying. Even though mean could also be a good option for such an indicator, is very sensitive to the sudden changes and could easily forget the history of the distribution. Therefore median is better than the mean, but still necessary to look for an unbiased statistical quantifier to estimate the amount of potentiated as a response to presynaptic and postsynaptic spike pair which can represent the history as well as the sudden changes to the weight distribution effectively. The other main drawback we see in our approach is the use of bins to chunk the process at synapses. The size of the bin might be a possible constraint that also limits the

Spike-Timing-Dependent Plasticity and Short-Term Plasticity Jointly Control the Excitation of Hebbian Plasticity
without Weight Constraints in Neural Networks

41

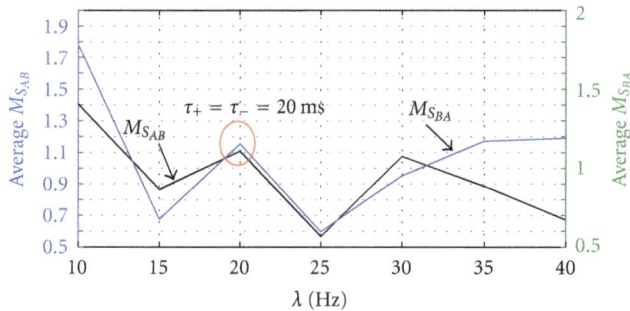

FIGURE 10: Distribution of the median of synaptic weights of neurons at different mean firing rates. Figure 10 illustrates the average of the weight medians calculated after both neurons were stabilized after applying Poisson inputs with mean rate λ. $M_{S_{AB}}$ is the average of the medians value of all the synapses of postsynaptic neuron A from presynaptic neuron B, and similarly $M_{S_{BA}}$ is the average of the medians value of all the synapses of postsynaptic neuron B to presynaptic neuron A. $\tau_+ = \tau_- = 20$ ms are STDP time constants for potentiation and depression.

performances of STDP on short-term dynamics. However, the model proposed in this research successfully balanced the synaptic excitation of the two neurons in an operational level without damaging their biological plausibility.

References

[1] D. O. Hebb, *The Organization of Behavior. The First Stage of the Perception: Growth of the Assembly*, John Wiley & Sons, New York, NY, USA, 1949.

[2] K. D. Miller and D. J. MacKey, "The role of constraints in Hebbian learning," *Neural Computation*, vol. 6, pp. 100–126, 1994.

[3] E. L. Bienenstock, L. N. Cooper, and P. W. Munro, "Theory for the development of neuron selectivity: orientation specificity and binocular interaction in visual cortex," *Journal of Neuroscience*, vol. 2, no. 1, pp. 32–48, 1982.

[4] L. F. Abbott and W. Gerstner, "Homeostasis and learning through spike-timing dependent plasticity," in *Presented at Summer School in Neurophzsics*, Les Houches, France, July 2004.

[5] G. J. Goodhill and H. G. Barrow, "The role of weight normalization in competitive learning," *Neural Computation*, vol. 6, no. 2, pp. 255–269, 1993.

[6] G. Q. Bi and M. M. Poo, "Synaptic modification by correlated activity: Hebb's postulate revisited," *Annual Review of Neuroscience*, vol. 24, pp. 139–166, 2001.

[7] L. F. Abbott and S. B. Nelson, "Synaptic plasticity: taming the beast," *Nature Neuroscience*, vol. 3, pp. 1178–1183, 2000.

[8] J. Lisman and N. Spruston, "Postsynaptic depolarization requirements for LTP and LTD: a critique of spike timing-dependent plasticity," *Nature Neuroscience*, vol. 8, no. 7, pp. 839–841, 2005.

[9] D. A. Butts and P. O. Kanold, "The applicability of spike dependent plasticity to development," *Frontier in Synaptic Neuroscience*, vol. 2, p. 30, 2010.

[10] R. Legenstein, C. Naeger, and W. Maass, "What can a neuron learn with spike-timing-dependent plasticity?" *Neural Computation*, vol. 17, no. 11, pp. 2337–2382, 2005.

[11] S. Song, K. D. Miller, and L. F. Abbott, "Competitive Hebbian learning through spike-timing-dependent synaptic plasticity," *Nature Neuroscience*, vol. 3, no. 9, pp. 919–926, 2000.

[12] M. C. W. van Rossum, G. Q. Bi, and G. G. Turrigiano, "Stable Hebbian learning from spike timing-dependent plasticity," *Journal of Neuroscience*, vol. 20, no. 23, pp. 8812–8821, 2000.

[13] M. C. W. van Rossum and G. G. Turrigiano, "Correlation based learning from spike timing dependent plasticity," *Neurocomputing*, vol. 38-40, pp. 409–415, 2001.

[14] T. Masquelier, R. Guyonneau, and S. J. Thorpe, "Spike timing dependent plasticity finds the start of repeating patterns in continuous spike trains," *PLoS ONE*, vol. 3, no. 1, Article ID e1377, 2008.

[15] F. Henry, E. Daucé, and H. Soula, "Temporal pattern identification using spike-timing dependent plasticity," *Neurocomputing*, vol. 70, no. 10–12, pp. 2009–2016, 2007.

[16] T. Tetzlaff, M. Helias, G. T. Einevoll, and M. Diesmann, "Decorrelation of neural-network activity by inhibitory feedback," *PLoS Computational Biology*, vol. 8, no. 8, Article ID e100259610, 2012.

[17] R. Gütig, R. Aharonov, S. Rotter, and H. Sompolinsky, "Learning input correlations through nonlinear temporally asymmetric Hebbian plasticity," *Journal of Neuroscience*, vol. 23, no. 9, pp. 3697–3714, 2003.

[18] J. S. Haas, T. Nowotny, and H. D. I. Abarbanel, "Spike-timing-dependent plasticity of inhibitory synapses in the entorhinal cortex," *Journal of Neurophysiology*, vol. 96, no. 6, pp. 3305–3313, 2006.

[19] Y. Luz and M. Shamir, "Balancing feed-forward excitation and inhibition via Hebbian inhibitory synaptic plasticity," *PLoS Computational Biology*, vol. 8, no. 1, Article ID e1002334, 2012.

[20] T. Branco and K. Staras, "The probability of neurotransmitter release: variability and feedback control at single synapses," *Nature Reviews Neuroscience*, vol. 10, no. 5, pp. 373–383, 2009.

[21] T. Branco, K. Staras, K. J. Darcy, and Y. Goda, "Local dendritic activity sets release probability at hippocampal synapses," *Neuron*, vol. 59, no. 3, pp. 475–485, 2008.

[22] R. S. Zucker and W. G. Regehr, "Short-term synaptic plasticity," *Annual Review of Physiology*, vol. 64, pp. 355–405, 2002.

[23] P. A. Appleby and T. Elliott, "Multispike interactions in a stochastic model of spike-timing-dependent plasticity," *Neural Computation*, vol. 19, no. 5, pp. 1362–1399, 2007.

[24] H. S. Seung, "Learning in spiking neural networks by reinforcement of stochastic synaptic transmission," *Neuron*, vol. 40, no. 6, pp. 1063–1073, 2003.

[25] A. M. Thomson, "Facilitation, augmentation and potentiation at central synapses," *Trends in Neurosciences*, vol. 23, no. 7, pp. 305–312, 2000.

[26] L. F. Abbott and W. G. Regehr, "Synaptic computation," *Nature*, vol. 431, no. 7010, pp. 796–803, 2004.

[27] W. Maass and A. M. Zador, "Dynamic stochastic synapses as computational units," *Neural Computation*, vol. 11, no. 4, pp. 903–917, 1999.

[28] G. G. Turrigiano, "Homeostatic plasticity in neuronal networks: the more things change, the more they stay the same," *Trends in Neurosciences*, vol. 22, no. 5, pp. 221–227, 1999.

[29] G. G. Turrigiano and S. B. Nelson, "Homeostatic plasticity in the developing nervous system," *Nature Reviews Neuroscience*, vol. 5, no. 2, pp. 97–107, 2004.

[30] S. D. Fernando, K. Yamada, and A. Marasinghe, "Observed stent's anti-Hebbian postulate on dynamic stochastic computational synapses," in *Proceedings of International Joint*

Conference on Neural Networks (IJCNN '11), pp. 1336–1343, San Jose, Calif, USA, 2011.

[31] S. Fernando, K. Ymamada, and A. Marasinghe, "New threshold updating mechanism to stabilize activity of Hebbian neuron in a dynamic stochastic 'multiple synaptic' network, similar to homeostatic synaptic plasticity process," *International Journal of Computer Applications*, vol. 36, no. 3, pp. 29–37, 2011.

[32] G. Q. Bi and M. M. Poo, "Synaptic modifications in cultured hippocampal neurons: dependence on spike timing, synaptic strength, and postsynaptic cell type," *Journal of Neuroscience*, vol. 18, no. 24, pp. 10464–10472, 1998.

[33] H. Markram, Y. Wang, and M. Tsodyks, "Differential signaling via the same axon of neocortical pyramidal neurons," *Proceedings of the National Academy of Sciences of the United States of America*, vol. 95, no. 9, pp. 5323–5328, 1998.

[34] G. G. Turrigiano, K. R. Leslie, N. S. Desai, L. C. Rutherford, and S. B. Nelson, "Activity-dependent scaling of quantal amplitude in neocortical neurons," *Nature*, vol. 391, no. 6670, pp. 892–896, 1998.

4

Random Bin for Analyzing Neuron Spike Trains

Shinichi Tamura,[1] Tomomitsu Miyoshi,[2] Hajime Sawai,[2] and Yuko Mizuno-Matsumoto[3]

[1] *NBL Technovator Co., Ltd., 631 Shindachimakino, Sennan City, Osaka 590-0522, Japan*
[2] *Department of Integrative Physiology, Graduate School of Medicine, Osaka University, Suita 565-0871, Japan*
[3] *Graduate School of Applied Informatics, University of Hyogo, Kobe 650-0047, Japan*

Correspondence should be addressed to Shinichi Tamura, tamuras@nblmt.jp

Academic Editor: Huiyan Jiang

When analyzing neuron spike trains, it is always the problem of how to set the time bin. Bin width affects much to analyzed results of such as periodicity of the spike trains. Many approaches have been proposed to determine the bin setting. However, these bins are fixed through the analysis. In this paper, we propose a randomizing method of bin width and location instead of conventional fixed bin setting. This technique is applied to analyzing periodicity of interspike interval train. Also the sensitivity of the method is presented.

1. Introduction

Bin width setting is always a problem, since it affects largely analyzed results. Neural spike train usually has time-varying characteristics. Therefore, data length of spike train in stationary state with the same characteristics is often limited. That is, the number of stable data is limited, and therefore there exists limitation in decreasing bin width to analyze more precisely. The more troublesome problem is that the results become different by how much to set the bin width or even the initial position.

Bin size has been determined to optimize some performance measure of time histogram [1, 2], time precision [3–5], information [6], rate estimation [7], and so forth. However, their bins are fixed after being optimized/determined. To avoid such troublesome problem, binless analysis methods are also used [8–10].

In this paper, we propose a method of setting various random bins. Random bin will be expected to decrease unfavorable effects up to the level of being neglectable. See the appendix section for preliminary easy explanation of the random bin.

2. Automutual Information of Spike-Interval Train

To analyze a spike train as a time sequence, there exist mainly 4 methods of (i) spectrum analysis [11] which includes sideband and therefore may be limited in precise time analysis, (ii) correlation [12] which reflects only linear relation, (iii) time histogram [1] whose precision may be limited by nonstationarity of the train, and (iv) information measure [6, 12, 13] which is expected to be possible to avoid such limitations. Automutual information method dealt in this paper belongs to (iv).

Mutual information (MI) is a measure of expressing common quantity of information between events A and B, as described by (1):

$$\mathrm{MI}(A;B) = \sum_{A}\sum_{B} P(A,B)\log_2\left[\frac{P(A,B)}{P(A)\cdot P(B)}\right] [\mathrm{bit}]. \quad (1)$$

More specifically, this is the difference between joint probability $P(A,B)$ and probability $P(A)\cdot P(B)$ in which A and B are assumed to be independent events. If A and B are indeed independent, they have no common information, and therefore the mutual information is zero. If we take an

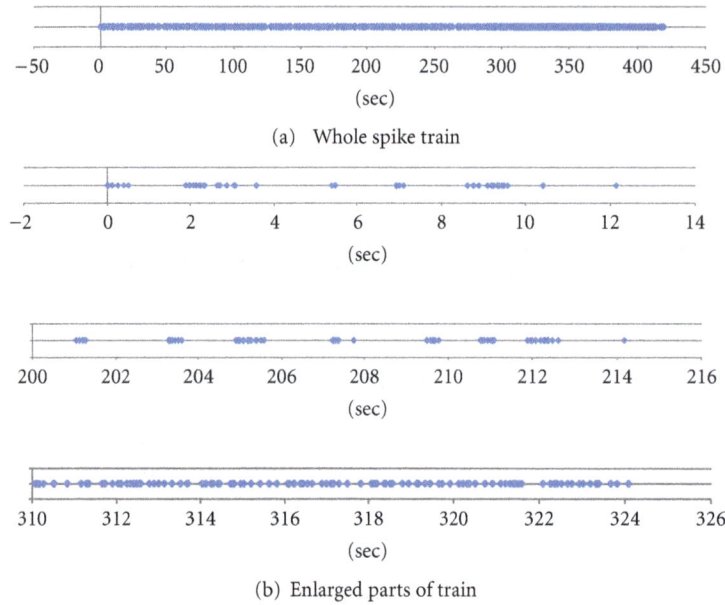

(a) Whole spike train

(b) Enlarged parts of train

FIGURE 1: Example of sorted spike train (Electrode No.16, light stimulated).

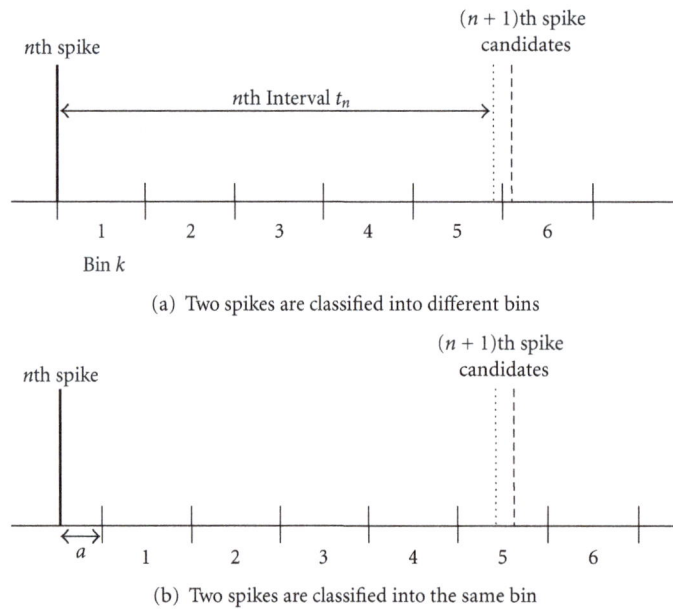

(a) Two spikes are classified into different bins

(b) Two spikes are classified into the same bin

FIGURE 2: Problem of discriminatability in discretizing analog interspike intervals. Almost the same intervals are treated differently by chance in the fixed bin setting.

inter-spike interval train as A, and one shifted by m intervals as B, mutual information becomes automutual information (AMI).

3. Spike Train

Figure 1 shows a spike train obtained from Electrode No.16 of V1 field of a rat with LED light stimulation of 30 ms duration at every 7 sec. This is a sorted data, which means it is processed by pattern recognition so as to catch only spikes from a specific neuron. Number of spikes is 1721

between 420 sec. Some enlarged parts of the train are shown in Figure 1(b).

To investigate the periodic characteristics, we calculated automutual information between interval-value train (A) and its shifted train (B) by m intervals.

4. Problem of Fixed Bin

Figure 2 shows how almost the same intervals are classified into different bins or the same bin almost by chance depending on where the border of the bins is in the case of

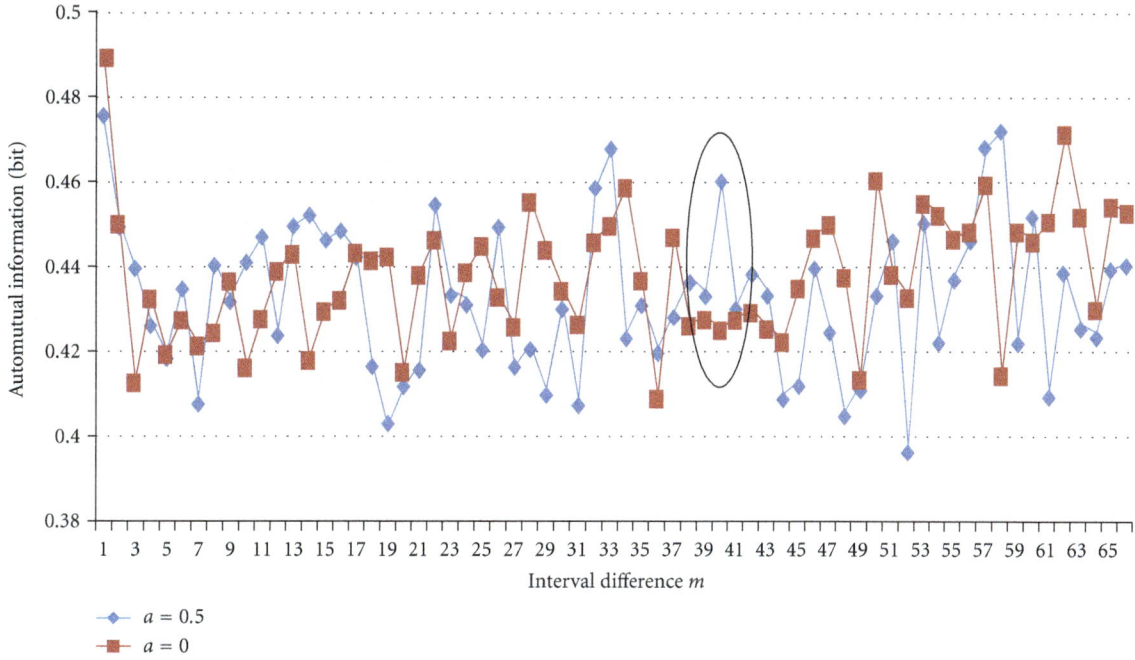

FIGURE 3: Difference of automutual information values for half-shift of bin position.

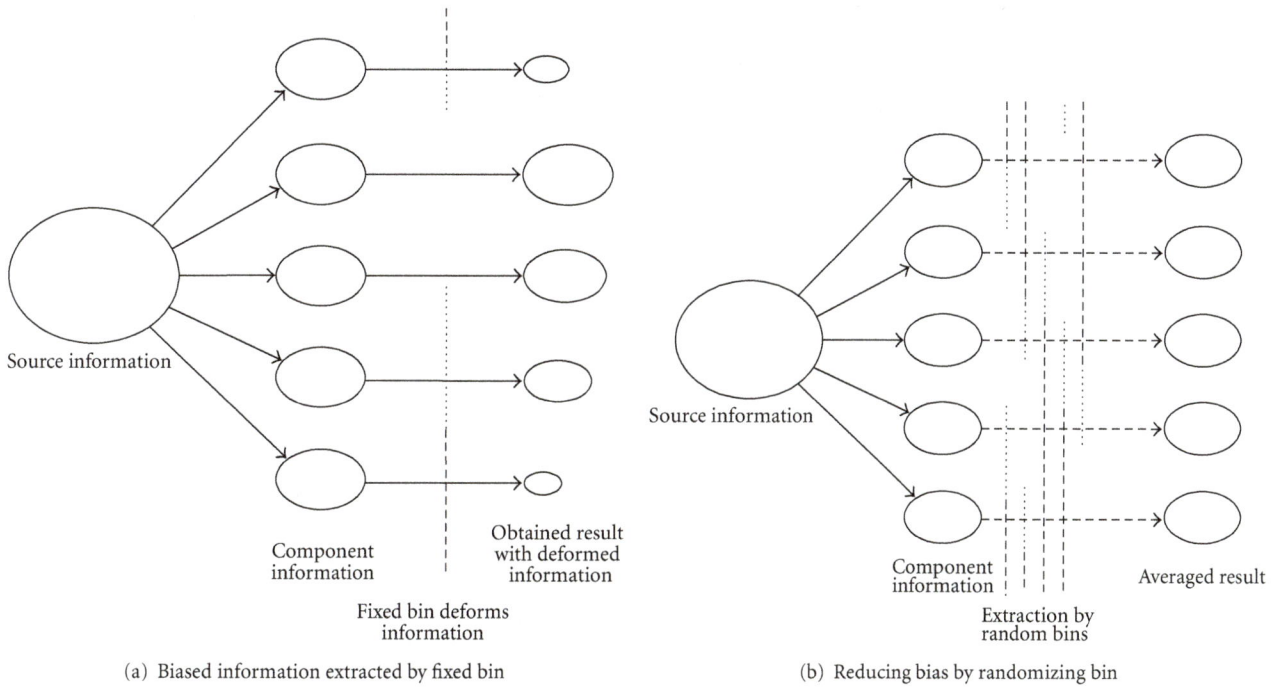

(a) Biased information extracted by fixed bin

(b) Reducing bias by randomizing bin

FIGURE 4: Reducing bias in obtained result by averaging randomly biased information.

conventional fixed bin setting. This affects the result of AMI calculation.

To show this, assume the bin is set as follows: number of bins is $K = 32$. The kth bin border ($k = 1, 2, \ldots, K$) is given by

$$b(k) = 0.01 \times 10^{(3.5 \times (k-a)/K)} = 0.01 \times 10^{(0.1094 \times (k-a))}. \quad (2)$$

In this paper, to be able to cover wide range of intervals, exponential bin setting is adopted differently from Figure 2. The bin is set as

Bin 1: 0-$b(1)$

Bin 2: $b(1)$-$b(2)$

$\ldots$

Bin 32: $b(31)$-$b(32)$.

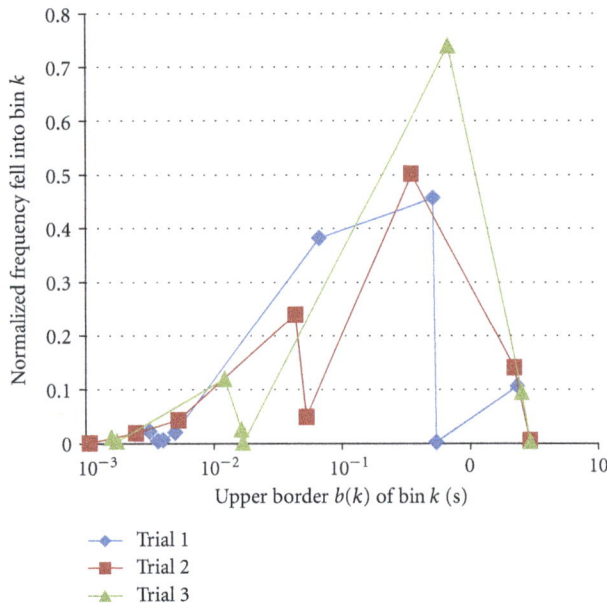

FIGURE 5: Random bin and rate fell into bin k. $K = 8$.

Data over $b(32)$ were neglected. Usually we set $a = 0$ which corresponds to the case of Figure 2(a). In order to check the problem of the fixed bin in this section, we compared in cases of $a = 0$ and $a = 0.5$. That is, in the latter case, the Bin borders are shifted by half as Figure 2(b). This may often happen since spike interval has lower bound by refractory period, and therefore bin setting at small interval values is nonsense.

Figure 3 shows two results of AMI calculation for the data of Figure 1 with shifted bin positions by half as shown in Figure 2. Their shapes are rather different. For example, at $m = 40$ curve of $a = 0.5$ has a peak, but $a = 0$ has not been as shown by a black ellipse. We can see that it is almost impossible to extract period components from the spike train by the fixed bin setting as it is.

5. Randomized Bin Setting

In order to suppress such instability, after having tried some methods including fluctuating initial position a of (2) and classifying an interval value into not one but adjacent two bins with weights, we decided finally to adopt a bin randomizing method, though it needs more computation time than the former.

First, 32 uniform random numbers between $[0, 32)$ are generated, rearranged in order from small to large, and they are substituted for k of (2). In a preliminary experiment, K was set 8, while it was set 32 in the main experiment, which was also extended to 128. Then we calculate one-trial AMI. Also start counting how many times one-trial AMI becomes the maximum at each m among, for example, 64 interval differences. This is a one trial with a random bin setting. We repeated these trials $N = 5,000$ to $500,000$ times and averaged to obtain final AMI. At the same time, we

also obtained normalized frequency of AMI becoming the maximum at each m.

This method will be explained in Figure 4; that is, fixed bin method (a) has some biased characteristics. If we generate random set of bin borders $\{b(k)\}$ as method (b), bias effects will be decreased by repeating many times.

In addition to the original data set (i) of spike trains from rat V1 field, we also prepared (ii) interval shuffled train (Shuf3/Shuf8) among successive 3 intervals or 8 intervals with sequentially shifting interval one by one and (iii) one repeated Shuf8 operation 2048 times (Shuf8–2048) or 256 times (Shuf8–256). Further we prepared (iv) three different randomly generated trains only having the same number of spikes with the original train but not the same interval distribution.

6. Experimental Results

6.1. Preliminary Experiment. Before starting the main experiment, we tried with a small size of $K = 8$ and $N = 5,000$. Examples of normalized frequency that fell into bins in three trials for original train shown in Figure 1 are shown in Figure 5.

Figure 6 shows changes of AMI and its frequency of taking local maximum at each interval difference m when shuffling the spike train. Generally speaking, by shuffling the train, AMI values do not decrease suddenly but gradually, since some rate of interval pairs moves in the same way with keeping the same relative interval difference. Large values of AMI and consequently large frequency of taking local maximum of Original train are often decreased by shuffling more as shown by black ellipses in Figure 6. Inversely, since total values of normalized frequency are 1, other new periodic components of AMI emerge/increase by shuffling, and consequently the rate of taking local maximum is also increased as shown by purple ellipses, though not completely.

Figure 7 shows an obtained scatter plot of AMI versus frequency of AMI value took local maximum for original train shown in Figure 1, its Shuf8, Shuf8–256, and Random trains. AMI curve has such characteristics that (i) AMI of original train usually takes the maximum at $m = 1$, since if a spike detected that time moves to front, preceding interval value is shortened and succeeding one is elongated; that is they have negative correlation relationship (low independency), and (ii) curve is sometimes inclined subtly. To cope to these at this stage, we took a local-maximum judgment separately at ranges of 1–4, 5–8, 9–16, 17–32, and 33–64 instead of maximum judgment at full range of 1–64. Therefore we see 4 outlier points of Original data most at right and 3 around horizontal axis in Figure 7. We can also see that Shuf8 points are almost overlapping on the Original ones, Shuf8–256 points are shifted to lower AMI values, and Random points shifted more. These are well separated. That is, the AMI with random bin method can well extract temporal information of the spike train.

6.2. Prefinal Experiment. To improve the result, we expanded number of bins to $K = 32$ and number of trials to

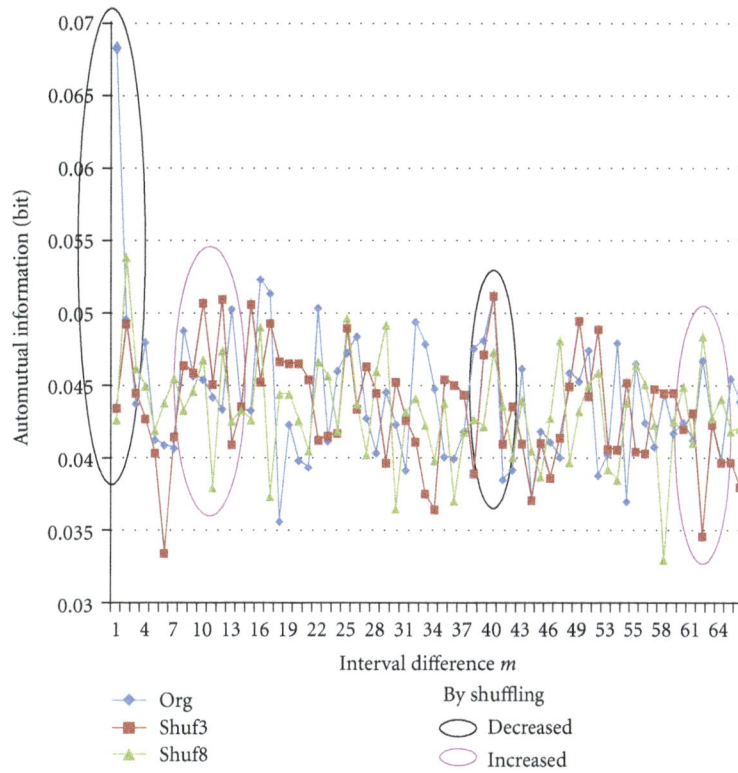

(a) Automutual information

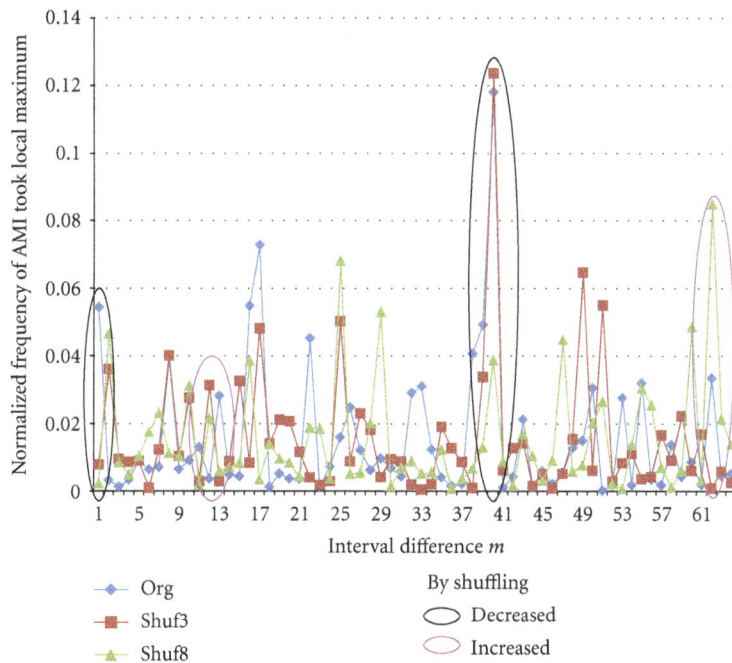

(b) Maximum-detection rate of AMI

FIGURE 6: Changes by shuffling spike train for $K = 8$ and $N = 5,000$. In black ellipses as examples, AMI values and its peak detection rate are decreased according to the train shuffled more. On the other hand, in purple ellipses, they are increased by disturbance of shuffling.

$N = 40,000$. When increasing K 4 times, since the probability of intervals falling into a bin decreases to 1/4, it may be reasonable also to increase N in this case 8 times. Examples of normalized frequency that fell into bins in four trials for original train shown in Figure 1 are shown in Figure 8. Figure 9 shows scatter plot of the obtained AMI versus frequency of AMI value took the local maximum of the Original train shown in Figure 1 with its Shuf8, Shuf8–256

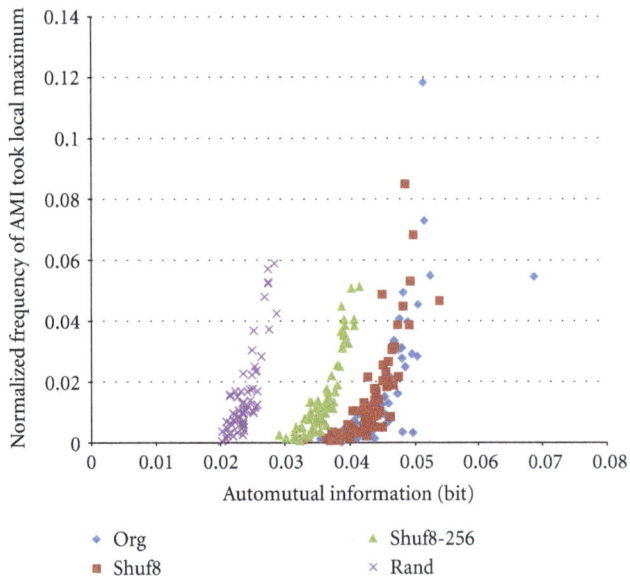

FIGURE 7: Scatter plot of automutual information versus probability of AMI that took local maximum. $K = 8$ and $N = 5,000$.

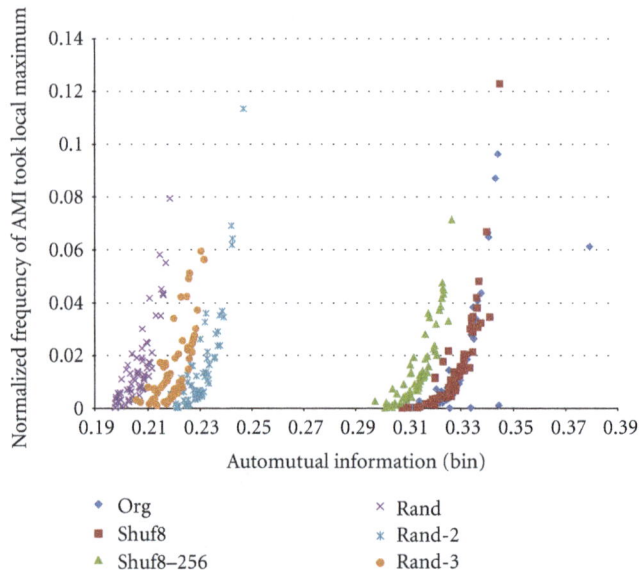

FIGURE 9: Scatter plot of automutual information versus probability of AMI that took local maximum. $K = 32$ and $N = 40,000$.

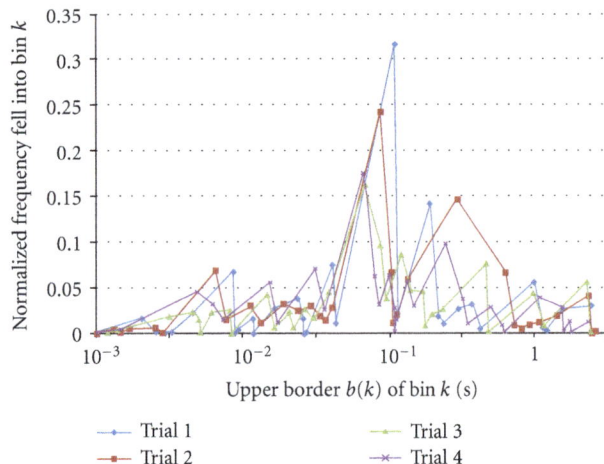

FIGURE 8: Examples of rate of intervals fell into bin k.

times, and three different Random trains. We can see that compared with Figure 7 the scatter plot converged more. Note that random trains have some divergence within trains.

6.3. Main Experiment. Increasing the number of trials more to $N = 500,000$, we obtained almost the same results as $N = 40,000$ but more improved than $N = 5,000, K = 8$ (8Bins). These are shown in Figure 10. We can see that N seems to have reached plateau already at 40,000. In this case, by suppressing the AMI value at $m = 1$ to 0, we could determine more fairly the maximum of AMI value through all ranges of 64 interval differences. Then, we could obtain a final scatter plot of Figure 11, where we can see clear one-to-one correspondence between AMI and maximum-detection frequency than Figure 9.

AMI curves in Figure 10(a) seem rather flat. Contrary to this, rate curve of AMI taking the maximum in Figure 10(b) appears more sensitive or too much sensitive to the periodicity. Essentially, however, they have the same information.

7. Sensitivity Check

7.1. By Test Train Only. The average of inter-spike interval of the train No.16 of Figure 1 was $\tau =$ (measuring time)/(number of spikes) $= 0.244$ sec. We generated a base train with constant interval of 0.244 sec. That is, the base train is

$$\mathbf{B} = (t_1, t_2, \ldots, t_{1721}) = (\tau, \tau, \ldots, \tau). \qquad (3)$$

Then, test trains were generated by adding periodic component such that

$$t_n = \begin{cases} (1+s)\tau & \text{if } n = P \times i, i = 1, 2, 3, \\ \tau & \text{otherwise,} \end{cases} \qquad (4)$$

where P is a period of the test component and s is its amplitude.

Figure 12(a) shows obtained AMIs for test trains with $P = 27$. It shows sharp peaks at $m = 27$, just corresponding to P as well as the 2nd peaks twice at interval difference $m = 54$. Figure 12(b) shows their peak values at $m = 27$ with extending range of s more than (a). This is a sensitivity of the proposed method.

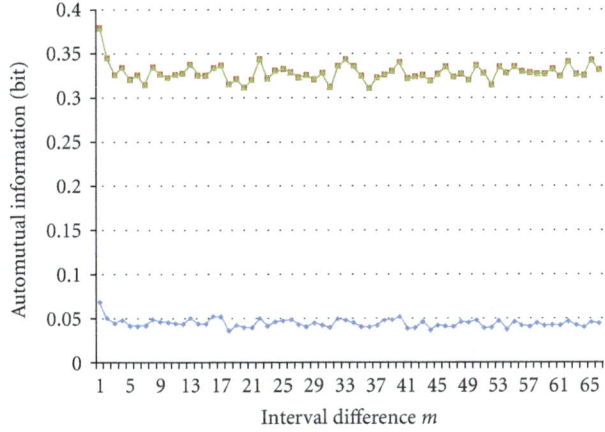

(a) Automutual information

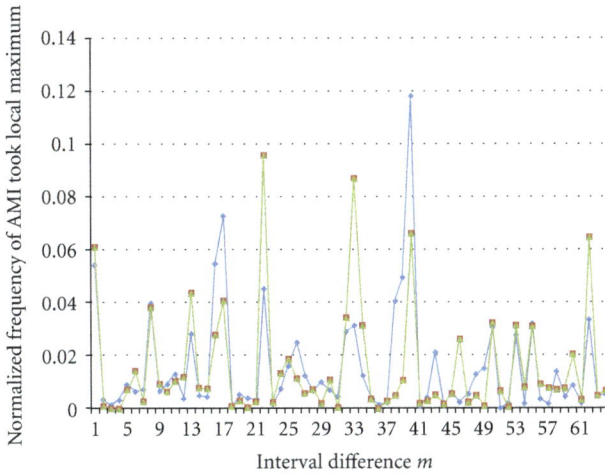

(b) Normalized frequency of AMI took the local maximum

FIGURE 10: Improvement of AMI (automutual information) calculation and local maximum detection by increasing numbers of bins and trials. 32-Bin 40k Trial and 32-Bin 500 k curves are almost the same and overlapping.

7.2. Test Train Added to Real Train. Figure 13 shows the results of test train with several amplitudes added to No.16 train. Test train is

$$t_n = s\tau \quad \text{if } n = P \times i \text{ where } i = 1, 2, 3, \ldots$$

$$0 \qquad \text{otherwise,} \tag{5}$$

where periodicity $P = 27$. We can see that in the Original train there exists low periodic component at $m = 27$. Then by adding test train with amplitude of more than 30% of average interval τ, periodic component appears; that is in No.16 train, there exist many periodic components with amplitude of several ten % of τ.

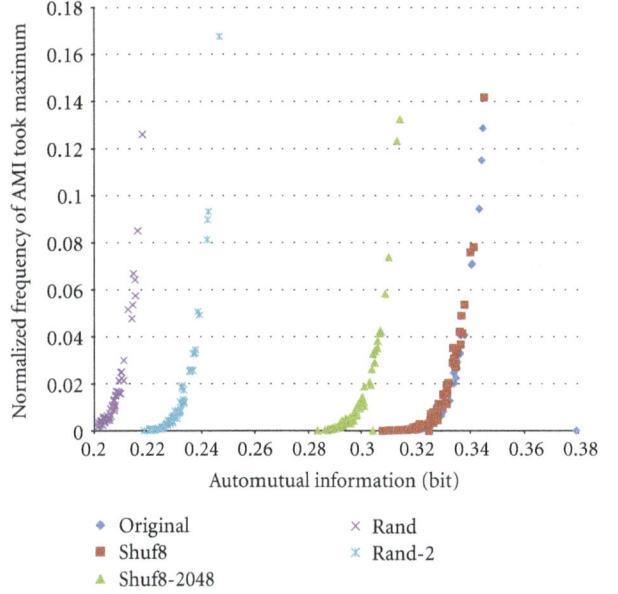

FIGURE 11: Scatter plot of automutual information versus probability of AMI that took the maximum for $K = 32$ and $N = 500,000$ with improving the maximum detection by suppressing AMI of $m = 1$. Electrode No.16, light stimulated.

8. Low Periodicity Train

The train from Electrode No.16 with light stimulation shown above is the one mostly showing its deep structure in the sense that AMI values of Shuf8–2048 are clearly lower than that of Org. This means that characteristics including periodicity are disturbed by interval shuffling. However, this is not always the case. An example of results of commonly typical (ordinary) train of nonstimulated spontaneous response of No.2 Electrode is shown in Figure 14 where characteristics of shuffled train (Shuf8, Shuf8–256) are not so different from original one (Org) but have larger AMI values than that of artificially generated Random trains. Number of spikes in this No.2 train is 729.

9. Extension to 128 Bins

We tried to extend the number of bins to 128 for some cases, though computation time takes several times compared with $K = 32$ cases. Figure 15 shows two examples of rate of spike intervals fell into the random bins. Figure 16 shows AMI, and Figure 17 shows the normalized frequency of AMI took the maximum of the spike train from the Electrode No.2 with the light stimulations and $K = 128$, $N = 20,000$. We can see in this case that peak of AMI showing periodicity is sharp at $P = 28$, and it disappears after interval shuffling and in random sequence.

10. Discussion

In the calculation of AMI, P(A,B) is estimated from the target data. As a result, it works as a learning effect. Consider an

(a) AMIs for test trains

(b) Sensitivity for test train

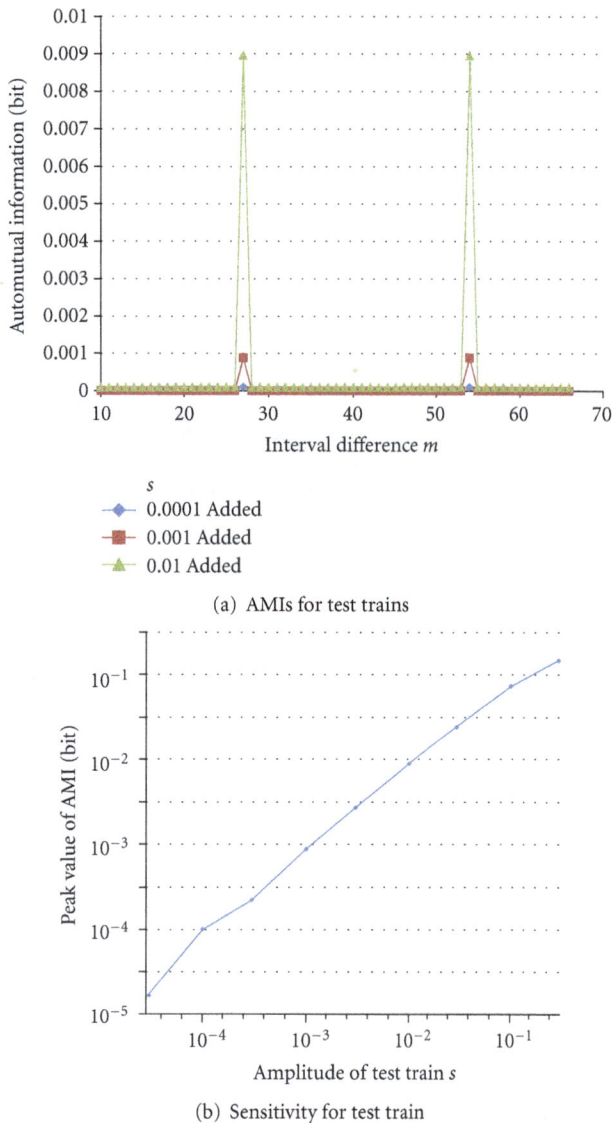

FIGURE 12: Sensitivity of AMI for various amplitudes of periodic test trains.

(a) Automutual information

(b) Rate of AMI took maximum

FIGURE 13: Results of test train added to No.16 train.

ultimate case with only two intervals t_1 and t_2 from three spikes, where we can estimate the future t_2 (generally t_{n+m}) perfectly with mutual information $\log_2 K$ if we know $t_1(t_n)$. Therefore, the smaller number of spikes we have, the more we can estimate future, and the higher mutual information we have between the present and future. Inversely, the larger number of spikes we have, the smaller level of the average AMI values we obtain. Figure 18 shows the relation between AMI level and number of spikes in a train of our experiments. There may be some theoretical relationship between these. However, we have not analyzed enough yet. Instead in the experiment, we expanded K values up to 128 and can see that we can obtain higher AMI level which means we are able to estimate more accurate future interval values by increasing K value. However, it is also true that since the number of spikes is limited, we cannot increase K value unlimitedly to estimate well $P(A,B)$.

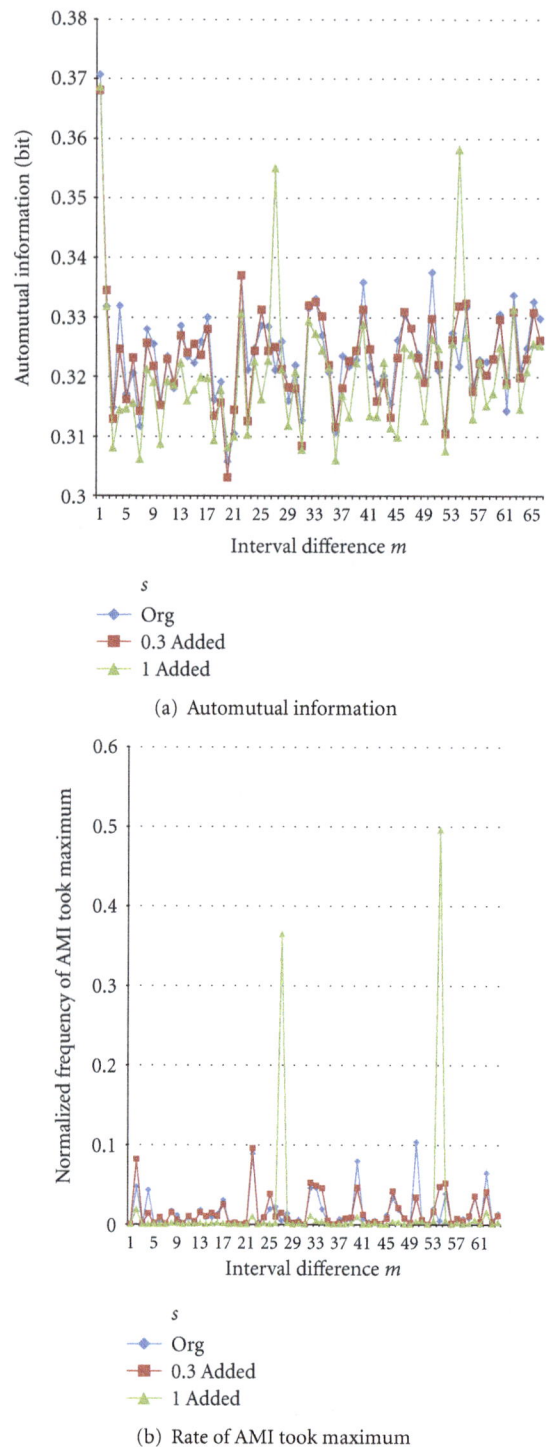

From viewpoint of circuit theory, each periodicity corresponds to a specific circuit excited by a trigger input. Then, by analyzing the interspike interval sequence, it may be possible to get known the participating circuit shape or structure. Through such analysis, it may become possible to analyze the information storage and communication mechanisms in the brain.

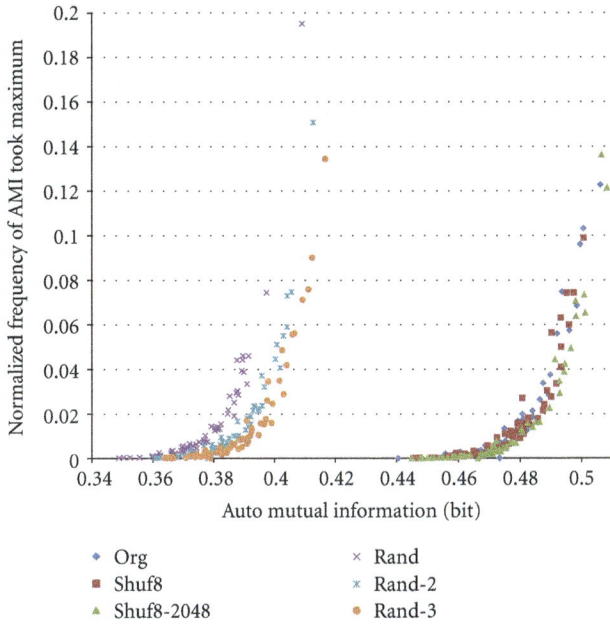

FIGURE 14: Example of typical scatter plot of common (ordinary) trains (nonstimulated; Electrode No.2 sorted) of $K = 32$ and $N = 40,000$ with improved maximum detection.

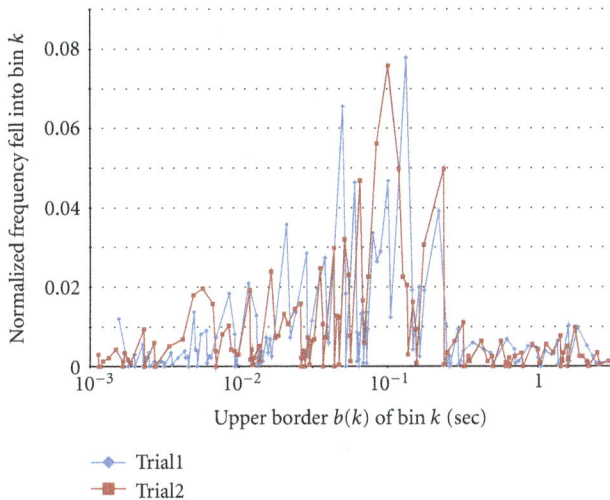

FIGURE 15: Two examples of rate of intervals fell into bin k when $K=128$.

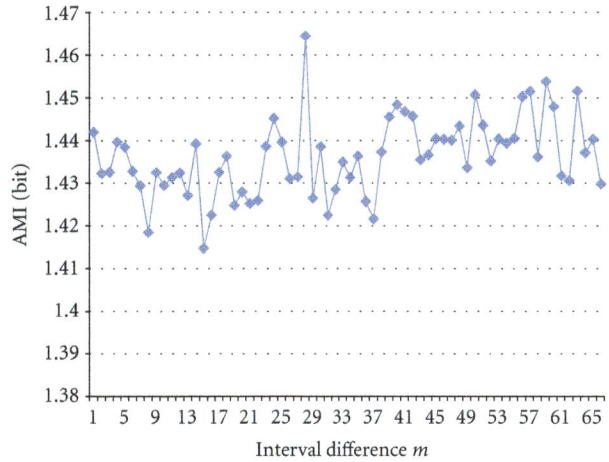

FIGURE 16: AMI of spike train No.2 with light stimulation and $K = 128$, $N = 20,000$.

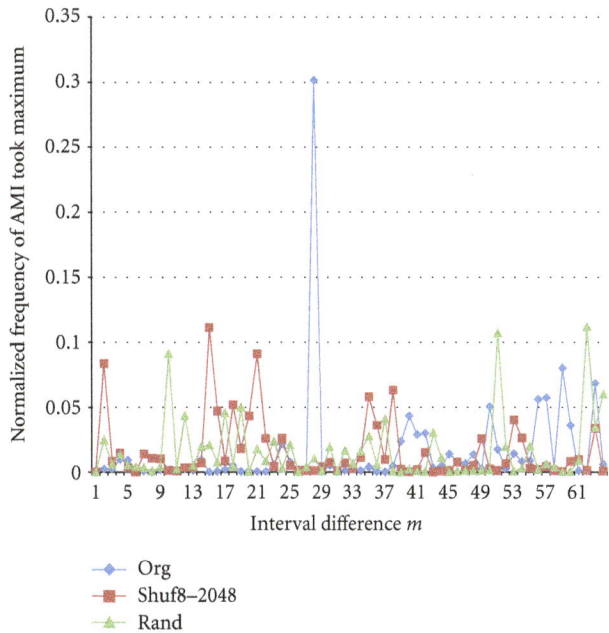

FIGURE 17: Normalized frequencies of AMI of Electrode No.2 with light stimulation took the maximum, interval shuffled (Shuf8-2048), and randomly generated (Rand).

Problem of the proposed method is computation time. Presently software is written in Basic interpreter language (BASICw32), and it takes 8 hours with 2.4 GHz i5 CPU of note PC to calculate 40k trials when $K = 32$ for train data with 1721 intervals. This may be possible to reduce to one severalth by using compiler language.

11. Conclusion

Sizes and positions of time bins have been usually fixed. It often causes effect to precision and stability of the results. In this paper, we proposed a bin randomizing method to avoid such troubles. As an analyzing method, we used automutual information, which has merits of (i) detectability of even non-monotonic relation than correlation (ii) since the AMI is calculated based on not the absolute time but the appearance order and independence relationship between train intervals, it can cope with nonstationarity such as expand and contract of the spike interval combined with the flexibility of randomized bin, and it has more (iii) direct precise analysis than spectrum analysis and able to cope with nonstationarity.

Demerit of the proposed method is the long computation time. However, as a postprocessing of the spike train, these are not severe demerits.

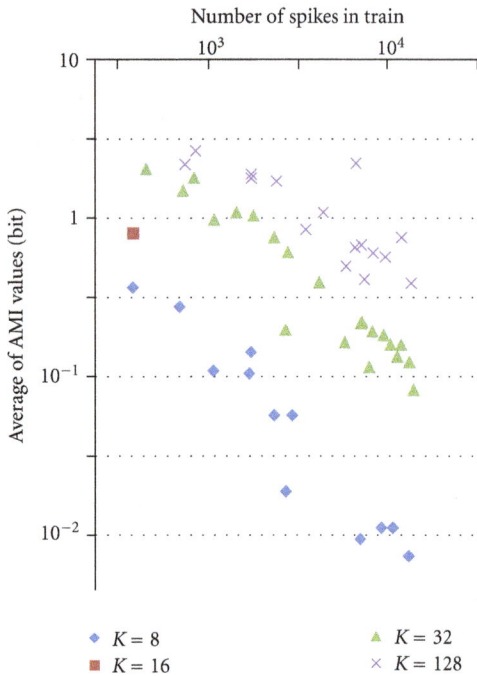

FIGURE 18: AMI level versus number of spikes. $N = 5,000$.

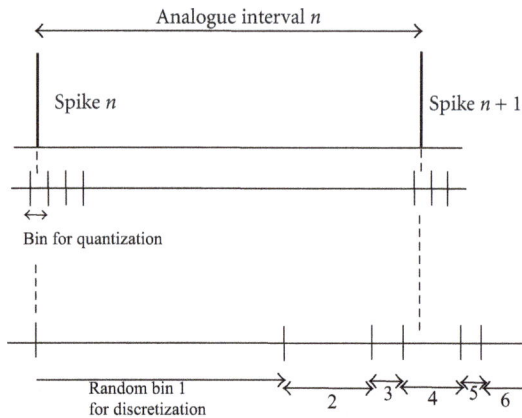

FIGURE 19: Conventional fixed bin for quantization (upper) and proposed random bin for time discretization (lower) to calculate mutual information. In this figure case, interval n is categorized to Random bin 4.

It is shown that there exists an almost one-to-one monotonic relation between AMI value and rate of AMI value, takes the maximum through many trials of random bin generation.

Though mainly we treated a problem of obtaining automutual information, the proposed method of random bin can also apply not only to the spike analysis but also to other problems of other fields.

Appendix

A. Preliminary Explanation of Random Bin Proposing

If the codes for communication are generated repeatedly by circulating pulses in a loop circuit, we can observe a sequence with a period. Then two time intervals separated with some number p of intervals will differ in their lengths reflecting local appearance patterns in the code such as 11, 101,10 01, and 10001 as well as subtle physical transmission-time differences between different cell connections. Therefore, if we calculate mutual information between time lengths of the two intervals, we may be able to estimate the period of the code. Conventionally in this calculation, time bin size is first fixed, and analogue time lengths of intervals are next quantized according to this fixed bin size as shown in Figure 19 upper part. Typically bin size (width) is set to around time fluctuation (uncertainty) of spike position or small so as no more one spike falls in a bin. However, the bin size determined beforehand affects the results considerably. Therefore, we cannot trust any more the results obtained through this fixed bin size.

To cope with this problem, we propose here a method of discretizing analogue time intervals by random time-bins as shown in Figure 19 lower part for calculating the mutual information in each trial. These time bins are different trial by trial. Some time, the time bins are well determined and can discriminate different time intervals, and some time not. As an average, the mutual information will show the true mutual information not affected by the bin or each bin size.

Finally, the mutual information between interval n and interval $n + m$ is shown as an automutual information (AMI) graph by changing m (interval difference).

Acknowledgments

The authors thank K. Nakamura and A. Kumamoto of graduate student of University of Hyogo for supporting the experiments. This work was supported in part by the Grant-in-Aid for Scientific Research of Exploratory Research 21656100 and Scientific Research (A) 22246054 of Japan Society for the Promotion of Science.

References

[1] T. Omi and S. Shinomoto, "Optimizing time histograms for non-Poissonian spike trains," *Neural Computation*, vol. 23, no.12, pp. 3125–3144, 2011.

[2] H. Shimazaki and S. Shinomoto, "A method for selecting the bin size of a time histogram," *Neural Computation*, vol. 19, no.6, pp. 1503–1527, 2007.

[3] D. Endres and M. Oram, "Feature extraction from spike trains with Bayesian binning: "Latency is where the signal starts"," *Journal of Computational Neuroscience*, vol. 29, no. 1-2, pp. 149–169, 2010.

[4] D. Endres, J. Schindelin, P. Földiák Peter, and M. W. Oram, "Modelling spike trains and extracting response latency with

Bayesian binning," *Journal of Physiology Paris*, vol. 104, no. 3-4, pp. 128–136, 2010.

[5] S. Louis, C. Borgelt, and S. Grün, "Complexity distribution as a measure for assembly size and temporal precision," *Neural Networks*, vol. 23, no. 6, pp. 705–712, 2010.

[6] D. D. Alan, "Estimating neuronal information: logarithmic binning of neuronal inter-spike intervals," *Entropy*, vol. 13, no. 2, pp. 485–501, 2011.

[7] H. Shimazaki and S. Shinomoto, "Kernel bandwidth optimization in spike rate estimation," *Journal of Computational Neuroscience*, vol. 29, no. 1-2, pp. 171–182, 2010.

[8] A. R. C. Paiva, I. Park, and J. C. Príncipe, "A comparison of binless spike train measures," *Neural Computing and Applications*, vol. 19, no. 3, pp. 405–419, 2010.

[9] P. B. Kruskal, J. J. Stanis, B. L. McNaughton, and P. J. Thomas, "A binless correlation measure reduces the variability of memory reactivation estimates," *Statistics in Medicine*, vol. 26, no. 21, pp. 3997–4008, 2007.

[10] B. Schrauwen and J. V. Campenhout, "Linking non-binned spike train kernels to several existing spike train metrics," *Neurocomputing*, vol. 70, no. 7-9, pp. 1247–1253, 2007.

[11] M. Rivlin-Etzion, Y. Ritov, G. Heimer, H. Bergman, and I. Bar-Gad, "Local shuffling of spike trains boosts the accuracy of spike train spectral analysis," *Journal of Neurophysiology*, vol. 95, no. 5, pp. 3245–3256, 2006.

[12] A. Scaglione, G. Foffani, G. Scannella, S. Cerutti, and K. A. Moxon, "Mutual information expansion for studying the role of correlations in population codes: how important are autocorrelations?" *Neural Computation*, vol. 20, no. 11, pp. 2662–2695, 2008.

[13] S. Ito, M. E. Hansen, R. Heiland, A. Lumsdaine, A. M. Litke, and J. M. Beggs, "Extending transfer entropy improves identification of effective connectivity in a spiking cortical network model," *PLoS ONE*, vol. 6, no. 11, Article ID e27431, 2011.

5

Emergent Central Pattern Generator Behavior in Gap-Junction-Coupled Hodgkin-Huxley Style Neuron Model

Kyle G. Horn,[1,2] **Heraldo Memelli,**[2,3] **and Irene C. Solomon**[2]

[1] *Program in Neuroscience, Stony Brook Universty, SUNY, Stony Brook, NY 11794-5230, USA*
[2] *Department of Physiology and Biophysics, Stony Brook Universty, SUNY, Stony Brook, NY 11794-8661, USA*
[3] *Department of Computer Science, Stony Brook Universty, SUNY, Stony Brook, NY 11794-4440, USA*

Correspondence should be addressed to Irene C. Solomon, irene.solomon@stonybrook.edu

Academic Editor: Daoqiang Zhang

Most models of central pattern generators (CPGs) involve two distinct nuclei mutually inhibiting one another via synapses. Here, we present a single-nucleus model of biologically realistic Hodgkin-Huxley neurons with random gap junction coupling. Despite no explicit division of neurons into two groups, we observe a spontaneous division of neurons into two distinct firing groups. In addition, we also demonstrate this phenomenon in a simplified version of the model, highlighting the importance of afterhyperpolarization currents (I_{AHP}) to CPGs utilizing gap junction coupling. The properties of these CPGs also appear sensitive to gap junction conductance, probability of gap junction coupling between cells, topology of gap junction coupling, and, to a lesser extent, input current into our simulated nucleus.

1. Introduction

Central pattern generators (CPGs) correspond to neural regions that spontaneously generate oscillatory behavior in the absence of patterned input. In both invertebrates and vertebrates, they appear to play a critical role in the formation of repeated oscillatory behaviors, including activities such as walking, swimming, heartbeating, and breathing [1–4]. Because of their roles in cardiac and respiratory function, CPGs may be considered vital for basic survival across much of the animal kingdom.

Originally, the oscillatory behaviors seen in locomotion were presumed to be generated through reflexes alone. An ever-growing body of evidence, however, suggests that both locomotor [2, 5] and respiratory oscillatory activities [6, 7] are generated centrally in spinal cord and brainstem regions, respectively, since these behaviors occur in the absence of descending cortical drive and sensory input. Modulation of CPG activity, however, is necessary for adapting locomotor and breathing patterns to ever-changing environmental conditions. Because of this, both the locomotor [8] and the respiratory [9, 10] systems exhibit a great deal of plasticity

in the face of changing conditions and, therefore, should be viewed as dynamic rhythm generating devices.

Traditionally, reciprocal synaptic inhibition between two neuronal populations (or two groups of neuronal populations, or even two individual neurons [11]) is seen as the standard method of generating CPG behavior in both biological and computational systems. Originally proposed by Brown [12], this style of CPG appears in biological models of lamprey [13] and stick insect locomotion [14]. It also appears in simulated salamander [15] and mammalian locomotion models [16] as well as in leech heart [17] and as a component in more complex models of respiratory activity [18]. This form of CPG is often referred to as the half-center model and is a prominent model for robotic locomotion controllers [19, 20].

While half-center CPGs typically focus on synaptic inhibition, recent work indicates that gap junction coupling may also play a role in locomotor patterns [21] and respiratory patterns in both amphibians [22] and mammals [23–26]. Gap junctions are a prominent mechanism for neuronal synchrony in the brainstem, cerebellum, and neocortex as well as among motoneurons, glia, and retinal cells [27–29], though they can also produce complex asynchronous patterns [30]. Gap junction proteins and functional gap junction coupling have also been demonstrated in numerous neurons associated with central respiratory control, including hypoglossal and phrenic motoneurons [25, 31] and the pre-Bötzinger complex [23, 32]. A combination of gap junctions and synaptic inhibition may also be responsible for synchrony in some neuronal populations [33], and even if gap junctions are not responsible for generating a mutually inhibitory connection, inhibitory currents coupled by gap junctions could easily play such a role.

Although hard-wired reciprocal synaptic inhibition may be easy to identify physiologically, we propose that this same style of inhibition can spontaneously form in a single pool of gap-junction-coupled neurons, mutually inhibiting one another via their slow afterhyperpolarization (sAHP). The slow sAHP following the action potential can be modified directly through calcium-gated potassium channels (called either I_{AHP} or I_{SK}), which are found in many neurons including motoneurons [34, 35] and are known to play an important role in burst frequency modulation. Since neurons would dynamically assign themselves to one of the two "half-centers," changes to gap junctions, I_{AHP}, or inputs alone could modify how individual neurons align their firing. This would produce a highly dynamic modifiable half-center CPG capable of adapting to the rapid demands of locomotion or respiration.

Here, we present two biologically realistic models of gap-junction-coupled neurons that exhibit multiple output rhythms typical of half-center CPGs. Unlike standard half-center CPG models, however, we have one pool of ubiquitous neurons with random gap junction coupling that are still able to output two or more distinct phase-shifted rhythms.

2. Models

The full model is the hypoglossal motoneuron model developed by Purvis and Butera [36] and modified with gap junction coupling from Perez Velazquez and Carlen [37]. Despite the model's seeming specificity, many of the ion channels contained in this model also feature prominently in a variety of other neurons and motoneurons [35].

The reduced model is a combination of the simple spiking Izhikevich [38] $I_{Na} + I_K$ model, which contains only a sodium, potassium, and leak current, with I_{Ca}, $[Ca^{2+}]_i$, I_{SK}, and gap junction coupling. The model itself for neuron i, where $i = [1, 2, \ldots, N]$, is as follows:

$$C\frac{dV_i}{dt} = I_{\text{input},i}(t) - \overline{g}_{\text{leak}}(V_i - E_{\text{leak}})$$

$$- \overline{g}_{Na} m_{\text{inf}}(V_i)(V_i - E_{Na}) - \overline{g}_K n_i(V_i - E_K)$$

$$- \overline{g}_{Ca} p_i(V_i - E_{Ca}) - \overline{g}_{AHP} z_i(V_i - E_K)$$

$$- \sum_{j=1}^{N} \overline{g}_{\text{gap},i,j}\left(V_i - V_j\right),$$

$$m_{\text{inf}}(V) = \frac{1}{1 + e^{(-26.5-V)/(14.5/\ln(5/3))}},$$

$$\frac{dn_i}{dt} = \frac{n_{\text{inf}}(V_i) - n_i}{\tau_n(V_i)},$$

$$n_{\text{inf}}(V) = \frac{1}{1 + e^{(-20-V)/5}}, \qquad \tau_n(V) = 3,$$

$$\frac{dp_i}{dt} = \frac{p_{\text{inf}}(V_i) - p_i}{\tau_p(V_i)},$$

$$p_{\text{inf}}(V) = \frac{1}{1 + e^{(-40-V)/5}}, \qquad \tau_p(V) = \frac{6}{1 + e^{(55+V)/2}} + 0.5,$$

$$\frac{dz_i}{dt} = \frac{z_{\text{inf}}(V_i) - z_i}{\tau_z(V_i)},$$

$$z_{\text{inf}}\left(\left[Ca^{2+}\right]_i\right) = \frac{1}{1 + \left(0.003/\left[Ca^{2+}\right]_i\right)^5}, \qquad \tau_z(V) = 10,$$

$$\frac{d\left[Ca^{2+}\right]_{i,i}}{dt} = -0.0005\, I_{Ca} - 0.04\left[Ca^{2+}\right]_{i,i},$$

$$I_{Ca} = \overline{g}_{Ca} p_i(V_i - E_{Ca}).$$

$$(1)$$

With the following parameters:

$$\overline{g}_{\text{leak}} = 0.38\,\mu S, \qquad \overline{g}_{Na} = 1.28\overline{3}\,\mu S, \qquad \overline{g}_K = 1.8\,\mu S,$$

$$\overline{g}_{Ca} = 0.08\,\mu S, \qquad \overline{g}_{AHP} = 0.5\,\mu S,$$

$$\overline{g}_{\text{gap},i,j} = \begin{cases} g_{\text{gap}} & \text{if neurons } i \text{ and } j \text{ share a gap junction,} \\ 0\,\mu S & \text{if neurons } i \text{ and } j \text{ are not connected,} \end{cases}$$

$$E_{\text{leak}} = -80\,\text{mV}, \qquad E_{Na} = 60\,\text{mV},$$

$$E_K = -80\,\text{mV}, \qquad E_{Ca} = 80\,\text{mV},$$

$$N = 100, \qquad dt = 0.02\,\text{ms}, \qquad C = 0.04\,\text{nF}.$$

$$(2)$$

Changes to default parameters in specific simulations will be given in the description of each simulation run.

3. Simulations and Analysis

Simulations were performed using the Python programming language (http://www.python.org/) accompanied by the Scientific Tools for Python package (http://www.scipy.org/). Speed-critical code was written in C++ and plugged into Python using the C-Extensions for Python library (http://www.cython.org/). Plotting was done via the matplotlib library (http://www.matplotlib.sourceforge.net).

Numerical integration of both the full and reduced motoneuron models was done via Euler's method. Both models were also tested over a variety of dt settings, and against RK4, the most common Runge-Kutta method [39], to ensure sufficient numerical accuracy by Euler's method.

Simulated neurons were deemed part of a singular cohesive firing group based on clustering via the Expectation-Maximization algorithm [40], with the number of clusters determined by the following method: first, the voltage of all neurons was summed together at every point in time. Next, this summed signal was smoothed via six passes of a 4 ms square wave moving average filter. Finally, this signal was normalized, local maxima were determined by observing when the derivative passed through zero, and an extremely low threshold of 6% of the total signal was used to discard spurious local optima that sometimes appeared when only a few neurons were active. Because firing neurons have significant depolarizations both before and after firing, analysis of summed voltage traces proved superior in identifying firing groups in comparison to the popular information theoretic "jump" method [41]. In addition, no groups were deemed to exist if the average of all voltage traces failed to reach a maximum value of 3 mV (out of a possible ~21 mV), implying that fewer than 15% of all spikes were well aligned in the largest group.

For the full model, all simulations were all performed with a 500 ms duration square wave input current of 0.5 ± 0.35 nA, where the variation corresponds to white noise changing at every time step. Since increasing or decreasing white noise amplitude did not noticeably produce distinct changes in the model, other levels of white noise were not further considered. The current was applied to each neuron beginning at 100 ± 10 ms into the simulation and ending at 600 ± 5 ms. Variation in input current start and end times simulates the varied delays in stimulation produced by axonal variation of input drive. Gap junctions were opened at 300 ms for the demonstration of behavior in the full model, at 125 ms when reducing SK conductance in the full model, and at 200 ms when examining plane topology in the full model. Gap junctions were closed for all full models at 700 ms. Gap junction connections were made at random, with each cell having a probability of 25% of connecting to any other cell in the full model without plane topology, and 50% of connecting to any cell within a radius of 5 in the model with topology. Neighboring cells in the plane topology were all evenly spaced in a square lattice, with nondiagonal neighbors at a distance of 1. The precise gap junction connectivity generated to demonstrate the behavior in the full model was reused in the simulations with SK conductance in the full-model, so that results could be directly compared. In the full-model simulation with SK conduction reduction, maximal conductance began at 0.3 μS. Starting at 300 ms, the conductance dropped in a linear value until hitting 0.0 μS at 500 ms, where it remained at zero for the duration of the simulation. All full simulations lasted 700 ms, had a gap junction conductance setting of 0.0005 μS, and consisted of 100 neurons.

For the reduced model, simulations were performed using a constant square wave impulse, beginning at 100 ± 200 ms, where negative start times imply a start time of zero and ending at 1400 ms. Input current for reduced SK conductance simulations was set to 0.08 nA and for plane topology simulations was set to 0.1 nA. Gap junction conductance g_{gap} in plane topology simulations was set to 0.003 μS. Gap junctions in all reduced simulations remained open for the complete duration of the simulation. Gap junction connectivity in all figures illustrating multiple runs

with varying parameters was performed using the same set of gap junction connections to ensure that results are not a byproduct of different gap junction connectivity. The reduced model gap junction conductance versus input current experiments and the connectivity in the reduced model gap junction conductance versus input current simulations also used the same gap junction connections to ensure comparability between generated results. The total simulation time for all reduced model runs was 1500 ms.

In our theoretical treatment of gap junction connection probability and connectivity radius, the cutoffs for each firing group were chosen based on the final column of Figure 7(a). An average total gap junction current below 0.09 nA produced ungrouped firing, between 0.09 nA and 0.1 nA produced three groups, between 0.1 nA and 0.3 nA produced two groups, and above 0.3 nA produced one group. The constant g, the conductance of each gap junction, was set to 0.003 μS, to match the conductance settings in the accompanying simulations. The voltage differential constant dV, which is arbitrarily defined, was set to $0.\overline{3}$ V so that $\sum_{j=1}^{N_i} g \, dV = 1/N_T$, where the total number of neurons in the simulation equals N_T. This ensures the maximum for average total gap junction current under an infinitely large radius and all-to-all connectivity yields 1.0 nA of average total gap junction current.

4. Results

4.1. Half-Center-Like Behavior. In the full model, following the opening of gap junctions, two distinct phase-shifted signals reminiscent of a traditional half-center CPG could be generated in a single nucleus with ubiquitous connectivity. An example of this behavior is shown in Figure 1, where opening the gap junctions shifted a fairly asynchronous firing pattern amongst the 100 neurons into two distinct neuronal firing groups. To highlight this division into two firing groups, the neurons were color coded according to their group affiliation (Figure 1(b)), and a raster plot of their spiking behavior was generated (Figure 1(c)). In general, one group was often better aligned than the other, and upon further investigation, the randomness and ubiquity of connectivity actually fostered conditions encouraging one slightly larger group to act as a "driver" for the smaller "follower" group. On average, connectivity between and within the firing groups is given in Table 1, with the probability of two neurons being gap junction coupled set to C, and the number of neurons in firing group i being S_i.

While this implies that both groups are sending roughly an equal amount of conductance between one another, as would be expected by bidirectional gap junction coupling, it masks a more important property. Since each group has a very different degree of interconnectedness, the amount of incoming drive in relation to internal drive is markedly different (see Table 2).

As $S_1 \gg 1$ and $S_2 \gg 1$, the ratio of internal connections in group 1 is identical to the ratio of connections in group 2 received from group 1. Regardless the size of each spontaneously formed group, the larger group always

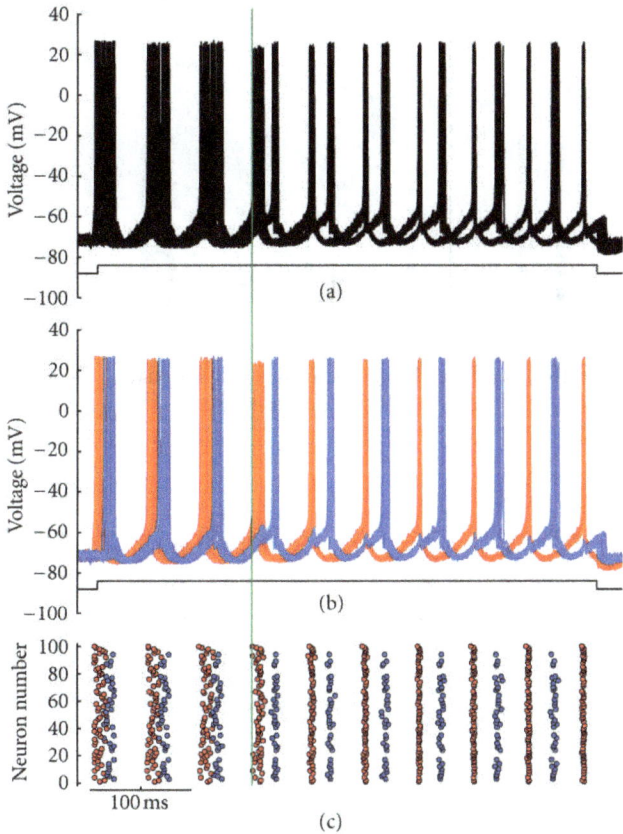

FIGURE 1: Example of half-center-like behavior in the full model. (a) Plot of voltage traces showing 40 of the 100 neurons. (b) Each voltage trace shown in (a) has been colored based on its neuron group affiliation. (c) A raster plot showing spikes for all 100 neurons, also colored based on neuron group affiliation. Vertical line corresponds to the opening of gap junctions.

TABLE 1: Total connections.

	From group 1	From group 2
Connections to group 1	$C(S_1^2 - S_1)/2$	$CS_1S_2/2$
Connections to group 2	$CS_1S_2/2$	$C(S_2^2 - S_2)/2$

TABLE 2: Percent connections.

	From group 1	From group 2
% Connections to group 1	$(S_1 - 1)/(S_1 + S_2 - 1)$	$(S_2)/(S_1 + S_2 - 1)$
% Connections to group 2	$(S_1)/(S_1 + S_2 - 1)$	$(S_2 - 1)/(S_1 + S_2 - 1)$

in input current were less predictive of a trend in the number of firing groups.

4.2. Behavior Requirements. We hypothesized that I_{AHP} is a prime candidate for the necessary reciprocal inhibition between firing groups. Unlike other currents, I_{AHP} is solely gated by intracellular calcium. This delays the onset of the current until shortly after the spike ends and maintains the level of inhibition high until intracellular calcium is pumped out. Thus, the I_{AHP} is long lasting and easily modified by altering intracellular calcium influx. This would also serve to inhibit neighboring electrically coupled cells firing shortly after a spike but have less of an effect on neighboring cells firing concurrently. Without I_{AHP}, the afterhyperpolarization and relative refractory period of the neuron are very brief. While it may be hypothetically plausible to use the relative refractory period generated by I_K or other comparatively brief negative currents for reciprocal inhibition, this is not observed in our models.

To explore the necessity and sufficiency of I_{AHP} on this behavior in our models, we first examine our reduced model. This model itself takes a simple spiking model ($I_{Na} + I_K$) and extends it with the minimal requirements for a functional I_{AHP} channel: voltage-gated calcium currents (I_{Ca}), intracellular calcium concentration ($[Ca^{2+}]_i$), and the AHP current (I_{AHP}). The fact that this behavior is observed in the reduced model is a demonstration that, with the right parameters, the addition of a functional I_{AHP} current to a model is sufficient to produce more than one half-center like firing groups. However, since proper parameter setting is required, it is not sufficient alone to merely add a functional I_{AHP} current and expect half-center like CPG behavior.

Based on our models, when I_{AHP} is gradually removed, the behavior ceases. In the full model, at sufficiently low I_{AHP}, the two firing groups merge into one (Figure 4). Greater detail is shown based on simulations from the reduced model, where changes to I_{AHP} were compared with the primary determinant of group size: gap junction conductance (Figure 5). In both the full and reduced models, multiple firing groups could not be maintained in the absence of I_{AHP}. Trivially, without the existence of gap junctions, neurons would be completely uncoupled and, thus, incapable of synchronizing.

receives more internal than external stimulation and extends more excitability to drive the smaller group. Thus, based on probability alone, a similar topology is consistently observed regardless of network size. This does not exclude the possibility of having two equally sized groups, which would be expected to have more balanced dynamics.

While the main focus of the current investigation is on half-center like CPGs, it should be noted that generating more than two groups is entirely feasible. In the reduced model, for example, not only one or two centers could be generated, but also occasionally three centers were observed (Figure 2).

Furthermore, even without altering patterns of connectivity, modifying conductance through gap junctions and/or input current into the system had the potential to shift the firing patterns between 1, 2, and 3 independent groups (Figure 3). In this case, gap junction conductance was seen to exert a greater influence than that of input current in determining the number of firing groups (Figure 3), with higher gap junction conductance being associated with fewer groups and lower gap junction conductance being associated with more groups. While input current was capable of producing a shift between different firing behaviors, changes

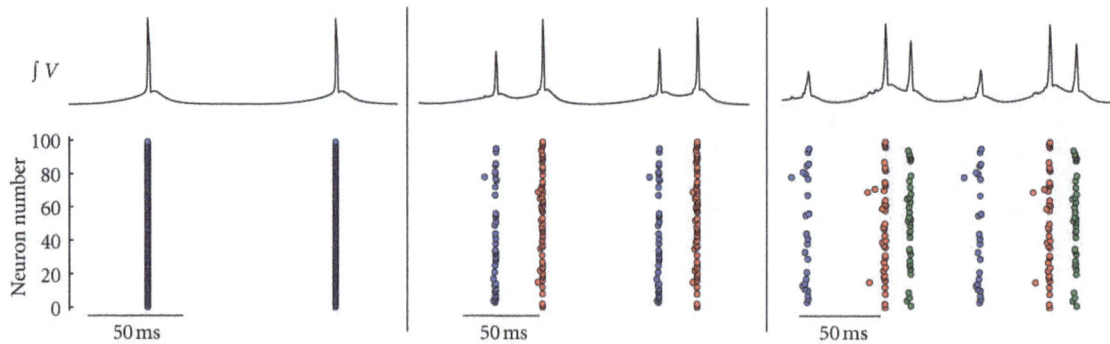

FIGURE 2: Examples of firing behaviors observed in the reduced model. One, two, and three firing groups through summation of all 100 neuronal voltage traces over time (upper panel) and raster plots of all spikes colored by group affiliation (lower panel).

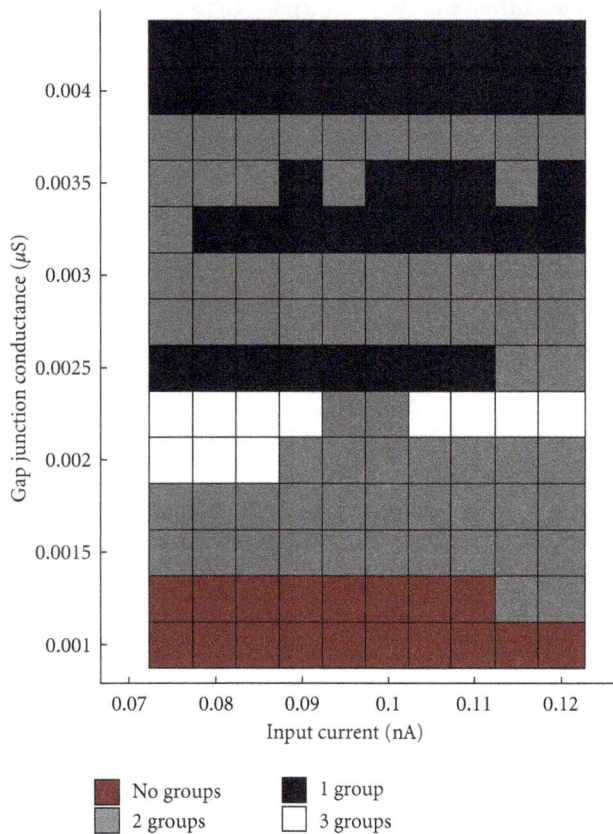

FIGURE 4: Effects of reduction and removal of I_{AHP} (full model). Simulation demonstrating that when $\bar{g}_{AHP}$ is reduced linearly, the two firing groups merge into one group before $\bar{g}_{AHP}$ reaches zero. This behavior can be observed in both (a) the voltage traces (shown for 40 of the 100 neurons) and (b) a raster plot of spikes for all 100 neurons. Both are color coded based on their firing group affiliation prior to the merging of the firing groups.

FIGURE 3: Gap junction conductance versus input current (reduced model). Multiple simulation runs examining the effects of changing gap junction conductance and input current on the formation firing groups.

4.3. Topology. Up until now, all of our models have used random connectivity without any concern for spatial placement of the cells. Since slice preparations are commonly used to study CPGs of the spinal cord [42], and the slice itself often has a thickness (350–600 μm) within the range of the dendritic span of motoneurons (250–700 μm) [43], where gap junctions primarily form between dendrites and/or somas, we opted to orient neurons along a two-dimensional plane as a first approximation to this layout to begin to

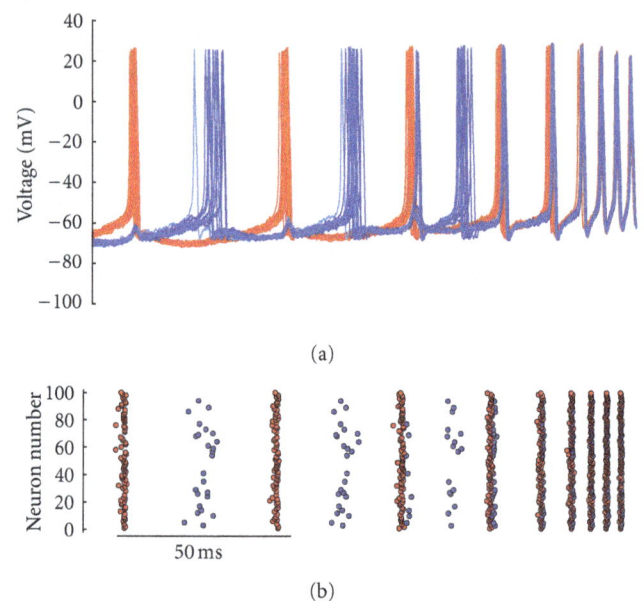

explore the effects of spatial connectivity on half-center like CPG group formation. While one might predict that neurons in each firing group would clump together into two massive nuclei, this is not the case. Instead, neuronal groups tended to form a mottled appearance, with clumps from each group equally interspersed (Figure 6). This configuration would ensure that each neuron would be exposed to some members of each firing group, thus loosely preserving connectivity reminiscent of a topology-free model. Figure 6 also reveals that one neuron spent five firing cycles with the firing group color coded in red before joining the firing group color coded in blue. This suggests that unlike in a traditional half-center CPG where group allegiance is a hard-set property of each

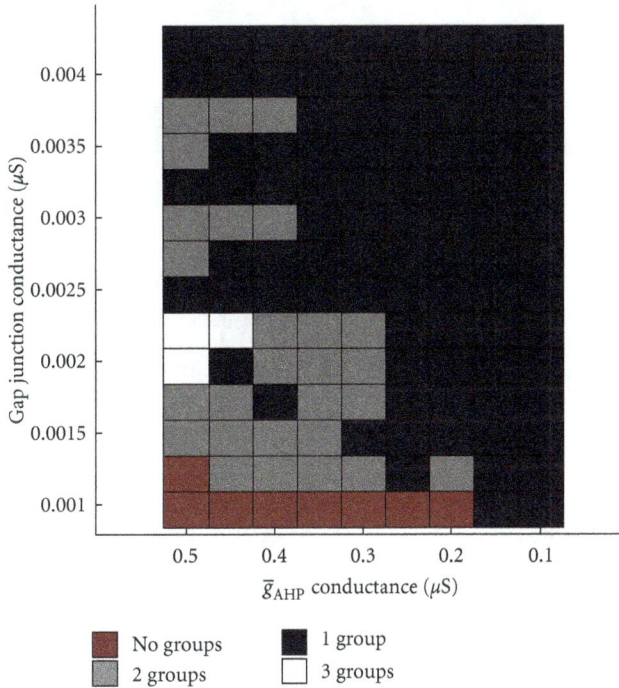

FIGURE 5: Gap junction conductance versus $\bar{g}_{AHP}$ conductance (reduced model). When multiple simulations are performed with varying gap junction conductances and values for $\bar{g}_{AHP}$, the incidence of having more than one firing group is abolished before reaching $\bar{g}_{AHP} = 0$. Note that the leftmost column of this figure corresponds to the input current = 0.08 nA column in Figure 3.

neuron, in some rare instances, a neuron may straddle the fence and move between two groups.

Finally, using the reduced model, we explored the parameter space to further examine the exact role that topology plays (Figure 7(a)). We found that the pronounced inverse relationship between connectivity probability and connectivity radius is best explained by overall gap junction current. We reasoned that by imposing a radius of connectivity, the number of available neurons to form gap junctions would be decreased, and the net current entering all gap junctions for each individual cell would be decreased. If we assume that every pair of cells has an identical voltage differential dV, probability C of connecting to other neurons within a radius R, and a gap junction conductance g, and also assume a plane of neurons of infinite size and some total current "window" where two firing groups can exist, we can generate a theoretical prediction of when two firing groups will form (Figure 7(b)). If we represent the number of neurons within a radius R of neuron i to be N_i neurons, and the total number of neurons to be N_T, then the equation for average total conductance is as follows:

$$\langle I_{gap,total} \rangle = \frac{\sum_{i=1}^{N_T} I_{gap,i,total}}{N_T} = \frac{\sum_{i=1}^{N_T} \left(C \sum_{j=1}^{N_i} g dV \right)}{N_T}. \quad (3)$$

5. Discussion

In this study, we have elucidated the properties of a novel type of CPG that oftentimes bear resemblance to the traditional half-center CPG depicted in the literature. Unlike most traditional models of CPGs, we have presented a more dynamic entity that allocates group membership on the fly and can modify its firing properties through a variety of different biological parameters. This property itself has a number of benefits and drawbacks.

It cannot be stressed enough that the dynamic nature of these CPGs would be especially beneficial for either the generation or modification of locomotor or respiratory central patterns. In contrast to our model, most models of CPGs incorporate static group affiliation, which alone may not be able to produce the sorts of dynamically changing locomotor and respiratory patterns seen in nature. In both locomotion and respiration, adaptation of rhythms to both external environmental changes and descending cortical commands may be more difficult in a simpler CPG, which may lack the requisite complexity required to describe the wealth of patterns that humans and other mammals are capable of exhibiting in these two activities. Moreover, CPGs formed through gap junctions can alter group affiliation without relying explicitly on changes in gap junction coupling. With this in mind, some of the rarer behaviors seen in the current model, including the more exotic three firing group behavior, might be easy for a biological system to generate and maintain as long as the initial state of the system is within the vicinity of the correct set of parameters. Evolutionarily speaking, it would also be easier to create a CPG that itself had no explicit wiring, but could rely on random connectivity to self-organize.

Though this novel CPG has many admirable traits, some inherent properties of these systems may make them harder to tune or more difficult to find biologically. The volatile nature of a system that drastically changes behavior with small changes in parameters could open such neural systems up to a plethora of neurological disorders. While we offer no strong hypotheses regarding known disorders that might stem from such a disruption, known disorders with errant or absent patterns certainly come to mind: spastic gait, persistent muscle spasms, and the sudden loss of breathing implicated in SIDS. Furthermore, because such systems can exist in a singular nucleus with otherwise ubiquitous physiological properties, the only way to identify such systems experimentally would be to observe them while active, rather than through simple histology alone.

5.1. Fast Pattern Generation. At first, the speed of oscillations in our proposed CPG may appear surprisingly fast, commonly ranging between 10 and 30 Hz. However, these speeds are not uncommon for many smaller animals in locomotion, respiration, and associated behaviors. For example, depending on its speed, the American Cockroach (*Periplaneta americana*) commonly has a stride frequency between 20 and 25 Hz, which at the highest speeds is often quadrupedal or bipedal, and only a few Hz below wing beat frequency [44]. In addition, the wing beat frequency of the

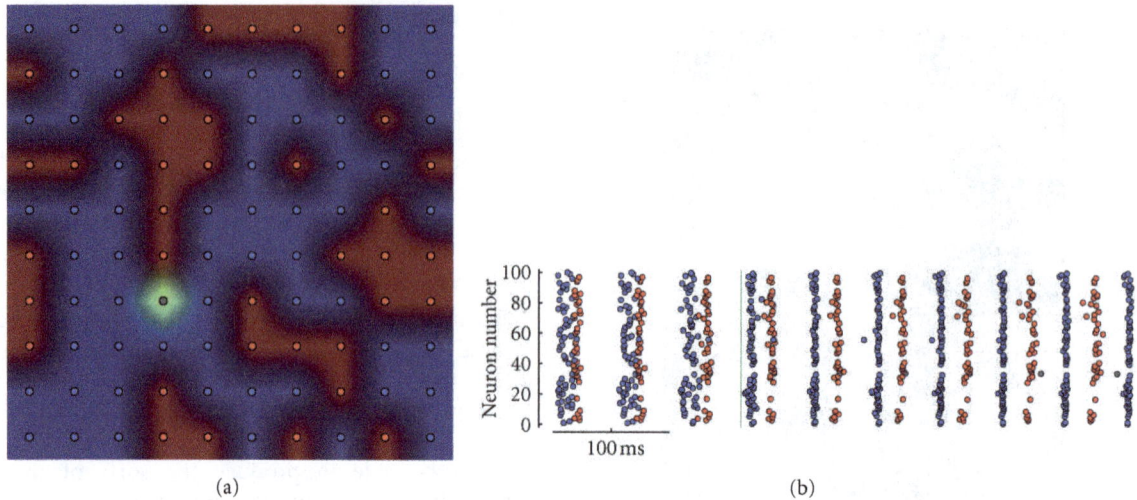

(a)   (b)

FIGURE 6: Mottled topology (full model). Demonstrations of a simulation run with a planar topology and a connectivity radius of 5. In (a), each neuron is plotted in its correct topological orientation and colored based on its group alignment. In (b), a raster plot confirms that these neurons are indeed split into two firing groups. The one grey/green neuron began as a member of the red firing group, but later transitioned to the blue group after several cycles of firing.

ruby-throated hummingbird (*Archilochus colubris*) nearly doubles this at 53 ± 3 Hz. [45]. At slightly lower frequencies, but still within this range, are whisking behaviors in rats (at 6–12 Hz) [46] and mice (at 19 ± 7 Hz) [47] as well as sniffing behaviors in rats (at ∼8 Hz) [48] and mice (at ∼12 Hz) [49], both of which are commonly associated with respiratory events. Finally, hypoglossal motoneurons, the motoneuron corresponding to our full model, have been observed with steady-state frequencies as low as 9.7 ± 3 Hz and as high as 70+ Hz [50]. While all the aforementioned frequencies are within one standard deviation of the lowest frequencies observed in hypoglossal motoneurons, slow motoneurons in the cockroach (*Blaberus discoidalis*) leg can fire even slower, at 2–5 Hz [51]. Given the broad range of behaviors that fall within the firing frequencies observed in both the model and biological neurons, we believe that our proposed CPG represents a reasonable approach for modeling a variety of locomotor and respiratory behaviors.

5.2. CPG or Downstream Modifier? While we have focused on pattern generation, our proposed CPG could also just act as easily as a pattern modification nucleus. Rather than using multiple firing groups to drive two separate sets of muscles, for example, it could be used to take a singular signal and double or triple its period. Considering the ability to rapidly swap between one and two firing groups, this could be especially useful in locomotor transitions, such as moving from walking to running, or in cortical processes used to keep track of time.

5.3. Limitations. As is the case with all modeling, a reasonable computational model does not necessarily imply the existence of a biological correlate. In addition, we are limited to the sets of parameters explicitly examined, and, therefore, we cannot rule out that additional exotic effects may not exist for this style of CPG. While we have tried to keep

biological plausibility in the forefront, the lack of plasticity in our models could, in theory, lead us to miss key properties of these systems.

5.4. Inhibitory Synapses. While the current models stress a role for I_{AHP} and gap junctions, similar outcomes in terms of emergent topology should be obtainable via synaptic connections. Recall that the most important property in our plane topology networks was total current flowing between cells. Thus, a benefit of inhibitory synapses would be that inhibition comes from neuronal firing rather than electrophysiological afterhyperpolarization. This would take the focus away from ion channels controlling the sAHP and instead focus on mechanisms of firing. Subsequently, this may still include channels associated with I_{AHP}, as they are strongly associated with modifications to firing frequency [35], but other currents, such as the hyperpolarization-activated current (I_h), may end up playing a more substantive role under inhibitory synapses, as observed by Marder and Bucher [2].

While synaptic connectivity may favor distally located cells in a CPG, gap junction coupling would certainly emphasize less broadly spaced neurons. This would make gap-junction-based CPGs more amenable to our CPG model, where connectivity is more dense and arbitrary, than spaced and planned out meticulously. The latter of which would likely be more favorable to an inhibitory synaptic design, as axons have reasonably high target specificity. Thus, synaptic-based versus gap-junction-based CPGs may favor different styles of design.

5.5. Emergent Connectivity. It is easy to get into the habit of viewing the brain as a large hard-wired microchip; however, such analogies neglect the lively self-organized properties of the brain that could lead to unique emergent behaviors, like those demonstrated by our proposed CPG.

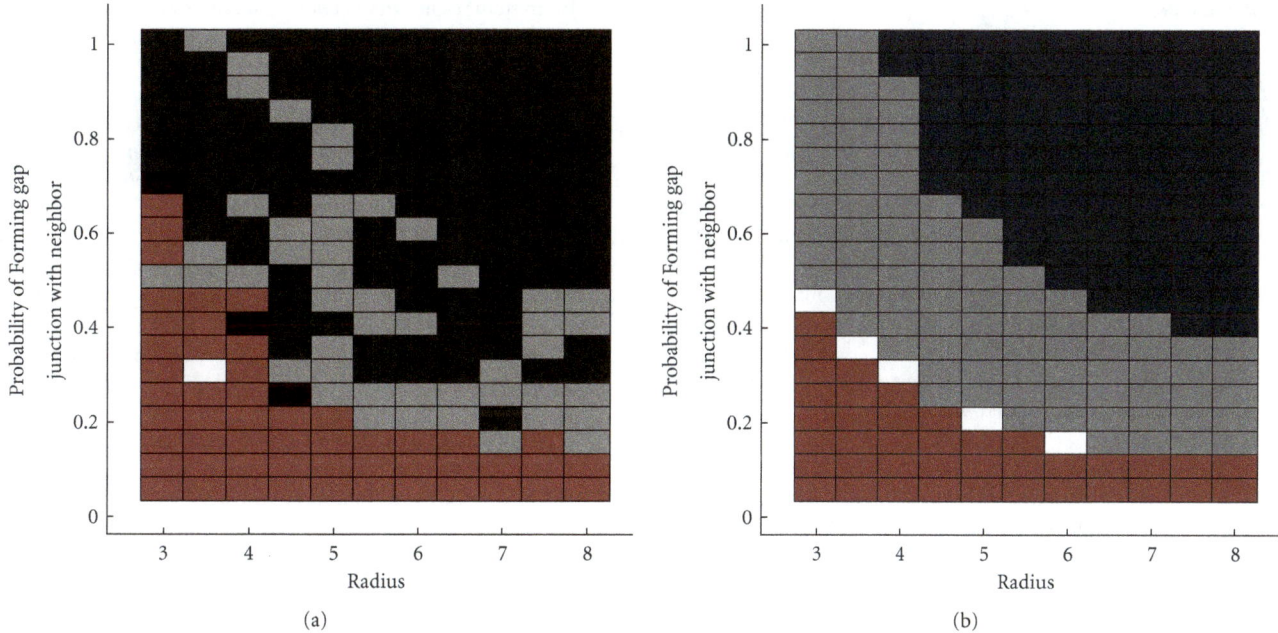

(a)   (b)

FIGURE 7: Connectivity topology versus radius (reduced and theory). (a) Multiple simulations are performed varying both the probability of forming a gap junction with a neighbor within a given radius and the radius itself. The results from this series of simulations resembles (b) a plot based on the average of the total hypothetical conductance received by each neuron.

While highly deterministic behaviors can arise from arbitrary connectivity, "dynamic" should not necessarily be conflated with "chaotic." Even in the most volatile systems, the laws of probability can impose order and, in some cases, may yield a more stable design paradigm than the one that is seemingly more orderly. Our simulations with random gap junction coupling clearly illustrate this by demonstrating that regardless of the size of each group, the connectivity always scales to ensure that one group acts as the leader and the other follows in suit. Such design principles are simple to implement, even if they are not obvious at first.

5.6. Connectivity versus Radius: Simulation and Theory. Our theoretical predictions reasonably capture the various regions in parameter space where zero, one, two, and three firing groups typically appear (Figures 7(a) and 7(b)). Certainly the predictions do not perfectly match with the simulations, but the dynamic nature of these models makes absolute knowledge of when a certain number of firing groups appear difficult to define. Moreover, because the cutoff conductances need to be specified in our theoretical predictions for when each set of firing groups would appear, it cannot be seen as a predictive theory so much as an explanatory one. If one knew roughly under what total conductances certain firing groups would form, then that alone would be sufficient to detail a great deal of how many firing groups would be likely to appear in other simulation runs with different parameters.

5.7. The Benefits of Full and Reduced Models. By using both full and reduced style models, we have been able to gain the benefits of both. With the full model, we have a demonstration of our style of CPG in a model that has, otherwise, been deemed accurate to the biology. Clearly, the plethora of channels in a realistic neuron does not disrupt the ability to form multiple neuronal groups. Simultaneously, with our reduced model, we have stripped down the neuron to just the bare essentials to demonstrate the minimum required to achieve the desired behavior. Additionally, we gain the speed benefits traditionally associated with reduced models, which permits sufficient simulation runs for the many parameter versus parameter plots generated as part of this study.

6. Conclusions

The formation and modulation of centrally generated signals for essential behaviors remains an exciting and open field. Simply knowing where CPGs might lie in the brain may not be enough to fully understand exactly how they create their signals. Likewise, there may exist useful methods for pattern generation not used in nature that may still be useful for robotics and artificial prostheses. Thus, it is of great importance to explore the theoretical possibilities, so that we can provide biological experimentalists and roboticists ample paradigms from which to choose. While we have demonstrated a novel type of CPG, relying less on deliberate connections and containing no synapses at all, there may still exist other unique methods for generating these signals that are not yet addressed by the CPG literature or the current study. Thus, it is important to keep an open mind about the possibilities of how a CPG might form.

Acknowledgment

This work was supported by NIH Grants NS049310 and HL063175.

References

[1] R. L. Calabrese, "Half-center oscillators underlying rhythmic movements," *Nature*, vol. 261, pp. 146–148, 1995.

[2] E. Marder and D. Bucher, "Central pattern generators and the control of rhythmic movements," *Current Biology*, vol. 11, no. 23, pp. R986–R996, 2001.

[3] J. L. Feldman and C. A. Del Negro, "Looking for inspiration: new perspectives on respiratory rhythm," *Nature Reviews Neuroscience*, vol. 7, no. 3, pp. 232–242, 2006.

[4] A. J. Ijspeert, "Central pattern generators for locomotion control in animals and robots: a review," *Neural Networks*, vol. 21, no. 4, pp. 642–653, 2008.

[5] D. M. Wilson, "The central nervous control of flight in a locust," *The Journal of Experimental Biology*, vol. 38, pp. 471–490, 1961.

[6] J. C. Rekling and J. L. Feldman, "Prebotzinger complex and pacemaker neurons: hypothesized site and kernel for respiratory rhythm generation," *Annual Review of Physiology*, vol. 60, pp. 385–405, 1998.

[7] N. Koshiya and J. C. Smith, "Neuronal pacemaker for breathing visualized in vitro," *Nature*, vol. 400, no. 6742, pp. 360–363, 1999.

[8] C. R. Gordon, W. A. Fletcher, G. Melvill Jones, and E. W. Block, "Adaptive plasticity in the control of locomotor trajectory," *Experimental Brain Research*, vol. 102, no. 3, pp. 540–545, 1995.

[9] D. D. Fuller, K. B. Bach, T. L. Baker, R. Kinkead, and G. S. Mitchell, "Long term facilitation of phrenic motor output," *Respiration Physiology*, vol. 121, no. 2-3, pp. 135–146, 2000.

[10] J. L. Feldman, N. V. Neverova, and S. A. Saywell, "Modulation of hypoglossal motoneuron excitability by intracellular signal transduction cascades," *Respiratory Physiology and Neurobiology*, vol. 147, no. 2-3, pp. 131–143, 2005.

[11] A. A. Sharp, F. K. Skinner, and E. Marder, "Mechanisms of oscillation in dynamic clamp constructed two-cell half-center circuits," *Journal of Neurophysiology*, vol. 76, no. 2, pp. 867–883, 1996.

[12] T. G. Brown, "On the nature of the fundamental activity of the nervous centres, together with an analysis of the conditioning of rhythmic activity in progression, and a theory of the evolution of function in the nervous system," *The Journal of Physiology*, vol. 48, no. 1, p. 18, 1914.

[13] S. Grillner and P. Wallen, "Central pattern generators for locomotion, with special reference to vertebrates," *Annual Review of Neuroscience*, vol. 8, no. 1, pp. 233–261, 1985.

[14] U. Bässler and A. Büschges, "Pattern generation for stick insect walking movements—multisensory control of a locomotor program," *Brain Research Reviews*, vol. 27, no. 1, pp. 65–88, 1998.

[15] A. J. Ijspeert, A. Crespi, D. Ryczko, and J. M. Cabelguen, "From swimming to walking with a salamander robot driven by a spinal cord model," *Science*, vol. 315, no. 5817, pp. 1416–1420, 2007.

[16] D. A. McCrea and I. A. Rybak, "Modeling the mammalian locomotor CPG: insights from mistakes and perturbations," *Progress in Brain Research*, vol. 165, pp. 235–253, 2007.

[17] R. L. Calabrese, J. D. Angstadt, and E. A. Arbas, "A neural oscillator based on reciprocal inhibition," *Perspectives in Neural Systems and Behavior*, vol. 10, pp. 33–50, 1989.

[18] J. C. Smith, A. P. L. Abdala, H. Koizumi, I. A. Rybak, and J. F. R. Paton, "Spatial and functional architecture of the mammalian brain stem respiratory network: a hierarchy of three oscillatory mechanisms," *Journal of Neurophysiology*, vol. 98, no. 6, pp. 3370–3387, 2007.

[19] K. Nakada, T. Asai, T. Hirose, and Y. Amemiya, "Analog current-mode CMOS implementation of central pattern generator for robot locomotion," in *Proceedings of the IEEE International Joint Conference on Neural Networks (IJCNN '05)*, pp. 639–644, August 2005.

[20] M. F. Simoni and S. P. DeWeerth, "Two-dimensional variation of bursting properties in a silicon-neuron half-center oscillator," *IEEE Transactions on Neural Systems and Rehabilitation Engineering*, vol. 14, no. 3, pp. 281–289, 2006.

[21] O. Kiehn and M. C. Tresch, "Gap junctions and motor behavior," *Trends in Neurosciences*, vol. 25, no. 2, pp. 108–115, 2002.

[22] L. Broch, R. D. Morales, A. V. Sandoval, and M. S. Hedrick, "Regulation of the respiratory central pattern generator by chloride-dependent inhibition during development in the bullfrog (Rana catesbeiana)," *Journal of Experimental Biology*, vol. 205, no. 8, pp. 1161–1169, 2002.

[23] J. C. Rekling, X. M. Shao, and J. L. Feldman, "Electrical coupling and excitatory synaptic transmission between rhythmogenic respiratory neurons in the preBötzinger complex," *The Journal of Neuroscience*, vol. 20, no. 23, p. RC113, 2000.

[24] A. Urbani and O. Belluzzi, "Riluzole inhibits the persistent sodium current in mammalian CNS neurons," *European Journal of Neuroscience*, vol. 12, no. 10, pp. 3567–3574, 2000.

[25] C. Bou-Flores and A. J. Berger, "Gap junctions and inhibitory synapses modulate inspiratory motoneuron synchronization," *Journal of Neurophysiology*, vol. 85, no. 4, pp. 1543–1551, 2001.

[26] I. C. Solomon, K. H. Chon, and M. N. Rodriguez, "Blockade of brain stem gap junctions increases phrenic burst frequency and reduces phrenic burst synchronization in adult rat," *Journal of Neurophysiology*, vol. 89, no. 1, pp. 135–149, 2003.

[27] W. T. Wong, J. R. Sanes, and R. O. L. Wong, "Developmentally regulated spontaneous activity in the embryonic chick retina," *Journal of Neuroscience*, vol. 18, no. 21, pp. 8839–8852, 1998.

[28] M. Galarreta and S. Hestrin, "A network of fast-spiking cells in the neocortex connected by electrical synapses," *Nature*, vol. 402, no. 6757, pp. 72–75, 1999.

[29] M. V. L. Bennett and R. S. Zukin, "Electrical coupling and neuronal synchronization in the mammalian brain," *Neuron*, vol. 41, no. 4, pp. 495–511, 2004.

[30] T. Zahid and F. K. Skinner, "Predicting synchronous and asynchronous network groupings of hippocampal interneurons coupled with dendritic gap junctions," *Brain Research*, vol. 1262, pp. 115–129, 2009.

[31] D. I. Cardone, T. J. Halat, M. N. Rodriguez, and I. C. Solomon, "Expression of gap junction proteins in cranial (hypoglossal) and spinal (phrenic) respiratory motor nuclei in rat," *Faseb Journal*, vol. 16, no. 5, pp. A810–A810, 2002.

[32] A. S. Chiang, Y. C. Liu, S. L. Chiu, S. H. Hu, C. Y. Huang, and C. H. Hsieh, "Differential expression of connexin26 and connexin32 in the pre-Bötzinger complex of neonatal and adult rat," *Journal of Comparative Neurology*, vol. 440, no. 1, pp. 12–19, 2001.

[33] G. Tamás, E. H. Buhl, A. Lörincz, and P. Somogyi, "Proximally targeted GABAergic synapses and gap junctions synchronize cortical interneurons," *Nature Neuroscience*, vol. 3, no. 4, pp. 366–371, 2000.

[34] J. G. McLarnon, "Potassium currents in motoneurones," *Progress in Neurobiology*, vol. 47, no. 6, pp. 513–531, 1995.

[35] J. C. Rekling, G. D. Funk, D. A. Bayliss, X. W. Dong, and J. L. Feldman, "Synaptic control of motoneuronal excitability," *Physiological Reviews*, vol. 80, no. 2, pp. 767–852, 2000.

[36] L. K. Purvis and R. J. Butera, "Ionic current model of a hypoglossal motoneuron," *Journal of Neurophysiology*, vol. 93, no. 2, pp. 723–733, 2005.

[37] J. L. Perez Velazquez and P. L. Carlen, "Gap junctions, synchrony and seizures," *Trends in Neurosciences*, vol. 23, no. 2, pp. 68–74, 2000.

[38] E. M. Izhikevich, *Dynamical Systems in Neuroscience : The Geometry of Excitability and Bursting*, The MIT Press, Cambridge, Mass, USA, 2007.

[39] W. H. Press, *Numerical Recipes : The Art of Scientific Computing*, Cambridge University Press, Cambridge, UK, 2007.

[40] A. P. Dempster, N. M. Laird, and D. B. Rubin, "Maximum likelihood from incomplete data via the EM algorithm," *Journal of the Royal Statistical Society*, vol. 39, no. 1, pp. 1–38, 1977.

[41] C. A. Sugar and G. M. James, "Finding the number of clusters in a dataset: an information-theoretic approach," *Journal of the American Statistical Association*, vol. 98, no. 463, pp. 750–763, 2003.

[42] J. C. Smith, H. H. Ellenberger, K. Ballanyi, D. W. Richter, and J. L. Feldman, "Pre-Botzinger complex: a brainstem region that may generate respiratory rhythm in mammals," *Science*, vol. 254, no. 5032, pp. 726–729, 1991.

[43] A. J. Berger, D. A. Bayliss, and F. Viana, "Development of hypoglossal motoneurons," *Journal of Applied Physiology*, vol. 81, no. 3, pp. 1039–1048, 1996.

[44] R. J. Full and M. S. Tu, "Mechanics of a rapid running insect: Two-, four- and six-legged locomotion," *Journal of Experimental Biology*, vol. 156, pp. 215–231, 1991.

[45] C. H. Greenewalt, "The wings of insects and birds as mechanical oscillators," *Proceedings of the American Philosophical Society*, vol. 104, no. 6, pp. 605–611, 1960.

[46] G. E. Carvell and D. J. Simons, "Abnormal tactile experience early in life disrupts active touch," *Journal of Neuroscience*, vol. 16, no. 8, pp. 2750–2757, 1996.

[47] J. Voigts, B. Sakmann, and T. Celike, "Unsupervised whisker tracking in unrestrained behaving animals," *Journal of Neurophysiology*, vol. 100, no. 1, pp. 504–515, 2008.

[48] S. L. Youngentob, M. M. Mozell, P. R. Sheehe, and D. E. Hornung, "A quantitative analysis of sniffing strategies in rats performing odor detection tasks," *Physiology and Behavior*, vol. 41, no. 1, pp. 59–69, 1987.

[49] D. W. Wesson, T. N. Donahou, M. O. Johnson, and M. Wachowiak, "Sniffing behavior of mice during performance in odor-guided tasks," *Chemical Senses*, vol. 33, no. 7, pp. 581–596, 2008.

[50] F. Viana, D. A. Bayliss, and A. J. Berger, "Repetitive firing properties of developing rat brainstem motoneurones," *Journal of Physiology*, vol. 486, part 3, pp. 745–761, 1995.

[51] J. T. Watson and R. E. Ritzmann, "Leg kinematics and muscle activity during treadmill running in the cockroach, Blaberus discoidalis: I. Slow running," *Journal of Comparative Physiology*, vol. 182, no. 1, pp. 11–22, 1998.

Why People Play: Artificial Lives Acquiring Play Instinct to Stabilize Productivity

Shinichi Tamura,[1] Shoji Inabayashi,[2] Waichi Hayakawa,[2] Takahiro Yokouchi,[2] Hiroshi Mitsumoto,[3] and Hisashi Taketani[4]

[1] NBL Technovator Co., Ltd., 631 Shindachimakino, Sennan 590-0522, Japan
[2] Image Processing Solutions Deptartment, Pacific Systems Corporation, 8-4-19 Tajima, Sakura-Ku, Saitama City 338-0837, Japan
[3] Osaka Electro-Communication University, 18-1 Hatsucho, Osaka, Neyagawa 572-8530, Japan
[4] Tsuyama National College of Technology, 624-1 Numa, Okayama, Ttsuyama 708-8509, Japan

Correspondence should be addressed to Shinichi Tamura, tamuras@nblmt.jp

Academic Editor: Yen-Wei Chen

We propose a model to generate a group of artificial lives capable of coping with various environments which is equivalent to a set of requested task, and likely to show that the plays or hobbies are necessary for the group of individuals to maintain the coping capability with various changes of the environment as a whole. This may be an another side of saying that the wide variety of the abilities in the group is necessary, and if the variety in a species decreased, its species will be extinguished. Thus, we show some simulation results, for example, in the world where more variety of abilities are requested in the plays, performance of the whole world becomes stable and improved in spite of being calculated only from job tasks, and can avoid the risk of extinction of the species. This is the good effect of the play.

1. Introduction

Life is a highly rationalized organization. There often exists hidden rationality even in an irrational action at a glance. It is often said that in the human organization the top a-third people work hard and draw the organization, but the bottom a-third are lazy and work as brakes for the development of the organization. However, if a new organization is made with only this bottom a-third, also new top a-third people are born from them as excellent constituents and draw the organization. Generally speaking, each individual has his or her own peculiar and unique abilities, and he or she exhibits the hidden ability when it is requested. Human society is composed of divided works and specialized individuals with various explicit and implicit abilities. There are some work in the literature dealing with artificial brain and mind [1–4]. However, behavior of various abilities embedded into mankind has not been dealt with.

We assume that human society is composed of individuals, business positions (tasks), and hobby world with various plays. Human society has an interaction with the outer world (environment) which requests to the human society various tasks. For example, if the density of CO_2 increases, a task to decrease or suppress it is requested to the human society from the environment. Further, if the internet becomes spread over the human society, a task to utilize it is requested from the environment although it is a product of human society and somehow artificial. Such environment varies from time to time. To cope with such environmental changes, human society must make their work or business system change by assigning to the tasks (positions) individuals with adequate abilities or reorganize them. However, each individual is not expected to have enough abilities which are requested in the tasks. In our model shown in Figure 1, lacking abilities are trained and their skill is increased in the plays.

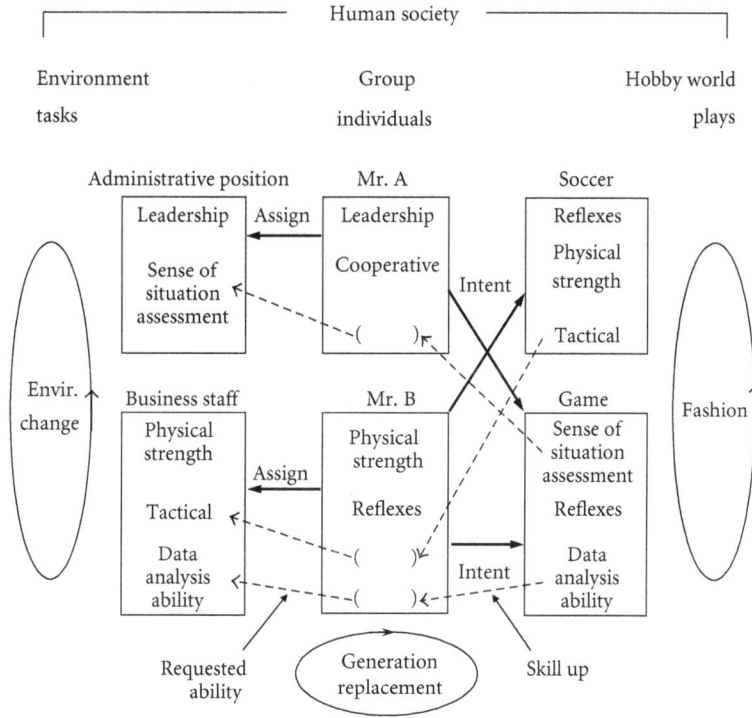

FIGURE 1: Gene world model with interaction to external environment. () shows lacking ability.

The play is often referred to as an opposite action to the diligence. However, like the play at housekeeping of the children, if we consider it as the training occasion of the social roles, the meaning of the play becomes largely different. Competitive playing sports may also be regarded as a hidden side of hot war to avoid destructive results by applying rules. The human society is requested to continuously cope with variously changing environment with its variety of abilities. At ordinary times most part of such abilities are hidden as they are, and a part of it is revealed as the play and training.

In this paper, using genetic algorithm we generate a group of artificial lives which are capable of coping with various environments which is a set of requested tasks. Under such situation, we like to show that the plays or hobbies are automatically generated as a necessary instinct or they are a natural result of evolution of the group of individuals in order to maintain the coping capability with various changes of the environment as a whole. This may be another side of that the wide variety of the abilities in the group is necessary and if the variety in a species decreased, its species will be extinguished.

This paper shows an attempt to generate an artificial mind which has two aspects of "diligent to his job" and "apt to run to pleasure." We show that the latter does not have a negative meaning but also has a positive meaning to cope with variously changing environment as a group and stabilize the world productivity.

2. Gene World Model

2.1. A Set of All Kinds of Human Abilities. In this paper we assume that human abilities are countable such as

$$a_1 = [\text{swift-of-foot}],$$

$$a_2 = [\text{patient}],$$

$$a_3 = [\text{skillful-in-calculation}],$$

$$a_4 = [\text{fair-spoken}],$$

$$a_5 = [\text{rich-in-leadership}], \quad (1)$$

$$a_6 = [\text{rich-in-curiosity}],$$

$$a_7 = [\text{rich-in-competitive-spirit}],$$

$$a_8 = [\text{aggressive}],$$

$$a_9 = [\text{interest-in-fishing}].$$

The set of such abilities is defined as

$$A = \{a_1, a_2, \ldots, a_{NA}\}, \text{ (Ability set).} \quad (2)$$

Null element may be included in A to make the following expressions simple, by which the sizes of some of the following sets may be made fixed by filling with it.

2.2. Environment. An environment (outer world) *Et* at time *t* is expressed as a set

$$Et = \{T_{t1}, \ldots, T_{tnt}\}; \quad t = 1, 2, \ldots, \text{ (Environment)} \tag{3}$$

of requested tasks in the environment, where tasks are, for example,

$$T_{\text{old-times } 1} = [\text{protect-from-wild-animals}],$$
$$T_{\text{recent-ages } 1} = [\text{publish-book}],$$
$$T_{\text{recent-ages } 2} = [\text{open-a-store}], \tag{4}$$
$$T_{\text{recent-ages } 3} = [\text{negotiate-with-other-company}],$$
$$T_{\text{today } 1} = [\text{set-computer-business-up}].$$

The environment changes from time to time.

2.3. Task. Each task *Ti* is expressed by a set of requested abilities to accomplish it as

$$Ti = \{a_{i1}, a_{i2}, \ldots, a_{ini}\}; \quad i = 1, \ldots, N_T, \text{ (Task)}. \tag{5}$$

For example,

$$T_1 (= \text{task of } [\text{protect-from-wild-animals}])$$
$$= \{\text{quick-motion}, \text{cool-mind}, \text{good ear}, \ldots\}.$$
$$T_2 (= \text{task of} [\text{political-action}]) \tag{6}$$
$$= \{\text{fair-spoken}, \text{planning-ability}, \ldots\}.$$

3. Play

On the other hand, there is a set *Pt* of all plays (hobbies) including self-development and volunteer activities to which people begin or join voluntarily at time *t*. That is,

$$Pt = \{R_{t1}, \ldots, R_{tmt}\}; \quad t = 1, 2, \ldots, \text{ (Play set)}. \tag{7}$$

For example,

$$Pt = \{\text{soccer}, \text{go-game}, \text{marathon},$$
$$\text{reading}, \text{travel}, \text{fishing}, \text{volunteer-activity}, \ldots\};$$
$$t = \text{today}. \tag{8}$$

Each play also requests a set of abilities:

$$Ri = \{a_{i1}, a_{i2}, \ldots, a_{imi}\}; \quad i = 1, \ldots, N_R, \text{ (Play)}. \tag{9}$$

For example,

$$R_1 (= \text{play of } [\text{fishing}])$$
$$= \{\text{interest-in-fishing}, \text{driving car}, \text{patient}, \ldots\}, \tag{10}$$
$$R_2 (= \text{play of } [\text{go-game}])$$
$$= \{\text{interest-in-go-game}, \text{pattern recognition}, \tag{11}$$
$$\text{spatial reasoning}, \text{tactical sense}, \ldots\}.$$

Abilities of [interest-in-*] are always necessary to begin the play of *. The set *Pt* of plays may also change from time to time according to the fashion. Only the difference between the task and the play is that the task is passively given or controlled by the environment, and the play is sought voluntarily or begun by urge and hard to stop.

3.1. Individual. Each individual *Ik* has his abilities and their skills:

$$Ik = \{a_{k1}, \ldots, a_{knI}; s_{k1}, \ldots, s_{knI}; c_k\};$$
$$k = 1, 2, \ldots, N_I, \text{ (Individual)}, \tag{12}$$

where *s* is a corresponding skill and *p* is a rate of spending time for the play where the hidden as well as the explicit abilities are trained. The set of individuals is generated and evolved by GA algorithm. Skills are raised by the experience in the job task or in the play but not inherited. The abilities and the ratio *c* are inherited across the generations. When one has more than two hobbies, time ratio *c* may be divided among them, or only the most fitted one is selected as shown in Section 3.4 and applied in the experiment.

3.1.1. Formulation. A simple example of the world model is shown in Figure 1. The outside environment is always changing. It is the origin of the evolution. In this section, some basic formulas of the simulation are given.

3.2. Skill. We assume here the skill is given by

$$s_{ki} = 0.5 + \sum_t \left[\alpha_{ki}^T \times (1 - c_k) + \alpha_{ki}^P \times c_k \right] \text{ (Skill)} \tag{13}$$

which is roughly proportional to the experienced time (years, or unit years); the ability a_{ki} was used in the task or the play. c_k and $1\text{-}c_k$ are rates of time used for the play and the task, respectively. For simplified example, in case of one working full time in weekdays and spending only for play full time in weekend, $c_k = 2/7$ and $1\text{-}c_k = 5/7$ by neglecting another time, for example, housekeeping. The value 0.5 is an initial skill when the individual has no experience. This means that one can do even unexperienced task to some extent using other means. α_{ki}^T takes 1 when the ability a_{ki} is used in the task and 0 when not used. α_{ki}^P is that of the play. The value of skill will be upper limited by their life-span automatically.

3.2.1. Assignment. When the environment Et is given, an individual most fitted to the task (see Section 3.3) among

(a) Ability spectrum of (i) requested and (ii) generated

(b) (iii) Performance and (iv) variance of ability distribution. Vertical axis is arbitrary unit

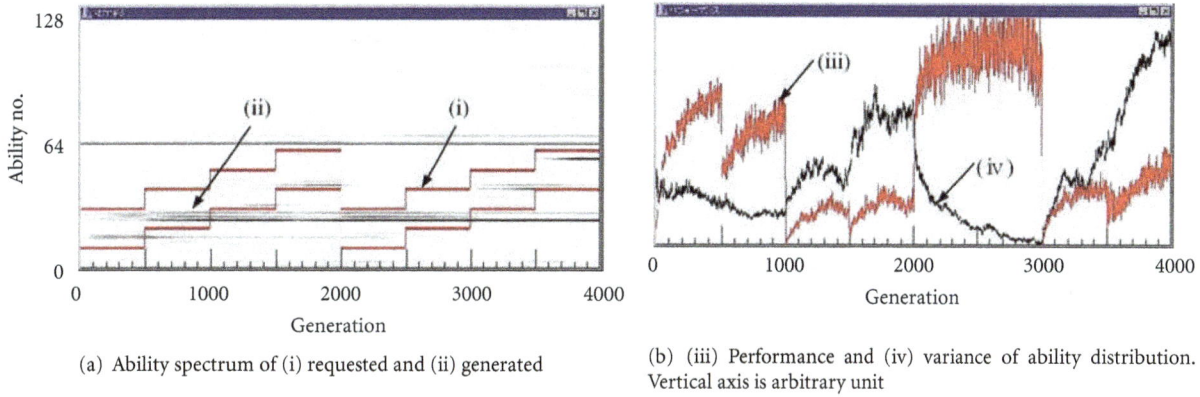

FIGURE 2: Typical behavior of the world with 64 kinds of less complicated plays each of which requires no ability.

randomly selected unassigned Ne (e.g., five) individuals is assigned to the task. Individuals not assigned to any task are called "window-side-folks" and direct his energy to the play with the time rate of

$$\min[1, b \times c_k]; \quad b \geq 1 \ (\text{e.g.}, b = 2). \tag{14}$$

Then the skills used implicitly as well as explicitly in the play will be increased as much.

3.3. Fitness to Task.
The fitness of the individual $Ik = \{a_{k1}, \ldots, a_{knI}; s_{k1}, \ldots, s_{knI}; c_k\}$ to the task $Ti = \{a_{i1}, a_{i2}, \ldots, a_{ini}\}$ is given by sum of the skills used in the task as

$$f_{ik} = (1 - c_k) \times \sum_{j=1}^{n_i} s_{kj} \ (\text{fitness}) \tag{15}$$

such that $a_{ij} = a_{kl}$ for some l in $\{1, 2, \ldots, n_I\}$.

If there are two same abilities in an individual, its skill is counted twice, and so on. That is, for example, if one has three genes of the same ability of [swift-of-foot], he may have three times of the skill of the [swift-of-foot]. On the other hand, if he has no such ability, s_{kl} is counted as 0.5 (min). Though this process may be more nonlinear, for the ease of modeling it was made linear.

3.4. Fitness to Play.
Also for the plays we can consider the same fitness, and each individual selects the most fitted Ri among the set of plays specified by ability of [interest-in-*] and practicing it. If one has no such ability, he will spend the leisure time with rate c idly; that is, his abilities are not trained in this time.

3.5. Performance of the Whole World.
Performance of the whole world is given by

$$Q = \sum_{i=1}^{NT} f_i^* \ (\text{performance}), \tag{16}$$

where f_i^* is f_{ik} of an individual Ik who is practically assigned to the task Ti. Note that hobbies are not evaluated at all in this model.

3.6. Evolution of Individuals.
Genetic algorithm [5, 6] is employed to make the evolution of the set of individuals. At every year (or unit year), some percentages of the individuals with fitness in order from the lowest or randomly to some extent are selected and erased. Then, those with the highest fitness are multiplied.

3.7. Genotype.
In order to simulate the artificial humans we must map the model to genotype. It may be more simple to represent the individual $(Ik' = \{a_{k1}, \ldots, a_{knI}; c_k\})$ by a sequence of binary numbers with $(\log_2 N_A \times n_I +$ bit length of $c_k)$ bits. Since skills are not inherited, $\{s_{k1}, \ldots, s_{knI}\}$ is not included in the Ik' different from Ik.

The environment is given from the outside to the group of individuals. It requests the tasks. On the other hand, the plays are generated as a result of the evolved genes. To do so, some specific binary sequences on the gene should be interpreted as [interest-in-*] with which he or she is represented as being fond of that kind of play and begin to do it. The set of trainable abilities in each play is given also from the outside.

4. Simulation and Experimental Results

We have developed a simulation software.

4.1. Parameters.
The initial parameters of the simulation are $N_A = 128$ (number of kinds of ability), $N_I = 120$ (number of populations), $N_T = 100$ (number of job tasks; 20 are unemployed), $n_i = 16$ (number of abilities required in each job tasks), $n_i = 16$ (number of abilities each person has), $N_R = 64$ (number of kinds of play), $m_i = 0$–64 (number of abilities required in each play), $b = 2$, $c_k = [7\text{bits}]$, and the rate of mutation is 0.05. Zone of the ability code number more than 64 (i.e., 65–128) is allocated to the abilities of interest-in-*. Initial genes are set randomly.

4.2. Stepwise Environmental Change.
Figure 2 shows a typical behavior of the world according to the generation lapse with 64 kinds of less complicated plays each of which requires no ability. In Figure 2(a), the vertical axis represents

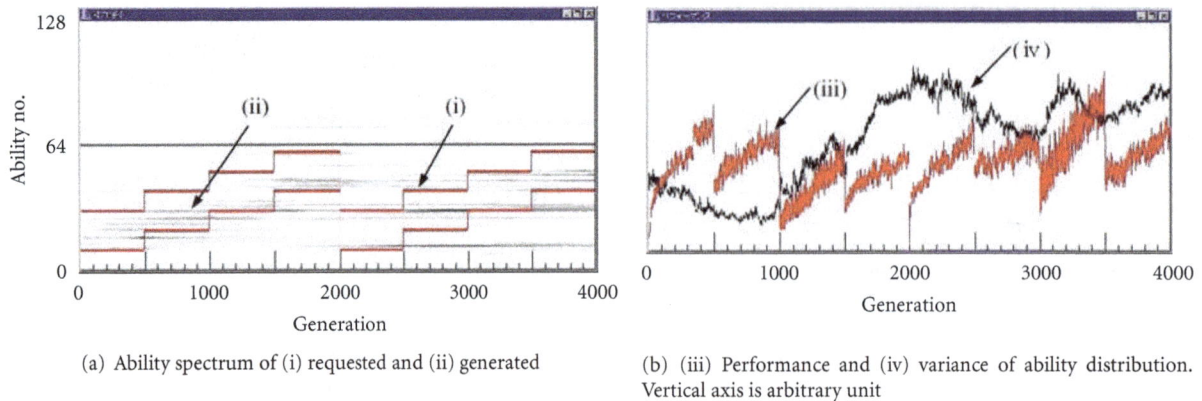

(a) Ability spectrum of (i) requested and (ii) generated

(b) (iii) Performance and (iv) variance of ability distribution. Vertical axis is arbitrary unit

FIGURE 3: Typical behavior of the world with 64 kinds of highly complicated plays each of which requires 64 abilities.

the code number of the ability which is expressed by seven bits in the gene. The environment requests abilities between two lines (i) with band width 20 and changing stepwise according to the generation. Requested 16 ($=n_i$) abilities for each task are selected from this band of 20 randomly. Then, various abilities (ii) in the group of the artificial humans become to appear. That is, (ii) represents a spectrum of ability set embedded in the artificial mankind and changing along the time. When the generated ability distribution as a result of the evolution matches well to the requested abilities, the performance (iii) of the world becomes high. When the environment change effects largely in the world, the performance of the world becomes almost zero and has a risk of extinction of the species as in the 1000th, 1500th, and 3000th generation. On the other hand, the change at the 2000th generation will be likened to the Industrial Revolution, where the performance of the world is raised up in spite of the sudden change of the environment (requested tasks) since the abilities in artificial mankind fit well to it. The curve (iv) shows the variance of the ability distribution (ii). We can see that when the ability distribution (ii) matches well to the environment (i) with small variance, the performance becomes very high as in between the 2000th and 3000th generations. Abilities between no 64 and no 128 that correspond to {interested-in-*} in Figure 2(a) are dispersed and thin, since the abilities of {interested-in-*} do not contribute to the performance.

Figure 3 shows a typical behavior of the world with 64 kinds of highly complicated plays each of which requires 64 abilities. We can see that the performance of the world is fairly steady compared with Figure 2. The variance curve is also steady than Figure 2.

We can see that in the world where more varieties of abilities are requested in the plays, or various kinds of plays are possible and popular as in the advanced countries, the performance of the whole world in spite of calculated only from tasks becomes more stable by the continuous skill up in the plays. This is the effect of the play.

In the first and the second simulations of Figures 2 and 3, the abilities of {interested-in-*} were only for plays. We have made the third simulation where we have defined the ability of {interested-in-task*}. The ability of {interested-in-task*}

is not always necessary to begin the task*. When individuals are assigned to tasks, the ability of {interested-in-task*} is used as an extra ability. Then, the performance is higher than those who don't have it as much as its skill. Thus, his skill is raised up more since he has more chance to be employed or promoted more, and it makes more chance to get good job, and so on.

Besides the abilities with code numbers 65–128 are allotted to plays one-to-one, each task is allotted randomly to one of the code numbers of 65–128. That is, each the code number between 65 and 128 is allotted to one play and 0, 1, or 2, ... tasks. An experimental result is shown in Figure 4, where the performance is more stable than Figure 2 and the risk of the extinction seems decreased. This may be because individuals having interest-in-task ability have more chance to fall in a good circulation of being selected and polish Their abilities. However, note that the stability is still lower than the world with the complicated plays of Figure 3 in this case.

We can say that there is a discrimination between the tasks as the duty one and the voluntary one with interest. This is the effect of the interest in this task. We can see that the improvement of the ability by which they performed the task voluntarily contributes to the variance of the ability. We can find that there are common parts between the {interested-in-task*} and the voluntary play, for example, the baseball is played as a profession and as a play (hobby). So we may be able to say this is the effect of the play.

4.3. Random Environmental Change. In order to see the difference by complexity of the plays from statistical point of view, we performed 100 trials of the simulation under the random environmental change shown in Figure 5.

Results are shown in Figure 6. Comparing performance and variance between worlds with different complexity of the play for the same environmental change, there are not so much differences at a glance in (a) and (b) of Figure 2. Therefore, the mean differences are calculated as Table 1. We can see that when the plays request more number of abilities, performance of the world becomes higher, and spectrum of genes within the world becomes more spread compared with the world with only simple plays. That is, there become people that have more wide range of abilities and talents.

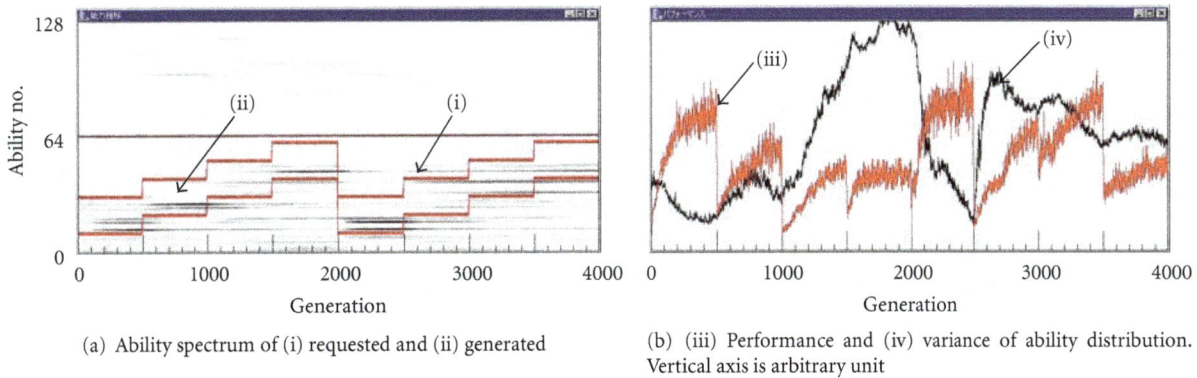

(a) Ability spectrum of (i) requested and (ii) generated

(b) (iii) Performance and (iv) variance of ability distribution. Vertical axis is arbitrary unit

FIGURE 4: Behavior of the world with 64 kinds of less complicated plays requiring no ability as Figure 2 but abilities of interest-in-task added.

TABLE 1: Improvement of means of performance and variance from simple play world. Values are arbitrary unit in Figure 6.

Abilities for play	Performance	Variance
0	—	—
8	18	39.26
16	35	38.57
32	14	41.61

Consequently, in the play-rich world, though productivity is not always higher, for sudden environmental changes drop of the productivity is small and the productivity after the drop is kept higher than in the simple play world as seen at around 2700 generation and 3400 generation in Figure 6. Though these differences do not look so large, they are accumulated and make a large difference.

Note that in the simulation, abilities are fixed beforehand. In real situations, however, they are newly generated to adapt more to the varying environment. Introducing this mechanism to the simulation, it will make the simulation more realistic. For example, in both worlds in Figure 6, we see that the variances after 3000 generation are decreasing though the performances are still raising and they show the limit of evolution by being too much adapted to the environment. These limits will be removed by the above improvement.

5. Discussion

Generally speaking, the simulation of GA is not always stable, but sometimes it happens that even when the performance is growing up steadily the evolution becomes bad suddenly. This is partly because of the limited number of individuals in the above simulation.

Though we have not tried fully yet, it is expected that a group of individuals with wide variety of abilities as a whole is generated by the effect of varying environment. In other words, required wide variety of abilities to cope with variously changing environment are embedded distributively in the group. As a result each individual comes to have his own personality and uniqueness. Individuals constitute the

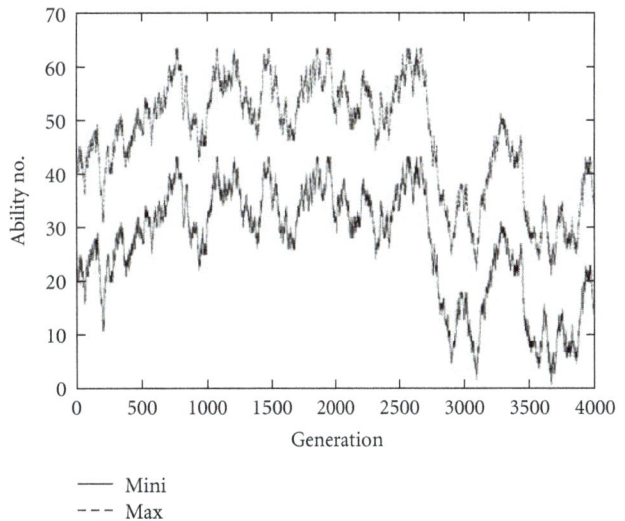

FIGURE 5: Random environmental change. Range of 20 between upper and lower curves is the requested abilities in each generation.

world (group or independent country) by sharing tasks required in the present environment. However, the whole abilities embedded in the group distributively are not only the ones required in the present environment but also for its possible changes. Some parts of such abilities embedded into an individual will appear as plays or hobbies. Thus, in spite of only evaluating the fitness of abilities of the individual to the changing environment, not only the variety of the abilities for the tasks but also the variety of the abilities for plays come to appear. As a result, three types of individuals will be generated. The first type of the individual is fitted to his present task, and this type of individual is happy only in his job. The second one is happy in his play rather than in his job. The third one is their intermediate type. Essentially, it may be nonsense to discriminate the abilities for the job tasks and for the plays, although in this paper we discriminated them for convenience in the simulation.

The present environment requests not the whole abilities of the group of humans but only the part. The unrequested abilities will appear as the play in the hobby world and

(a) World with simple plays requesting no ability

(b) World with complex plays requesting 32 abilities

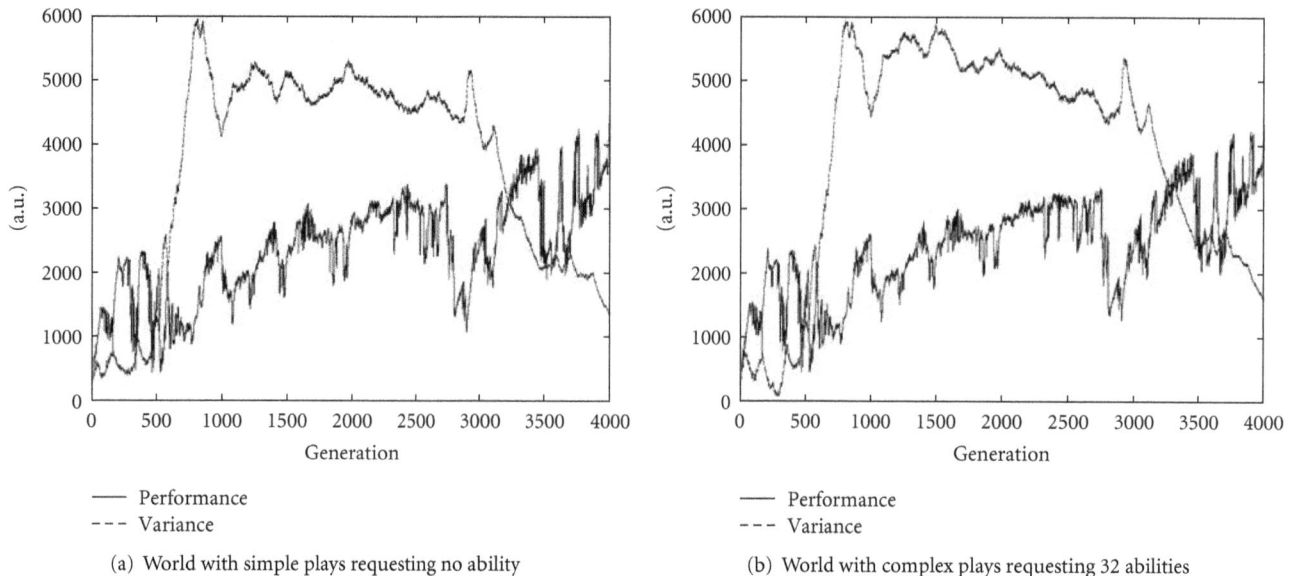

FIGURE 6: Average of 100 trials of the world. Lower dark curve is the performance and upper thinner curve is the variance of the gene spectrum.

prepare the next chance to be requested from the environment by being trained in the play.

Though the paper deals with a world of mankind, the frame of the research will also be applicable to various worlds with struggle for existence, such as animal world and business world.

6. Conclusion

In this paper, we have proposed a gene world model of a group of artificial humans capable of coping with various environments, where individuals with mind of plays or hobbies are generated. We have shown in a simulation that in the world, where more variety of abilities are requested in the plays, the performance of the whole worlds in spite of calculated only from tasks becomes stable. This is the effect of the play. Thus, we may be able to say that we could confirm that the role of plays will be partly some kind of training for implicit or explicit abilities which will be useful for unexpected changes of the environment. It can be regarded as play expresses a margin of ability.

Further improvement along the following items will be necessary.

(1) The system behavior is rather dependent on how the environment changes in speed and range. Therefore, we must investigate what type of changes will generate what type of artificial lives.

(2) Structure of the model of the task is flat of only a personal job level. It might be needed to construct a task system with a hierarchical structure such as a company composed of individuals.

(3) It may be difficult in the real world to discriminate the job tasks as the duty one and the hobbies as the voluntary one with pleasure. It may be a way not

discriminating [interested-in-play*] and [interested-in-task*].

(4) In the real world, evaluation of the world is made not only by the job tasks but by including the plays.

(5) "Variance" may not be adequate since it depends on how to allot the number to the ability. Therefore, entropy may be more adequate to express the diversity of the gene.

(6) Though in the simulation, abilities in the mankind are fixed beforehand in the real situation they are evolved to adapt more to the changing environment. Then, the artificial lives will evolve more drastically.

Acknowledgments

This work was partially supported by the Grant-in-Aid for Scientific Research of Exploratory Research 12878060, 21656100, and Scientific Research (A) 22246054 of Japan Society for the Promotion of Science, Japan.

References

[1] S. Franklin, *Artificial Minds*, The MIT Press, Cambridge, Mass, USA, 1995.

[2] M. A. Arbib, Ed., *The Handbook of Brain Theory and Neural Networks*, MIT Press, Cambridge, Mass, USA, 1998.

[3] Masanori Sugisaka and Y. G. Zhang, "A review to the International Symposium on Artificial Life and Robotics (AROB)," in *Proceedings of the 5th International Symposium on Artificial Life and Robotics (AROB '00)*, 2000.

[4] D. Ai, X. Ban, S. Zhang, and W. Wang, "Cognitive modeling of artificial fish learning and memory," in *Proceedings of the 12th International Symposium on Artificial Life and Robotics (AROB '07)*, pp. 280–283, January 2007.

[5] D. E. Goldberg, *Genetic Algorithm in Search, Optimization, and Machine Learning*, Addison-Wesley, Reading, Mass, USA, 1989.

[6] M. Mitvhell, *An Introduction to Genetic Algorithms*, MIT Press, Boston, Mass, USA, 1998.

7

CUDAICA: GPU Optimization of Infomax-ICA EEG Analysis

Federico Raimondo,[1] **Juan E. Kamienkowski,**[2]
Mariano Sigman,[2] **and Diego Fernandez Slezak**[1]

[1] *Departamento de Computación, Pabellón I, Ciudad Universitaria, C1428EGA Ciudad Autonoma de Buenos Aires, Argentina*
[2] *Laboratory of Integrative Neuroscience, Physics Department, University of Buenos Aires, Buenos Aires, Argentina*

Correspondence should be addressed to Diego Fernandez Slezak, dfslezak@dc.uba.ar

Academic Editor: Hujun Yin

In recent years, Independent Component Analysis (ICA) has become a standard to identify relevant dimensions of the data in neuroscience. ICA is a very reliable method to analyze data but it is, computationally, very costly. The use of ICA for online analysis of the data, used in brain computing interfaces, results are almost completely prohibitive. We show an increase with almost no cost (a rapid video card) of speed of ICA by about 25 fold. The EEG data, which is a repetition of many independent signals in multiple channels, is very suitable for processing using the vector processors included in the graphical units. We profiled the implementation of this algorithm and detected two main types of operations responsible of the processing bottleneck and taking almost 80% of computing time: vector-matrix and matrix-matrix multiplications. By replacing function calls to basic linear algebra functions to the standard CUBLAS routines provided by GPU manufacturers, it does not increase performance due to CUDA kernel launch overhead. Instead, we developed a GPU-based solution that, comparing with the original BLAS and CUBLAS versions, obtains a 25x increase of performance for the ICA calculation.

1. Introduction

Analysis of brain imaging data has two intrinsic difficulties: dealing with high volumes of data (and often high dimensional) and a usually low signal-to-noise ratio due to persistent artifacts. A significant number of methods have been developed, usually based on some form of dimensionality reduction of data, to cope with these difficulties. Multivariate statistical analysis for the separation of signals is a widely studied topic of great complexity because of the large number of sources and the low signal-to-noise ratio, inherent in this kind of signals. Specific approaches have been developed to separate the signals generated by the study of those sources that contribute only noise, such as principal component analysis (PCA) [1], factor analysis [2], and projection pursuit [3], among others.

Independent Component Analysis (ICA) [4–6] is one of the most effective methods for source separation and removal of noise and artifacts. The most emblematic example was the separation of audio sources in noisy environments [5]. In recent years, it has become a standard in brain imaging-electroencephalogram (EEG) [7–9], magnetoencephalogram (MEG) [10] and functional magnetic resonance imaging (fMRI) [11–13]. It has been used for the removal of artifacts arising from eye movements [14], but also for the selection of relevant dimensions of the data [15, 16]. In fact, the most popular open packages for EEG analysis—EEGLAB (http://sccn.ucsd.edu/eeglab/) and Fieldtrip (http://fieldtrip.fcdonders.nl/)—strongly rely on ICA.

Analyzing EEG, MEG, or fMRI with ICA does not come without a cost and requires a huge amount of computing power. For example, analyzing a typical single-subject EEG experiment (data from 132 channels, at 512 Hz of sampling rate, stored in single-precision, for a 1-hour experiment, which amounts to a total of 1.5 GB of data), typically takes around 12 hours on an Intel i5, 4 GB RAM, or 8.5 hours on an Intel i7, 16 GB RAM. Often one needs to look at the output of ICA before making a decision on how to proceed with the data, and this long-lasting processing heavily conditions the flexibility of analysis. More importantly, it makes this analysis completely prohibitive for online access of the data, for instance in Brain Computer Interface applications.

Current standard CPU hardware include different extensions of the basic instruction set architecture for vector processing support, for example, the Streaming SIMD Extensions (SSE) or Advanced Vector Extensions (AVX). These extensions enable the parallel execution of the same operations on multiple data, a key requirement for signal-processing. Although, no effective tools to operate with more than a few tens of floating point numbers simultaneously are available in current CPUs, which makes the calculation more efficient but it still lacks the key feature needed to get the results quick. In other words, processors of coming years will only improve slightly this performance.

One approach to solving this problem is using Beowulf parallel computing clusters [17]. The main drawback of this implementation is the communication overhead needed to synchronize the different compute nodes, since the memory is distributed over the nodes.

Here, we propose to use of the massive parallel processors included in the graphical units (GPU) for ICA calculations. Contrary to a Beowulf, this architecture has a common shared memory allowing a much faster parallelization of ICA algorithms. In addition, the low cost of GPUs makes this project available for virtually every user. We investigate the processing of EEG data series using CUDA: a parallel computing platform and programming model. CUDA extends C/C++ programming language, enabling the programmer to write a serial program (functions, also called *cuda kernels*) that executes in parallel across a set of threads operating over different memory positions [18].

Many implementations of ICA are available, for example, Infomax [5], SOBI [19], and FastICA [20]. These implementations use mostly linear algebra operations which are included in off-the-shelf optimized GPU standard libraries. FastICA has demonstrated that parallelization may be relatively straightforward replacing linear algebra routines by standard GPU libraries [21].

However, ICA of human EEG data is much better approximated by Infomax [7] enhanced which has made it a standard in EEG analysis. We show that simply replacing linear algebra routines with GPU libraries do not show better performance. Parallelizing Infomax requires an efficient optimization of GPU memory access and kernel dispatch, in order to obtain an increased performance. Here, we set to develop and implement CUDAICA, an optimization algorithm to increase processing time by a 25x factor, at almost no cost, without changing the original algorithm.

2. Independent Component Analysis

ICA was introduced in 1994 by Comon [22] independent. The concept of ICA can be seen as an extension of the PCA, where the linear transformation minimizes the statistic dependence between its components.

The following statistic model is assumed [23]:

$$x = My + v, \qquad (1)$$

where x, y, and v are random vectors with values in $\mathbb{R}$ or $\mathbb{C}$ with zero mean and finite covariance, M is a rectangular matrix with at most as many columns as rows and vector y has statistically independent components.

The problem set by ICA can be summarized as follows: given T samples of vector x, an estimation of matrix M is desired, and the corresponding samples from vector y. However, because of the presence of noise v, it is in general impossible to reconstruct the exact vector y. Since the noise v is assumed here to have an unknown distribution, it can only be treated as a nuisance, and the ICA cannot be devised for the noisy model above. Instead, it will be assumed that:

$$x = As, \qquad (2)$$

where s is a random vector whose components are maximizing statistical independence [22].

Both, EEGLAB and FieldTrip analysis software, use the Infomax algorithm [5] for estimation of independent components [24, 25]. Infomax is based on a neural network with three columns of neurons, each representing: (1) the original data (X); (2) the registered data (r); (3) the approximated independent data (Y). Each column of neurons combine linearly by matrices A and W.

The principle used by this algorithm is maximizing the mutual information that output Y of a neural network processor contains about its input X, defined as

$$I(Y, X) = H(Y) - H(Y \mid X), \qquad (3)$$

where $H(Y)$ is the entropy of output Y and $H(Y \mid X)$ is the entropy of the output that did not come from the input. In fact, $H(Y)$ is the differential entropy of Y with respect to some reference, such as the noise level or the accuracy of discretization of the variables in X and Y. Thus, only the gradient of information-theoretic quantities with respect to some parameter w is considered [5]. Then, the equation (3) can be differentiated, with respect to a parameter w as:

$$\frac{\partial}{\partial w} I(X, Y) = \frac{\partial}{\partial w} H(Y), \qquad (4)$$

because $H(X \mid Y)$ does not depend on w.

In the system (1), $H(X \mid Y) = v$. Whatever the level of the additive noise, maximization of the mutual information is equivalent to the maximization of the output entropy, because $(\partial/\partial w) H(v) = 0$. In consequence, for any invertible continuous deterministic mappings, the mutual information between inputs and outputs can be maximized by maximizing the entropy of the outputs alone.

The natural (or relative) gradient method simplifies considerably the method. The natural gradient principle [26, 27] is based on the geometric structure of parameters space and it is related to the relative gradient principle [28] that ICA uses.

Using this approach, the authors propose the following iteration of the gradient method to estimate the W matrix:

$$\Delta W \propto W - \tanh\left(\frac{Wx}{2}\right)(Wx)^T W. \qquad (5)$$

In summary, Infomax ICA consists of the following steps:

(1) $U = W \times \text{perm}(x)$ (where perm is a random permutation),

(2) $Y = -\tanh(U/2)$,

(3) $YU = Y \times U^T$,

(4) $YU = YU + I$,

(5) $W = l\text{rate} \times YU \times W + W$,

where $l$rate is the learning rate for each iteration of the method, generally lower than $1e^{-2}$. High values of $l$rate may lead to faster computation but a bad choice could destroy convergence. We propose a fast implementation using GPUs keeping the original algorithm intact, including the selection criteria for $l$rate.

3. Material and Methods

We first present, as a baseline measure, the performance of Infomax ICA implementations available. We then compare them with our development for Infomax ICA based on CUDA optimized for GPU processing.

3.1. Infomax ICA Implementations. Current implementations (C/C++ and Matlab) make use of Basic Linear Algebra Subprograms (BLAS). This library provides the standard routines for performing basic vector and matrix operations [29]. Several implementations of BLAS can be used to compute ICA: ATLAS, Intel MKL, and CUBLAS, among others.

The most popular implementation of these routines is ATLAS, a portable self-optimizing BLAS, included in most unix distributions [30]. Intel offers the Math Kernel Library (Intel MKL), a computing math library of highly optimized, extensively threaded math routines. NVIDIA offers CUBLAS, an implementation of BLAS on top of the NVIDIA CUDA driver for GPU optimization of linear algebra routines.

Generally, EEG equipment represents electrode data as a time series of single-precision floats. Nevertheless, all implementations of Infomax ICA use double-precision floats to ensure numerical stabilities and avoid numerical error propagation due to precision. Thus, a GPU with CUDA compute capability 1.3 (or greater) must be used, in order to support double-precision float operations.

3.2. Testing Hardware. Performance tests of ICA computation have been made on several hardware configurations. Three different CPUs where used: i7-2600, i5 430M, and Xeon E541a. For GPU technology, we used a Tesla C2070 and GTX 560. To compare the performance of our solution, we chose the high-end equipment: the Intel Core i7-2600 with 16 GB of RAM and the Nvidia Tesla C2070 video card. Some tests have been done with an Nvidia Quadro 4000 which showed a comparable performance to Tesla hardware.

Performance comparisons were made using real datasets with the number of channels varying from 32 to 256 in 32 channel steps. The amount of samples vary from 15 minutes experiments to 105 minutes in 15 minutes steps, with a sampling rate of 512 Hz.

3.3. Testing Datasets. The testing experiment consists of 70 trials where the subject freely explores between a set of dots distributed along the horizontal line. The trial starts with the participant fixating in a small dot in the bottom half of the screen, and the set of dots appear in the top half of the screen. Participants have 5 seconds to explore and typically perform 15 saccades in each trial. Participants are instructed to explore in random order and are free to blink. We purposely gave this instruction to have data contaminated by typical eye-movement and blinking artifacts. Simultaneous eye movement and EEG were recorded using eye tracker Eyelink 2K, SR-Research, and Active-Two EEG with 128-channel, Biosemi system.

4. Results

4.1. Infomax ICA Profiling. We performed a detailed analysis of all the operations involved in the execution of ICA using *callgrind*, a tool for sequential profiling and optimization [31]. We executed the method for different datasets and profiled the function calls, observing that BLAS routines dgemv and dgemm consumed more than 80% of total calculation time. For instance, dgemv and dgemm take, respectively, 47.9% and 40.78% of the total amount of time used to calculate ICA for a 136 channels dataset and 22528 samples, recorded at 512 Hz. Thus, we selected these as the most relevant candidates for GPU optimization. The dgemv and dgemm symbols correspond to BLAS functions for matrix-vector and matrix-matrix multiplication, respectively.

4.2. MKL Implementation. The easiest way to achieve significant performance increment in applications with substantial amounts of BLAS operations is using the available optimized libraries developed by the processor manufacturers, the Intel MKL Libraries. These libraries use multithreaded implementations of BLAS operations and exploit all features of installed processors. This upgrade is extremely simple since no code or compiler commands are needed to be modified. Instead, simply compiling and linking to MKL libraries provides a more efficient execution of Infomax ICA with multithreaded BLAS operations.

Using this multithreaded version, we compared the performance of MKL Infomax ICA versus the standard one, on a Intel Core i7. ATLAS implementation takes 1.5 seconds per step on the smallest experiment (32 channels, 15 minutes of experiment) and grows almost linearly as channels and total time increase, reaching 620 seconds per step on the biggest experiment (256 channels, 105 minutes of experiment), as shown in Figure 3(a). MKL implementation shows a significant increase of performance, showing a maximum of 142 seconds per step of experiment (see Figure 3(a)). We obtained a maximum speedup of 4.5 using the MKL libraries for the experiments with more channels (see Figure 3(b)).

4.3. CUDAICA Implementation. The time series nature of data, make GPU computing a very promising approach for EEG processing optimization. A naive and simple optimization can be done using CUBLAS—NVIDIA implementation of BLAS on top of the NVIDIA CUDA driver—replacing symbols dgemv and dgemm with CUBLAS symbols cublas-Dgemv and cublasDgemm, respectively. This approach was first used for a basic optimization of FastICA [21].

4.3.1. CUBLAS Approach. CUBLAS uses a different memory space than standard sequential calculations, that is, the video card RAM. Thus, before starting computation, data must be transferred from CPU memory to video memory. This results in a few additional calls to memory movement operations.

Performance measurements revealed that replacing the BLAS routines to their corresponding CUBLAS routines did not increase processing speed. The amount of time required to process the experiment was longer than ATLAS. In 32 and 64 channels, CUBLAS-implementation took 6.5x and 2.4x times longer than the ATLAS implementations. Only with 128 channels, the CUBLAS-implementation performance equals to ATLAS implementation, suggesting that if no effort wants to be invested in an ad-hoc solution, MKL is the best solution.

This lack of improvement is caused by the constant overhead of running the parallel kernels on GPU. Before starting each iteration, Infomax ICA creates a vector with a random permutation of the indexes, from 0 to $N - 1$, being N the number of samples in the dataset. Then, matrix W is multiplied by the column of x as indicated by the index in the random permutation vector using the corresponding BLAS matrix-vector operation. This results in as many matrix-vector multiplication as the numbers of samples in the dataset instead of a small number of matrix-matrix multiplications. The execution of each iteration results in an accumulated overhead that is only compensated with high dimensional datasets, where each step involves many operations.

As mentioned before, we observed matrix-vector and matrix-matrix multiplications consumes about 50% and 40% percent of the total amount of time required to compute Infomax, respectively. We performed performance tests comparing ATLAS against CUBLAS routines on matrix-matrix and matrix-vector operations using scenarios from Infomax algorithm. The results were conclusive: matrix-matrix operations were faster using CUBLAS, but matrix-vector operations were the bottleneck in this computation. Due to the actual dimensions of matrix W and the size of the column of x, ATLAS performed faster. The performance increase achieved with CUBLAS for the matrix-matrix operations (40% of total time) was opaqued by the overhead involved in execution of GPU functions and the lack of performance gain in the matrix-vector computation.

4.3.2. Hybrid Approach: CUDAICA. Based on this profiling, we implemented an hybrid solution of the method using CUDA (CUDAICA). The main objective of this

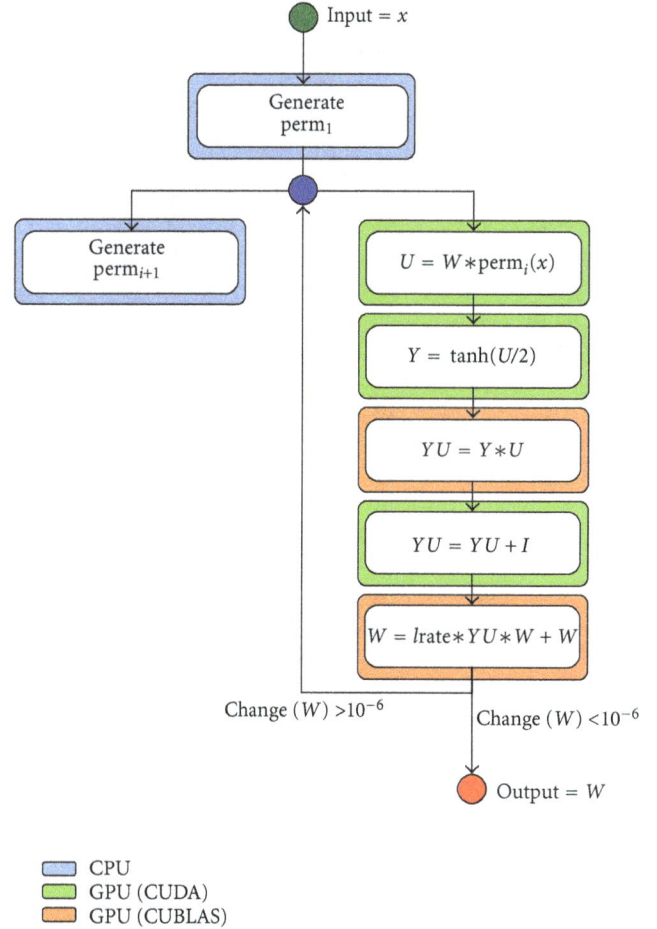

FIGURE 1: Flow diagram of hybrid implementation. We divide operations in three groups: CPU (blue box), GPU using standard CUBLAS libraries (orange box), and GPU using own CUDA implementation (green box).

implementation was to keep the original algorithm intact: reduce the computation time of the same operations. All the matrix-matrix operations are computed using CUBLAS and the remaining operations are solved by an ad-hoc implementation based on CUDA (see Figure 1).

Initially, all matrix and vectors are copied into the GPU device global memory. In the process, single precision floats—from the EEG time series—are converted to double precision. The optimizations applied to Infomax ICA consisted on two main optimizations: (1) optimizations on memory access and kernels executions in particular matrix operations ($U = W * \text{perm}(x), Y = -\tanh(U/2)$, and $YU = YU + I$) and (2) combining asynchronous execution in CPU and GPU to generate each step permutation perm_{i+1} of vector x, while GPU is computing step i.

Taking advantage of CUDA shared memory space, this optimization first copies the vector indexed by the permutation into shared memory. Once transferred, U is computed by an ad-hoc matrix-vector multiplication implementation. In the same kernel, as matrix U is being calculated, matrix Y is computed by computing $-\tanh(z/2)$, where z represents the values being computed in U. This combined operations reduce significantly the number of kernels launched and therefore the overhead involved in kernel initialization.

The amount of shared memory depends on GPU compute capability version. Since version 2.0, CUDA supports double-precision floating points operations and 48 KB of shared memory. These features allow to compute matrix-vector operations for dimensions up to 6144 elements at a time. By optimizing the algorithm implementation to use aligned memory, we obtained optimum memory access resulting in 128 Byte transactions to main memory without discarding any data.

Other significant improvement is made when computing $YU = YU + I$: instead of using the generic sum function, a specific CUDA kernel is used to modify only the elements in the diagonal of YU.

Random permutations can be performed by generating an index vector of random indexes in CPU while GPU computation is being performed. Then, when permutation is needed the random vector is used to index in the data matrix to locate the corresponding randomly permuted vector.

Combining these tweaks, we obtained an optimized version of Infomax ICA, with a processing time of less than 0.5 seconds per step in the 30-minute experiment with 32 channels (see Figure 2(a)).

In Figure 2(a) we show the performance comparison in a 30-minute experiment between CUDAICA and both Infomax implementations: ATLAS and MKL based. We observe an exponential growth of time per step (note de log scale in y-axis) in the MKL and CUDAICA implementations. ATLAS show a decrease of slope as channels increase, but always bigger than MKL and CUDAICA. Starting from 96 channels, MKL and CUDAICA show a similar slope, indicating an almost constant increase of performance of CUDAICA over MKL of 4.5x (Figure 2(b)). We obtained a maximum performance increase of CUDAICA versus the ATLAS implementation of more than 20x, for the experiments of 192 channels or more. Interestingly, we found that CUDAICA performed best versus MKL for 128 channels, the most typical configuration of our EEG setup.

All performance comparisons were performed in the best available hardware: i7 and Tesla C2070. In Figure 2(c), we show the time per step of the method in the different hardware available, with several channel configurations. As expected, we observe an almost linear increase of all runs as number of channels grow. Comparison between standard CPU and GPU processing shows a significant performance increase using any GPU (top-of-line Tesla or standard GTX560 card) against the non-GPU hardware tested.

4.4. Reliability of CUDAICA Relative to Previous Algorithms. The new optimization involves reordering of operations which may potentially affect the numerical stability of the calculations. Thus, we verified that CUDAICA produces results which are not distinguishable from previous implementations. For this purpose, we removed the random vector generation used in the first part of the method. After each step, we compared the output for the same inputs and the exact same result was achieved.

Then, we compared the output of the full algorithm using the original code (ATLAS based) and CUDAICA, executing both implementations and estimating the independent components of the same dataset. In Figure 3 we show the first 12 independent components calculated with both implementations (Figures 3(a) and 3(b)). As expected, executions show similar (but not identical) independent components, due to its random nature—that is, some ICs were in different order or with inverted weights. The first 3 components show almost the exact same behavior. As an example, in Figures 3(c) and 3(d) we plot the spectra and eye-movement artifact locked to saccade offset for IC1 and IC7-IC6 of both implementations. Difference are indistinguishable (correlation coefficient $R = 1$, $p < 0.0001$); we also show an amplified region where is possible to appreciate that there are, indeed, two curves in each plot in magenta (original code) and red (CUDAICA). We calculate the average trial-by-trial correlation between pairs of ICs of ATLAS and CUDAICA runs (Figure 3(e)). Almost all rows have a single white spot corresponding to the matched IC calculated by the other implementation. This spots are not necessarily aligned in the diagonal, as order could be switched and even weights could be inverted.

5. Discussion and Conclusions

ICA is one of the de-facto standard methods for source separation and removal of noise and artifacts. In neuroscience, it has been widely used for EEG [7, 8], fMRI [12, 14], and invasive electrophysiology [32].

In all these neuroimaging methods, technology has increased the data volume, improving spatial and temporal resolution. With current standards, analyzing data with ICA requires a vast, often intractable amount of computing power.

In practical EEG analysis, this computer power requirements impedes the rapid exploration of different methods since each implementation of ICA runs overnight or even taking more than one day. A rapid iteration and examination of different procedures becomes completely impractical. Even more, ICA is difficult to use for online access of the data. Over the last years there has been an exponential development of brain computing interfaces which require online access to the relevant dimensions of the data [13, 33–35].

ICA constitutes a formidable tool for finding relevant directions and BCI procedures [36] which use ICA to present participants with different components to determine which are easier to control is a timely necessity. For this, it is imperative to implement ICA at much faster speeds than is being implemented with current CPU and here we present a major advancement in this direction.

Our aim here was purely methodological: improve the speed of Infomax ICA by at least 10x. We performed a

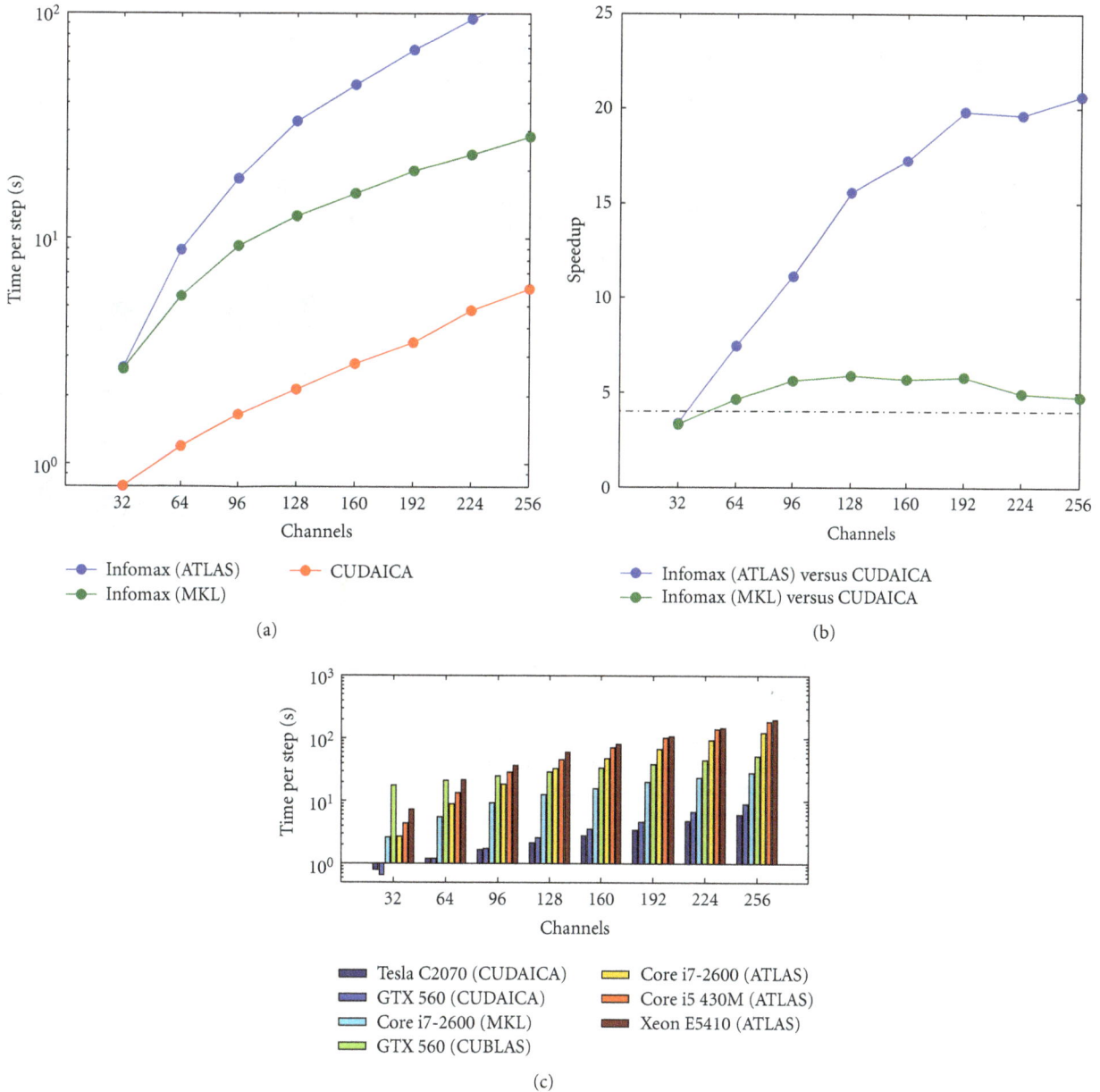

FIGURE 2: Execution time of Infomax step using ATLAS, MKL, and CUDAICA for 30-minute experiments (a). This short experiments may last more than 4 hours of computation time in the standard BLAS implementation with 256 channels. MKL shows significant performance improvement over BLAS, and CUDAICA behaves the best. Comparison between ATLAS and MKL versus CUDAICA (b). CUDAICA shows a maximum speedup of 4.7 over MKL, and more than 20x over standard ATLAS implementation. In (c), we show the time per step of different implementations of Infomax with several channel configurations, running under the available hardware.

detailed profiling and detected the bottleneck in the calculation of independent components, showing that vector-matrix and matrix-matrix operations take almost all computational time. Based on these results, we implemented an hybrid ad-hoc solution for GPU optimizations: CUDAICA. With this solution, we compared CUDAICA to the original BLAS (compiled with standard ATLAS and the optimized MKL libraries) and CUBLAS implementations. We observed a 25x performance increment using CUDAICA, over the standard ATLAS implementation, and 4.5x performance increment compared to the MKL implementation.

With this calculation time, a 128-electrodes EEG of 1-hour experiment would take 1500 seconds approximately to compute the independent components. This timing opens up new possibilities of the method, for instance for Brain Computer Interface applications, making possible to think of an experiment where Independent components may be calculated during the experiment and use them as a feedback feature.

CUDAICA was developed under the GNU General Public License, and is freely available from our wiki with a description of application features, FAQ and installation

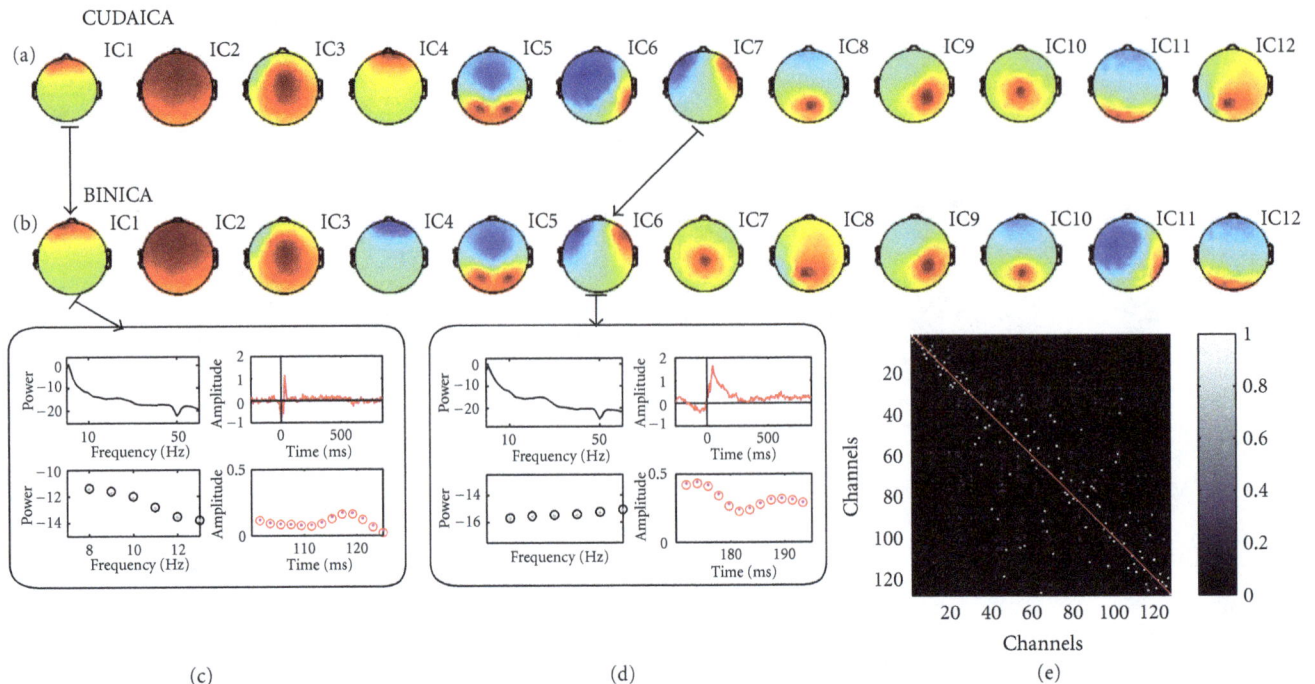

FIGURE 3: Comparison between CUDAICA and original code estimated components (ICs): (a, b) first twelve ICs of CUDAICA corresponds to the first twelve ICs, note that the order could be switched and the weights could be inverted (IC4). (c, d) Example of two pairs of components, IC1-IC1 and IC7-IC6, respectively. Both spectra (top-left panels) and eye-movement artifact locked to saccade offset (top-right panels) are indistinguishable by naked eye. We also show an amplified regions where it is possible to appreciate that there are two curves in each plot (bottom panels). Black and red lines: CUDAICA, and grey and magenta lines: Infomax ICA. (e) Average trial-by-trial correlation between pairs of ICs. Almost all rows have a single white spot corresponding to the matched IC calculated by the other implementation, and these spots are not necessarily aligned in the diagonal. Note that there are some subsets of components that do not have the correspondent ICs as these subsets correspond to a subset in other implementation.

instructions (http://calamaro.exp.dc.uba.ar/cudaica/doku.php?id=start). CUDAICA woks as a standalone application and integrates to the EEGLAB Toolbox adding an option to process ICA using CUDAICA, just like any other ICA implementation. It was designed for standard EEGLAB users, with no extra effort needed to run this implementation. It works under CUDA enabled hardware, that is, almost every modern graphic card, making CUDAICA widely available and easy to use.

Acknowledgment

This work was supported by the Nvidia Academic Partnership.

References

[1] E. Oja, "Principal components, minor components, and linear neural networks," *Neural Networks*, vol. 5, no. 6, pp. 927–935, 1992.

[2] H. H. Harman, *Modern Factor Analysis*, University of Chicago, 1976.

[3] J. H. Friedman, "Exploratory projection pursuit," *Journal of the American Statistical Association*, vol. 82, no. 397, pp. 249–266, 1987.

[4] A. Hyvärinen, E. Oja, and J. Karhuen, *Independent Component Analysis*, Wiley-Interscience, 2001.

[5] A. J. Bell and T. J. Sejnowski, "An information-maximization approach to blind separation and blind deconvolution.," *Neural computation*, vol. 7, no. 6, pp. 1129–1159, 1995.

[6] S. I. Amari, "Natural gradient works efficiently in learning," *Neural Computation*, vol. 10, no. 2, pp. 251–276, 1998.

[7] A. Delorme, T. Sejnowski, and S. Makeig, "Enhanced detection of artifacts in EEG data using higher-order statistics and independent component analysis," *NeuroImage*, vol. 34, no. 4, pp. 1443–1449, 2007.

[8] S. Makeig, T. P. Jung, A. J. Bell, D. Ghahremani, and T. J. Sejnowski, "Blind separation of auditory event-related brain responses into independent components," *Proceedings of the National Academy of Sciences of the United States of America*, vol. 94, no. 20, pp. 10979–10984, 1997.

[9] S. Makeig, A. J. Bell, T. P. Jung, and T. J. Sejnowski, "Independent component analysis of electroencephalographic data," in *Advances in Neural Information Processing Systems*, pp. 145–151, MIT Press, Cambridge, Mass, USA, 1996.

[10] R. Vigário, J. Särelä, V. Jousmäki, M. Hämäläinen, and E. Oja, "Independent component approach to the analysis of EEG and MEG recordings," *IEEE Transactions on Biomedical Engineering*, vol. 47, no. 5, pp. 589–593, 2000.

[11] V. Schöpf, C. H. Kasess, R. Lanzenberger, F. Fischmeister, C. Windischberger, and E. Moser, "Fully exploratory network

ICA (FENICA) on resting-state fMRI data," *Journal of Neuroscience Methods*, vol. 192, no. 2, pp. 207–213, 2010.

[12] V. D. Calhoun and T. Adali, "Unmixing fMRI with independent component analysis," *IEEE Engineering in Medicine and Biology Magazine*, vol. 25, no. 2, pp. 79–90, 2006.

[13] V. D. Calhoun, T. Eichele, and G. Pearlson, "Functional brain networks in schizophrenia: a review," *Frontiers in Human Neuroscience*, vol. 3, p. 17.

[14] T. P. Jung, S. Makeig, M. J. Mckeown, A. J. Bell, T. E. W. Lee, and T. J. Sejnowski, "Imaging brain dynamics component analysis," *Proceedings of the IEEE*, vol. 89, no. 7, pp. 1107–1122, 2001.

[15] S. Makeig, S. Debener, J. Onton, and A. Delorme, "Mining event-related brain dynamics," *Trends in Cognitive Sciences*, vol. 8, no. 5, pp. 204–210, 2004.

[16] S. Makeig and J. Onton, "ERP features and EEG dynamics: an ICA perspective," in *Oxford Handbook of Event-Related Potential Components*, Oxford University Press, New York, NY, USA, 2011.

[17] D. B. Keith, C. C. Hoge, R. M. Frank, and A. D. Malony, "Parallel ICA methods for EEG neuroimaging," in *Proceedings of the 20th International Parallel and Distributed Processing Symposium (IPDPS '06)*, p. 10, IEEE, 2006.

[18] J. Nickolls, I. Buck, M. Garland, and K. Skadron, "Scalable parallel programming with CUDA," *Queue*, vol. 6, no. 2, pp. 40–53, 2008.

[19] A. C. Tang, B. A. Pearlmutter, N. A. Malaszenko, D. B. Phung, and B. C. Reeb, "Independent components of magnetoencephalography: localization," *Neural Computation*, vol. 14, no. 8, pp. 1827–1858, 2002.

[20] A. Hyvarinen, "Fast and robust fixed-point algorithms for independent component analysis," *IEEE Transactions on Neural Networks*, vol. 10, no. 3, pp. 626–634, 1999.

[21] R. Ramalho, P. Tomás, and L. Sousa, "Efficient independent component analysis on a GPU," in *Proceedings of the 10th IEEE International Conference on Computer and Information Technology (CIT '10)*, pp. 1128–1133, July 2010.

[22] P. Comon, "Independent component analysis, A new concept?" *Signal Processing*, vol. 36, no. 3, pp. 287–314, 1994.

[23] A. Hyvärinen and E. Oja, "Independent component analysis: algorithms and applications," *Neural Networks*, vol. 13, no. 4-5, pp. 411–430, 2000.

[24] A. Delorme and S. Makeig, "EEGLAB: an open source toolbox for analysis of single-trial EEG dynamics including independent component analysis," *Journal of Neuroscience Methods*, vol. 134, no. 1, pp. 9–21, 2004.

[25] R. Oostenveld, P. Fries, E. Maris, and J. M. Schoffelen, "Field-Trip: open source software for advanced analysis of MEG, EEG, and invasive electrophysiological data," *Computational Intelligence and Neuroscience*, vol. 2011, Article ID 156869, 9 pages, 2011.

[26] S. Amari, "A new learning algorithm for blind signal separation," in *Advances in Neural Information Processing Systems*, 1996.

[27] S. Amari, "Neural learning in structured parameter spaces-natural Riemannian radient," in *Advances in Neural Information Processing Systems*, 1997.

[28] J. F. Cardoso and B. H. Laheld, "Equivariant adaptive source separation," *IEEE Transactions on Signal Processing*, vol. 44, no. 12, pp. 3017–3030, 1996.

[29] J. J. Dongarra, J. D. Croz, S. Hammarling, and I. Duff, "Set of level 3 basic linear algebra subprograms," *ACM Transactions on Mathematical Software*, vol. 16, no. 1, pp. 1–17, 1990.

[30] R. Clint Whaley, A. Petitet, and J. J. Dongarra, "Automated empirical optimizations of software and the ATLAS project," *Parallel Computing*, vol. 27, no. 1-2, pp. 3–35, 2001.

[31] J. Weidendorfer, "Sequential performance analysis with callgrind and kcachegrind," in *Tools for High Performance Computing*, pp. 93–113, Springer, 2008.

[32] G. D. Brown, S. Yamada, and T. J. Sejnowski, "Independent component analysis at the neural cocktail party," *Trends in Neurosciences*, vol. 24, no. 1, pp. 54–63, 2001.

[33] J. M. Carmena, M. A. Lebedev, R. E. Crist et al., "Learning to control a brain-machine interface for reaching and grasping by primates," *PLoS Biology*, vol. 1, no. 2, article e42, 2003.

[34] M. Laubach, J. Wessberg, and M. A. L. Nicolelis, "Cortical ensemble activity increasingly predicts behaviour outcomes during learning of a motor task," *Nature*, vol. 405, no. 6786, pp. 567–571, 2000.

[35] M. A. L. Nicolelis, L. A. Baccala, R. C. S. Lin, and J. K. Chapin, "Sensorimotor encoding by synchronous neural ensemble activity at multiple levels of the somatosensory system," *Science*, vol. 268, no. 5215, pp. 1353–1358, 1995.

[36] N. Xu, X. Gao, B. Hong, X. Miao, S. Gao, and F. Yang, "BCI competition 2003–data set IIb: enhancing P300 wave detection using ICA-based subspace projections for BCI applications," *IEEE Transactions on Biomedical Engineering*, vol. 51, no. 6, pp. 1067–1072, 2004.

8

Channel Identification Machines

Aurel A. Lazar and Yevgeniy B. Slutskiy

Department of Electrical Engineering, Columbia University, New York, NY 10027, USA

Correspondence should be addressed to Aurel A. Lazar, aurel@ee.columbia.edu

Academic Editor: Cheng-Jian Lin

We present a formal methodology for identifying a channel in a system consisting of a communication channel in cascade with an asynchronous sampler. The channel is modeled as a multidimensional filter, while models of asynchronous samplers are taken from neuroscience and communications and include integrate-and-fire neurons, asynchronous sigma/delta modulators and general oscillators in cascade with zero-crossing detectors. We devise channel identification algorithms that recover a projection of the filter(s) onto a space of input signals loss-free for both scalar and vector-valued test signals. The test signals are modeled as elements of a reproducing kernel Hilbert space (RKHS) with a Dirichlet kernel. Under appropriate limiting conditions on the bandwidth and the order of the test signal space, the filter projection converges to the impulse response of the filter. We show that our results hold for a wide class of RKHSs, including the space of finite-energy bandlimited signals. We also extend our channel identification results to noisy circuits.

1. Introduction

Signal distortions introduced by a communication channel can severely affect the reliability of communication systems. If properly utilized, knowledge of the channel response can lead to a dramatic improvement in the performance of a communication link. In practice, however, information about the channel is rarely available a priori and the channel needs to be identified at the receiver. A number of channel identification methods [1] have been proposed for traditional clock-based systems that rely on the classical sampling theorem [2, 3]. However, there is a growing need to develop channel identification methods for asynchronous nonlinear systems, of which time encoding machines (TEMs) [4] are a prime example.

TEMs naturally arise as models of early sensory systems in neuroscience [5, 6] as well as models of nonlinear samplers in signal processing and analog-to-discrete (A/D) converters in communication systems [4, 6]. Unlike traditional clock-based amplitude-domain devices, TEMs encode analog signals as a strictly increasing sequence of irregularly spaced times $(t_k)_{k \in \mathbb{Z}}$. As such, they are closely related to irregular

(amplitude) samplers [4, 7] and, due to their asynchronous nature, are inherently low-power devices [8]. TEMs are also readily amenable to massive parallelization [9]. Furthermore, under certain conditions, TEMs faithfully represent analog signals in the time domain; given the parameters of the TEM and the time sequence at its output, a time decoding machine (TDM) can recover the encoded signal loss-free [4, 5].

A general TEM of interest is shown in Figure 1. An analog multidimensional signal **u** is passed through a channel with memory that models physical communication links. We assume that the effect of this channel on the signal **u** can be described by a linear multidimensional filter. The output of the channel v is then mapped, or encoded, by a nonlinear asynchronous sampler into the time sequence $(t_k)_{k \in \mathbb{Z}}$. A few examples of samplers include asynchronous A/D converters such as the one based on an asynchronous sigma/delta modulator (ASDM) [4], nonlinear oscillators such as the van der Pol oscillator in cascade with a zero-crossing detector (ZCD) [6], and spiking neurons such as the integrate-and-fire (IAF) or the threshold-and-fire (TAF) neurons [9]. The above-mentioned asynchronous samplers incorporate the temporal dynamics of spike (pulse) generation and allow one

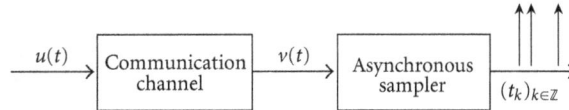

FIGURE 1: Modeling the channel identification problem. A known multidimensional signal $u(t)$, $t \in \mathbb{R}$, is first passed through a communication channel. A nonlinear sampler then maps the output v of the channel into an observable time sequence $(t_k)_{k \in \mathbb{Z}}$.

to consider, in particular for neuroscience applications, more biologically plausible nonlinear spike generation (sampling) mechanisms.

In this paper, we investigate the following *nonlinear* identification problem: given both the input signal **u** and the time sequence $(t_k)_{k \in \mathbb{Z}}$ at the output of a TEM, what is the channel filter? System identification problems of this kind are key to understanding the nature of neural encoding and processing [10–14], process modeling and control [15], and, more generally, methods for constructing mathematical models of dynamical systems [16].

Identification of the channel from a time sequence is to be contrasted with existing methods for rate-based models in neuroscience (see [10] for an extensive review). In such models the output of the system is taken to be its instantaneous response rate and the nonlinear generation of a time sequence is not explicitly modeled. Furthermore, in order to fit model parameters, identification methods for such models typically require the response rate to be known [17]. This is often difficult in practice since the same experiment needs to be repeated a large number of times to estimate the response rate. Moreover, the use of the same stimulus typically introduces a systematic bias during the identification procedure [10].

The channel identification methodology presented in this paper employs test signals that are neither white nor have stationary statistics (e.g., Gaussian with a fixed mean/ variance). This is a radical departure from the widely employed nonlinear system identification methods [10], including the spike-triggered average [18] and the spike-triggered covariance [19] methods. We carry out the channel identification using input signals that belong to reproducing kernel Hilbert spaces (RKHSs), and, in particular, spaces of bandlimited functions, that is, functions that have a finite support in the frequency domain. The latter signals are extensively used to describe sensory stimuli in biological systems and to model signals in communications. We show that for such signals the channel identification problem becomes mathematically tractable. Furthermore, we demonstrate that the choice of the input signal space profoundly effects the type of identification results that can be achieved.

The paper is organized as follows. In Section 2, we introduce three application-driven examples of the system in Figure 1 and formally state the channel identification problem. In Section 3, we present the single-input single-output (SISO) channel identification machine (CIM) for the finite-dimensional input signal space of trigonometric polynomials. Using analytical methods and simulations, we demonstrate that it is possible to identify the projection of the filter onto the input space loss-free and show that the SISO

CIM algorithm can recover the original filter with arbitrary precision, provided that both the bandwidth and the order of the input space are sufficiently high. Then, in Section 4, we extend our methodology to multidimensional systems and present multi-input single-output (MISO) CIM algorithms for the identification of vector-valued filters modeling the channel. We generalize our methods to classes of RKHSs of input signals in Section 5.1 and work out in detail channel identification algorithms for infinite-dimensional Paley-Wiener spaces. In Section 5.2 we discuss extensions of our identification results to noisy systems, where additive noise is introduced either by the channel or the asynchronous sampler. Finally, Section 6 concludes our work.

2. The Channel Identification Problem

We investigate a general I/O system comprised of a filter or a bank of filters (i.e., a linear operator) in cascade with an asynchronous (nonlinear) sampler (Figure 1). The I/O circuit belongs to the class of [Filter]-[Asynchronous Sampler] circuits. In general terms, the input to such a system is a vector-valued analog signal $\mathbf{u} = [u^1(t), u^2(t), \ldots, u^M(t)]^T$, $t \in \mathbb{R}$, $M \in \mathbb{N}$, and the output is a time sequence $(t_k)_{k \in \mathbb{Z}}$ generated by its asynchronous sampling mechanism. In the neural coding literature, such a system is called a time encoding machine (TEM) [4] as it encodes an unknown signal **u** into an observable time sequence $(t_k)_{k \in \mathbb{Z}}$.

2.1. Examples of Asynchronous SISO and MISO Systems. An instance of the TEM in Figure 1 is the SISO [Filter]-[Ideal IAF] neural circuit depicted in Figure 2(a). Here the filter is used to model the aggregate processing of a stimulus performed by the dendritic tree of a sensory neuron. The output of the filter v is encoded into the sequence of spike times $(t_k)_{k \in \mathbb{Z}}$ by an ideal integrate-and-fire neuron. Identification of dendritic processing in such a circuit is an important problem in systems neuroscience. It was first investigated in [20]. Another instance of the system in Figure 1 is the SISO [Filter]-[Nonlinear Oscillator-ZCD] circuit shown in Figure 2(b). In contrast to the first example, where the input was coupled additively, in this circuit the biased filter output v is coupled multiplicatively into a nonlinear oscillator. The zero-crossing detector then generates a time sequence $(t_k)_{k \in \mathbb{Z}}$ by extracting zeros from the observable modulated waveform at the output of the oscillator. Called a TEM with multiplicative coupling [6], this circuit is encountered in generalized frequency modulation [21].

An example of a MISO system is the [Filter]-[ASDM-ZCD] circuit shown in Figure 2(c). Similar circuits arise practically in all modern-day A/D converters and constitute important front-end components of measurement and

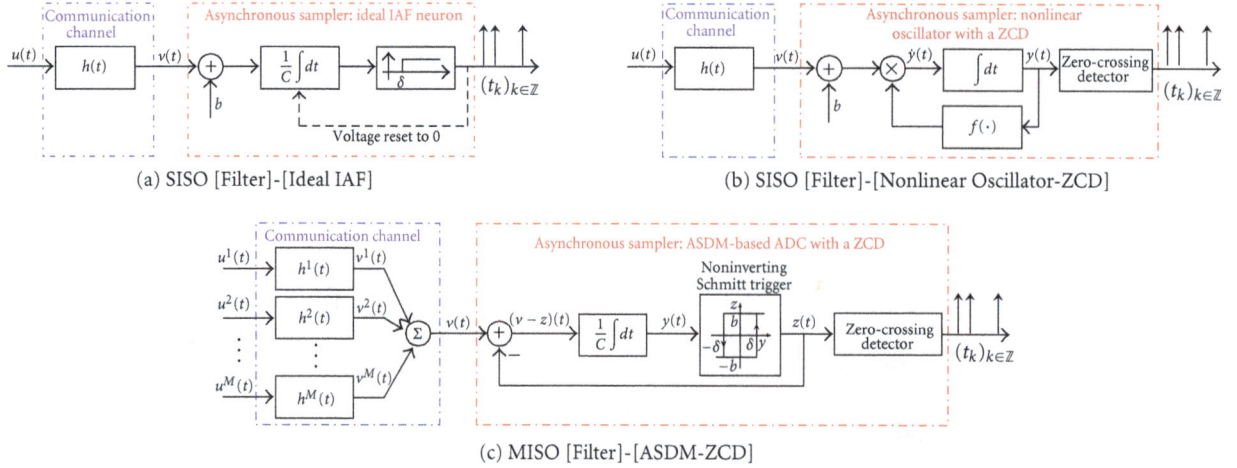

(a) SISO [Filter]-[Ideal IAF]

(b) SISO [Filter]-[Nonlinear Oscillator-ZCD]

(c) MISO [Filter]-[ASDM-ZCD]

FIGURE 2: Examples of systems arising in neuroscience and communications. (a) Single-input single-output model of a sensory neuron. (b) Single-input single-output nonlinear oscillator in cascade with a zero-crossing detector. (c) Multi-input single-output analog-to-discrete converter implemented with an asynchronous sigma-delta modulator. M liner filters model M (different) communication links.

communication systems. Each signal $u^m(t)$, $t \in \mathbb{R}$, $m = 1, 2, \ldots, M$, is transmitted through a communication channel and the effect of the channel on each signal is modeled using a linear filter with an impulse response $h^m(t)$, $t \in \mathbb{R}$, $m = 1, 2, \ldots, M$. The aggregate channel output $v(t) = \sum_{m=1}^{M} v^m(t) = \sum_{m=1}^{M} (u^m * h^m)(t)$, where $u^m * h^m$ denotes the convolution of u^m with h^m, is additively coupled into an ASDM. Specifically, $v(t)$ is passed through an integrator and a noninverting Schmitt trigger to produce a binary output $z(t) \in \{-b, b\}$, $t \in \mathbb{R}$. A zero-crossing detector is then used to extract the sequence of zero-crossing times $(t_k)_{k \in \mathbb{Z}}$ from $z(t)$. Thus, the output of this [Filter]-[ASDM-ZCD] circuit is the time sequence $(t_k)_{k \in \mathbb{Z}}$.

2.2. Modeling the Input Space.
We model channel input signals $u = u(t)$, $t \in \mathbb{R}$, as elements of the space of trigonometric polynomials $\mathcal{H}$ (see Section 5.1 for more general input spaces).

Definition 1. The space of trigonometric polynomials $\mathcal{H}$ is a Hilbert space of complex-valued functions

$$u(t) = \frac{1}{\sqrt{T}} \sum_{l=-L}^{L} u_l \exp\left(\frac{jl\Omega t}{L}\right), \quad t \in [0, T], \quad (1)$$

where $u_l \in \mathbb{C}$, Ω is the bandwidth, L is the order and $T = 2\pi L/\Omega$, endowed with the inner product $\langle \cdot, \cdot \rangle : \mathcal{H} \times \mathcal{H} \to \mathbb{C}$

$$\langle u, w \rangle = \int_0^T u(t)\overline{w(t)} dt. \quad (2)$$

Given the inner product in (2), the set of elements

$$e_l(t) = \frac{1}{\sqrt{T}} \exp\left(\frac{jl\Omega t}{L}\right), \quad l = -L, -L+1, \ldots, L, \quad (3)$$

forms an orthonormal basis in $\mathcal{H}$. Thus, any element $u \in \mathcal{H}$ and any inner product $\langle u, w \rangle$ can be compactly written as

$u = \sum_{l=-L}^{L} u_l e_l$ and $\langle u, w \rangle = \sum_{l=-L}^{L} u_l \overline{w_l}$. Moreover, $\mathcal{H}$ is a reproducing kernel Hilbert space (RKHS) with a reproducing kernel (RK) given by

$$K(s, t) = \sum_{l=-L}^{L} e_l(s)\overline{e_l(t)} = \frac{1}{T} \sum_{l=-L}^{L} \exp\left(\frac{jl\Omega}{L}(s-t)\right), \quad (4)$$

also known as a Dirichlet kernel [22].

We note that a function $u \in \mathcal{H}$ satisfies $u(0) = u(T)$. There is a natural connection between functions on an interval of length T that take on the same values at interval end-points and functions on $\mathbb{R}$ that are T-periodic: both provide equivalent descriptions of the same mathematical object, namely a function on a circle. By abuse of notation, in what follows u will denote both a function defined on an interval of length T and a function defined on the entire real line. In the latter case, the function u is simultaneously periodic with period T and bandlimited with bandwidth Ω, that is, it has a finite spectral support $\text{supp}(\mathcal{F}u) \subseteq [-\Omega, \Omega]$, where $\mathcal{F}$ denotes the Fourier transform. In what follows we will assume that $u_l \neq 0$ for all $l = -L, -L+1, \ldots, L$, that is, a signal $u \in \mathcal{H}$ contains all $2L + 1$ frequency components.

2.3. Modeling the Channel and Channel Identification.
The channel is modeled as a bank of M filters with impulse responses h^m, $m = 1, 2, \ldots, M$. We assume that each filter is linear, causal, BIBO-stable and has a finite temporal support of length $S \leq T$, that is, it belongs to the space $H = \{h \in \mathbb{L}^1(\mathbb{R}) \mid \text{supp}(h) \subseteq [0, T]\}$. Since the length of the filter support is smaller than or equal to the period of an input signal, we effectively require that for a given S and a fixed input signal bandwidth Ω, the order L of the space $\mathcal{H}$ satisfies $L \geq S \cdot \Omega/(2\pi)$. The aggregate channel output is given by $v(t) = \sum_{m=1}^{M} (u^m * h^m)(t)$. The asynchronous sampler maps the input signal v into the output time sequence $(t_k)_{k=1}^{n}$, where n denotes the total number of spikes produced on an interval $t \in [0, T]$.

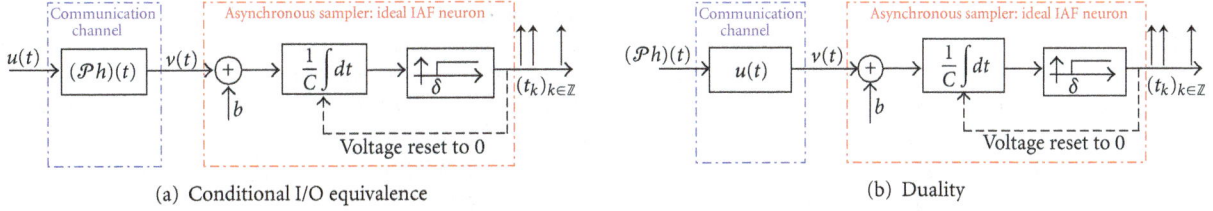

(a) Conditional I/O equivalence

(b) Duality

FIGURE 3: Conditional duality between channel identification and time encoding. (a) For all $u \in \mathcal{H}$, the [Filter]-[Ideal IAF] circuit with an input-filter pair (u, h) is I/O equivalent to a [Filter]-[Ideal IAF] circuit with an input-filter pair $(u, \mathcal{P}h)$. (b) The input-filter pair $(u, \mathcal{P}h)$ in channel identification is dual to the $(\mathcal{P}h, u)$ pair in time encoding.

Definition 2. A signal $\mathbf{u} \in \mathcal{H}^M$ at the input to a [Filter]-[Asynchronous Sampler] circuit together with the resulting output $\mathbb{T} = (t_k)_{k=1}^n$ of that circuit is called an input/output (I/O) pair and is denoted by $(\mathbf{u}, \mathbb{T})$.

We are now in a position to define the channel identification problem.

Definition 3. Let $(\mathbf{u}^i)$, $i = 1, 2, \ldots, N$, be a set of N signals from a test space $\mathcal{H}^M$. A channel identification machine implements an algorithm that estimates the impulse response of the filter from the I/O pairs $(\mathbf{u}^i, \mathbb{T}^i)$, $i = 1, 2, \ldots, N$, of the [Filter]-[Asynchronous Sampler] circuit.

Remark 4. We note that a CIM recovers the impulse response of the filter based on the knowledge of I/O pairs $(\mathbf{u}^i, \mathbb{T}^i)$, $i = 1, 2, \ldots, N$, and the sampler circuit. In contrast, a time decoding machine recovers an encoded signal $\mathbf{u}$ based on the knowledge of the entire TEM circuit (both the channel filter and the sampler) and the output time sequence $\mathbb{T}$.

3. SISO Channel Identification Machines

As already mentioned, the circuits under investigation consist of a channel and an asynchronous sampler. Throughout this paper, we will assume that the structure and the parameters of the asynchronous sampler are known. We start by formally describing asynchronous channel measurements in Section 3.1. Channel identification algorithms from asynchronous measurements are given in Section 3.2. Examples characterizing the performance of the identification algorithms are discussed in Section 3.3.

3.1. Asynchronous Measurements of the Channel Output. Consider the SISO [Filter]-[Ideal IAF] neural circuit in Figure 2(a). In this circuit, an input signal $u \in \mathcal{H}$ is passed through a filter with an impulse response (or kernel) $h \in H$ and then encoded by an ideal IAF neuron with a bias $b \in \mathbb{R}_+$, a capacitance $C \in \mathbb{R}_+$, and a threshold $\delta \in \mathbb{R}_+$. The output of the circuit is a sequence of spike times $(t_k)_{k=1}^n$ on the time interval $[0, T]$ that is available to an observer. This neural circuit is an instance of a TEM and its operation can be described by a set of equations

$$\int_{t_k}^{t_{k+1}} (u * h)(s)ds = q_k, \quad k = 1, 2, \ldots, n-1, \quad (5)$$

where $q_k = C\delta - b(t_{k+1} - t_k)$. Intuitively, at every spike time t_{k+1} the ideal IAF neuron is providing a measurement q_k of the signal $v(t) = (u * h)(t)$ on the time interval $[t_k, t_{k+1}]$.

Definition 5. The mapping of an analog signal $u(t)$, $t \in \mathbb{R}$, into an increasing sequence of times $(t_k)_{k \in \mathbb{Z}}$ (as in (5)) is called the *t*-transform [4].

Definition 6. The operator $\mathcal{P} : H \rightarrow \mathcal{H}$ given by

$$(\mathcal{P}h)(t) = \int_0^T h(s)\overline{K(s,t)}ds \quad (6)$$

is called the projection operator.

Proposition 7 (conditional duality). *For all $u \in \mathcal{H}$, a [Filter]-[Ideal IAF] TEM with a filter kernel h is I/O-equivalent to a [Filter]-[Ideal IAF] TEM with the filter kernel $\mathcal{P}h$. Furthermore, the CIM algorithm for identifying the filter kernel $\mathcal{P}h$ is equivalent to the TDM algorithm for recovering the input signal $\mathcal{P}h$ encoded by a [Filter]-[Ideal IAF] TEM with the filter kernel u.*

Proof. Since $u \in \mathcal{H}$, $u(t) = \langle u(\cdot), K(\cdot, t) \rangle$ by the reproducing property of the kernel $K(s, t)$. Hence, $(u * h)(t) \overset{(a)}{=} \int_{\mathbb{R}} h(w)u(t-w)dw \overset{(b)}{=} \int_0^T h(w) \int_0^T u(z)\overline{K(z, t-w)}dz\, dw \overset{(c)}{=} \int_0^T u(z) \int_0^T h(w)\overline{K(w, t-z)}dw\, dz \overset{(d)}{=} \int_0^T u(z)(\mathcal{P}h)(t-z)dz \overset{(e)}{=} (u * \mathcal{P}h)(t)$, where (a) follows from the commutativity of convolution, (b) from the reproducing property of the kernel K and the assumption that $\text{supp}(h) \subseteq [0, T]$, (c) from the equality $K(z, t-w) = K(w, t-z)$, (d) from the definition of $\mathcal{P}h$ in (6), and (e) from the definition of convolution for periodic functions [23]. It follows that on the interval $t \in [0, T]$, (5) can be rewritten as

$$\int_{t_k}^{t_{k+1}} (u * \mathcal{P}h)(s)ds \overset{(f)}{=} \int_{t_k}^{t_{k+1}} (\mathcal{P}h * u)(s)ds = q_k, \quad (7)$$

for all $k = 1, 2, \ldots, n-1$, where (f) comes from the commutativity of convolution. The right-hand side of (7) is the *t*-transform of a [Filter]-[Ideal IAF] TEM with an input $\mathcal{P}h$ and a filter that has an impulse response u. Hence, a TDM can identify $\mathcal{P}h$, given a filter-output pair $(u, \mathbb{T})$. $\qquad \square$

The conditional duality between time encoding and channel identification is visualized in Figure 3. First, we note the *conditional I/O equivalence* between the circuit in Figure 3(a) and the original circuit in Figure 2(a). The equivalence is conditional since $\mathscr{P}h$ is a projection onto a particular space $\mathcal{H}$ and the two circuits are I/O-equivalent only for input signals in that space. Second, identifying the filter of the circuit in Figure 3(a) is the same as decoding the signal encoded with the circuit in Figure 3(b). Note that the filter projection $\mathscr{P}h$ is now treated as the input to the [Filter]-[Ideal IAF] circuit and the signal u appears as the impulse response of the filter. Effectively, we have transformed the channel identification problem into a time decoding problem and we can use the TDM machinery of [5] to identify the filter projection $(\mathscr{P}h)(t)$ on $t \in [0, T]$.

3.2. Channel Identification from Asynchronous Measurements. Given the parameters of the asynchronous sampler, the measurements q_k of the channel output v can be readily computed from spike times $(t_k)_{k=1}^n$ using the definition of q_k ((5) for the IAF neuron). Furthermore, as we will now show, for a known input signal, these measurements can be reinterpreted as measurements of the channel itself.

Lemma 8. *There is a function $\phi_k(t) = \sum_{l=-L}^{L} \phi_{l,k} e_l(t) \in \mathcal{H}$, such that the t-transform of the [Filter]-[Ideal IAF] neuron in (7) can be written as*

$$\langle \mathscr{P}h, \phi_k \rangle = q_k, \tag{8}$$

and $\phi_{l,k} = \sqrt{T} \int_{t_k}^{t_{k+1}} \overline{u_l} \overline{e_l}(t) dt$ for all $l = -L, -L+1, \ldots, L$ and $k = 1, 2, \ldots, n-1$.

Proof. The linear functional $\mathcal{L}_k : \mathcal{H} \to \mathbb{R}$ defined by

$$\mathcal{L}_k(w) = \int_{t_k}^{t_{k+1}} (u * w)(s) ds, \tag{9}$$

where $w \in \mathcal{H}$, is bounded. Thus, by the Riesz representation theorem [22], there exists a function $\phi_k \in \mathcal{H}$ such that $\mathcal{L}_k(w) = \langle w, \phi_k \rangle$, $k = 1, 2, \ldots, n-1$, and $q_k = \mathcal{L}_k(\mathscr{P}h) = \int_{t_k}^{t_{k+1}} (u * \mathscr{P}h)(s) ds = \langle \mathscr{P}h, \phi_k \rangle$. Since $\phi_k \in \mathcal{H}$, we have $\phi_k(t) = \sum_{l=-L}^{L} \phi_{l,k} e_l$ for some $\phi_{l,k} \in \mathbb{C}$, $l = -L, -L+1, \ldots, L$. To find the latter coefficients, we note that $\phi_{l,k} = \langle \phi_k, e_l \rangle = \overline{\langle e_l, \phi_k \rangle} = \overline{\mathcal{L}_k(e_l)}$. By definition of $\mathcal{L}_k$ in (9), $\mathcal{L}_k(e_l) = \int_{t_k}^{t_{k+1}} (u * e_l)(t) dt = \int_{t_k}^{t_{k+1}} \int_0^T \sum_{i=-L}^{L} u_i e_i(s) e_l(t-s) ds\, dt = \sqrt{T} \int_{t_k}^{t_{k+1}} u_l e_l(t) dt$. $\square$

Since $q_k = \int_{t_k}^{t_{k+1}} (u * \mathscr{P}h)(s) ds = \langle v, \mathscr{P}1_{[t_k, t_{k+1}]} \rangle$, the measurements q_k are projections of $v = u * \mathscr{P}h$ onto $\mathscr{P}1_{[t_k, t_{k+1}]}$, $k = 1, 2, \ldots, n-1$. Assuming that u is known and there are enough measurements available, $\mathscr{P}h$ can be obtained by first recovering v from these projections and then deconvolving it with u. However, this two-step procedure does not work when the circuit is not producing enough measurements and one cannot recover v. A more direct route is suggested by Lemma 8, since the measurements $(q_k)_{k=1}^{n-1}$ can also be interpreted as the projections of $\mathscr{P}h$ onto ϕ_k, that is, $\langle \mathscr{P}h, \phi_k \rangle$, $k = 1, 2, \ldots, n-1$. A natural question then is how to identify $\mathscr{P}h$ directly from the latter projections.

Lemma 9. *Let $u \in \mathcal{H}$ be the input to a [Filter]-[Ideal IAF] circuit with $h \in H$. If the number of spikes n generated by the neuron in a time interval of length T satisfies $n \geq 2L + 2$, then the filter projection $\mathscr{P}h$ can be perfectly identified from the I/O pair $(u, \mathbb{T})$ as $(\mathscr{P}h)(t) = \sum_{l=-L}^{L} h_l e_l(t)$, where $\mathbf{h} = \mathbf{\Phi}^+ \mathbf{q}$ with $[\mathbf{q}]_k = q_k$ and $\mathbf{\Phi}^+$ denotes the Moore-Penrose pseudoinverse of $\mathbf{\Phi}$. The matrix $\mathbf{\Phi}$ is of size $(n-1) \times (2L+1)$ and its elements are given by*

$$[\mathbf{\Phi}]_{kl} = \begin{cases} u_l(t_{k+1} - t_k), & l = 0, \\ \dfrac{u_l L \sqrt{T} (e_l(t_{k+1}) - e_l(t_k))}{jl\Omega}, & l \neq 0. \end{cases} \tag{10}$$

Proof. Since $\mathscr{P}h \in \mathcal{H}$, it can be written as $(\mathscr{P}h)(t) = \sum_{l=-L}^{L} h_l e_l(t)$. Then from (8) we have

$$q_k = \langle \mathscr{P}h, \phi_k \rangle = \sum_{l=-L}^{L} h_l \overline{\phi_{l,k}}. \tag{11}$$

Writing (11) for all $k = 1, 2, \ldots, n-1$, we obtain $\mathbf{q} = \mathbf{\Phi}\mathbf{h}$ with $[\mathbf{q}]_k = q_k$, $[\mathbf{\Phi}]_{kl} = \overline{\phi_{l,k}}$ and $[\mathbf{h}]_l = h_l$. This system of linear equations can be solved for $\mathbf{h}$, provided that the rank $r(\mathbf{\Phi})$ of the matrix $\mathbf{\Phi}$ satisfies $r(\mathbf{\Phi}) = 2L + 1$. A necessary condition for the latter is that the number of measurements q_k is at least $2L+1$, or, equivalently, the number of spikes $n \geq 2L + 2$. Under this condition, the solution can be computed as $\mathbf{h} = \mathbf{\Phi}^+ \mathbf{q}$. $\square$

Remark 10. If the signal u is fed directly into the neuron, then $\int_{t_k}^{t_{k+1}} (u * \mathscr{P}h)(t) dt = \int_{t_k}^{t_{k+1}} u(t) dt$, for $k = 1, 2, \ldots, n-1$, that is, $(\mathscr{P}h)(t) = K(t, 0)$, $t \in \mathbb{R}$. In other words, if there is no processing on the input signal u, then the kernel $K(t, 0)$ in $\mathcal{H}$ is identified as the filter projection. This is also illustrated in Figure 7.

In order to ensure that the neuron produces at least $2L+1$ measurements in a time interval of length T, it suffices to have $t_{k+1} - t_k \leq T/(2L+2)$. Since $t_{k+1} - t_k \leq C\delta/(b-c)$ for $|v(t)| \leq c < b$, it suffices to have $C\delta < (b-c)T/(2L+2)$. Using the definition of $T = 2\pi L/\Omega$ and taking the limit as $L \to \infty$, we obtain the familiar Nyquist-type criterion $C\delta < \pi(b-c)/\Omega$ for a bandlimited stimulus $u \in \Xi$ [4, 20] (see also Section 5.1).

Ideally, we would like to identify the impulse response of the filter h. Note that unlike $h \in H$, the projection $\mathscr{P}h$ belongs to the space $\mathcal{H}$. Nevertheless, under quite natural conditions on h (see Section 3.4), $\mathscr{P}h$ approximates h arbitrarily closely on $t \in [0, T]$, provided that both the bandwidth and the order of the signal u are sufficiently large (see also Figure 9).

The requirement of Lemma 9 that the number of spikes n produced by the system in Figure 2(a) has to satisfy $n \geq 2L + 2$ is quite stringent and may be hard to meet in practice, especially if the order L of the space $\mathcal{H}$ is high. In that case we have the following result.

Theorem 11 (SISO channel identification machine). *Let $\{u^i \mid u^i \in \mathcal{H}\}_{i=1}^N$ be a collection of N linearly independent*

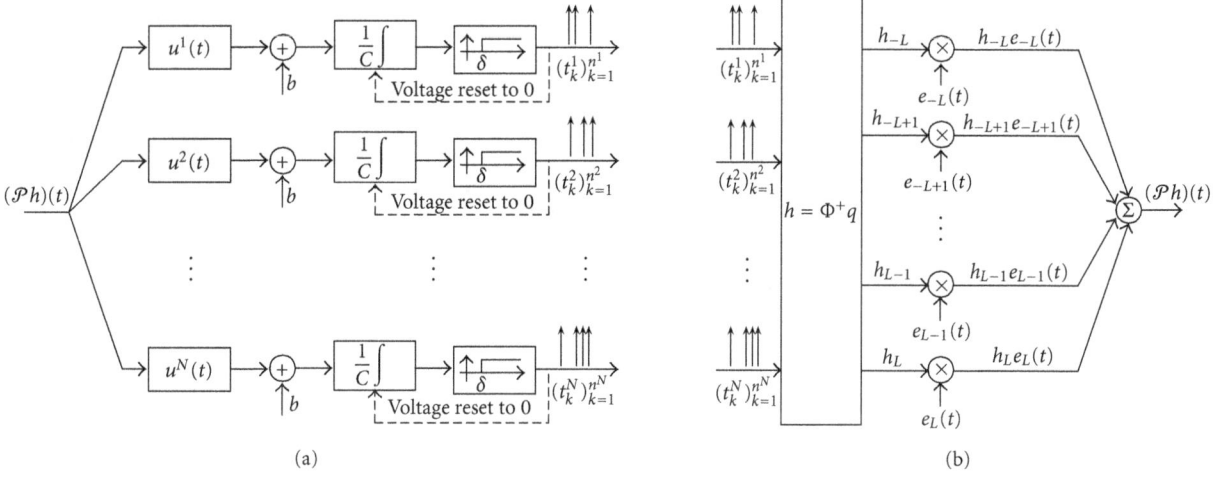

(a)

(b)

FIGURE 4: SISO CIM algorithm for the [Filter]-[Ideal IAF] circuit. (a) Time encoding interpretation of the channel identification problem. (b) Block diagram of the SISO channel identification machine.

stimuli at the input to a [Filter]-[Ideal IAF] circuit with $h \in H$. If the total number of spikes $n = \sum_{i=1}^{N} n^i$ generated by the neuron satisfies $n \geq 2L+N+1$, then the filter projection $\mathcal{P}h$ can be perfectly identified from a collection of I/O pairs $\{(u^i, \mathbb{T}^i)\}_{i=1}^{N}$ as

$$(\mathcal{P}h)(t) = \sum_{l=-L}^{L} h_l e_l(t), \qquad (12)$$

where $\mathbf{h} = \mathbf{\Phi}^+ \mathbf{q}$. Furthermore, $\mathbf{\Phi} = [\mathbf{\Phi}^1; \mathbf{\Phi}^2; \ldots; \mathbf{\Phi}^N]$, $\mathbf{q} = [\mathbf{q}^1; \mathbf{q}^2; \ldots; \mathbf{q}^N]$ and $[\mathbf{q}^i]_k = q_k^i$ with each $\mathbf{\Phi}^i$ of size $(n^i - 1) \times (2L+1)$ and $\mathbf{q}^i$ of size $(n^i - 1) \times 1$. The elements of matrices $\mathbf{\Phi}^i$ are given by

$$\left[\mathbf{\Phi}^i\right]_{kl} = \begin{cases} u_l^i \left(t_{k+1}^i - t_k^i\right), & l = 0, \\[2mm] \dfrac{u_l^i L\sqrt{T}\left(e_l\left(t_{k+1}^i\right) - e_l\left(t_k^i\right)\right)}{jl\Omega}, & l \neq 0, \end{cases} \qquad (13)$$

for all $k = 1, 2, \ldots, n-1$, $l = -L, -L+1, \ldots, L$, and $i = 1, 2, \ldots, N$.

Proof. Since $\mathcal{P}h \in \mathcal{H}$, it can be written as $(\mathcal{P}h)(t) = \sum_{l=-L}^{L} h_l e_l(t)$. Furthermore, since the stimuli are linearly independent, the measurements $(q_k^i)_{k=1}^{n^i-1}$ provided by the IAF neuron are distinct. Writing (5) for a stimulus u^i, we obtain

$$q_k^i = \langle \mathcal{P}h, \phi_k^i \rangle = \sum_{l=-L}^{L} h_l \overline{\phi_{l,k}^i}, \qquad (14)$$

or $\mathbf{q}^i = \mathbf{\Phi}^i \mathbf{h}$, with $[\mathbf{q}^i]_k = q_k^i$, $[\mathbf{\Phi}^i]_{kl} = \overline{\phi_{l,k}^i}$ and $[\mathbf{h}]_l = h_l$. Repeating for all $i = 1, \ldots, N$, we get $\mathbf{q} = \mathbf{\Phi}\mathbf{h}$ with $\mathbf{\Phi} = [\mathbf{\Phi}^1; \mathbf{\Phi}^2; \ldots; \mathbf{\Phi}^N]$ and $\mathbf{q} = [\mathbf{q}^1; \mathbf{q}^2; \ldots; \mathbf{q}^N]$. This system of linear equations can be solved for $\mathbf{h}$, provided that the rank $r(\mathbf{\Phi})$ of matrix $\mathbf{\Phi}$ satisfies $r(\mathbf{\Phi}) = 2L+1$. A necessary condition for the latter is that the total number $n = \sum_{i=1}^{N} n^i$ of spikes generated in response to all N signals satisfies $n \geq 2L+N+1$. Then, the solution can be computed as $\mathbf{h} = \mathbf{\Phi}^+ \mathbf{q}$.

To find the coefficients $\overline{\phi_{l,k}^i}$, we note that $\phi_{l,k}^i = \overline{\mathcal{L}_k^i(e_l)}$ (see Lemma 8). Hence, the result follows. □

The time encoding interpretation of the channel identification problem for a SISO [Filter]-[Ideal IAF] circuit is shown in Figure 4(a). The block diagram of the SISO CIM in Theorem 11 is shown in Figure 4(b). Note that the key idea behind the SISO CIM is the introduction of multiple linearly independent test signals $u^i \in \mathcal{H}$, $i = 1, 2, \ldots, N$. When the [Filter]-[Ideal IAF] circuit is producing very few measurements of $\mathcal{P}h$ in response to any given test signal u^i, we use more signals to obtain additional measurements. We can do so and identify $\mathcal{P}h$ because $\mathcal{P}h \in \mathcal{H}$ is fixed. In contrast, identifying $\mathcal{P}h$ in a two-step deconvolving procedure requires reconstructing at least one v^i. This is an ill-posed problem since each v^i is signal-dependent and has a small number of associated measurements.

3.3. Examples.

We now demonstrate the performance of the identification algorithms in Lemma 9 and Theorem 11. First, we identify a filter in the SISO [Filter]-[Ideal IAF] circuit (Figure 2(a)) from a single I/O pair when this circuit produces a sufficient number of measurements in an interval of length T. Second, we identify the filter using multiple I/O pairs for the case when the number of measurements produced in response to any given input signal is small. Finally, we consider the SISO [Filter]-[Nonlinear Oscillator-ZCD] circuit with multiplicative coupling (Figure 2(b)) and identify its filter from multiple I/O pairs.

3.3.1. SISO [Filter]-[Ideal IAF] Circuit, Single I/O Pair.

We model the dendritic processing filter using the causal linear kernel

$$h(t) = ce^{-\alpha t}\left[\frac{(\alpha t)^3}{3!} - \frac{(\alpha t)^5}{5!}\right], \qquad t \in [0, 0.1]\,\text{s}, \qquad (15)$$

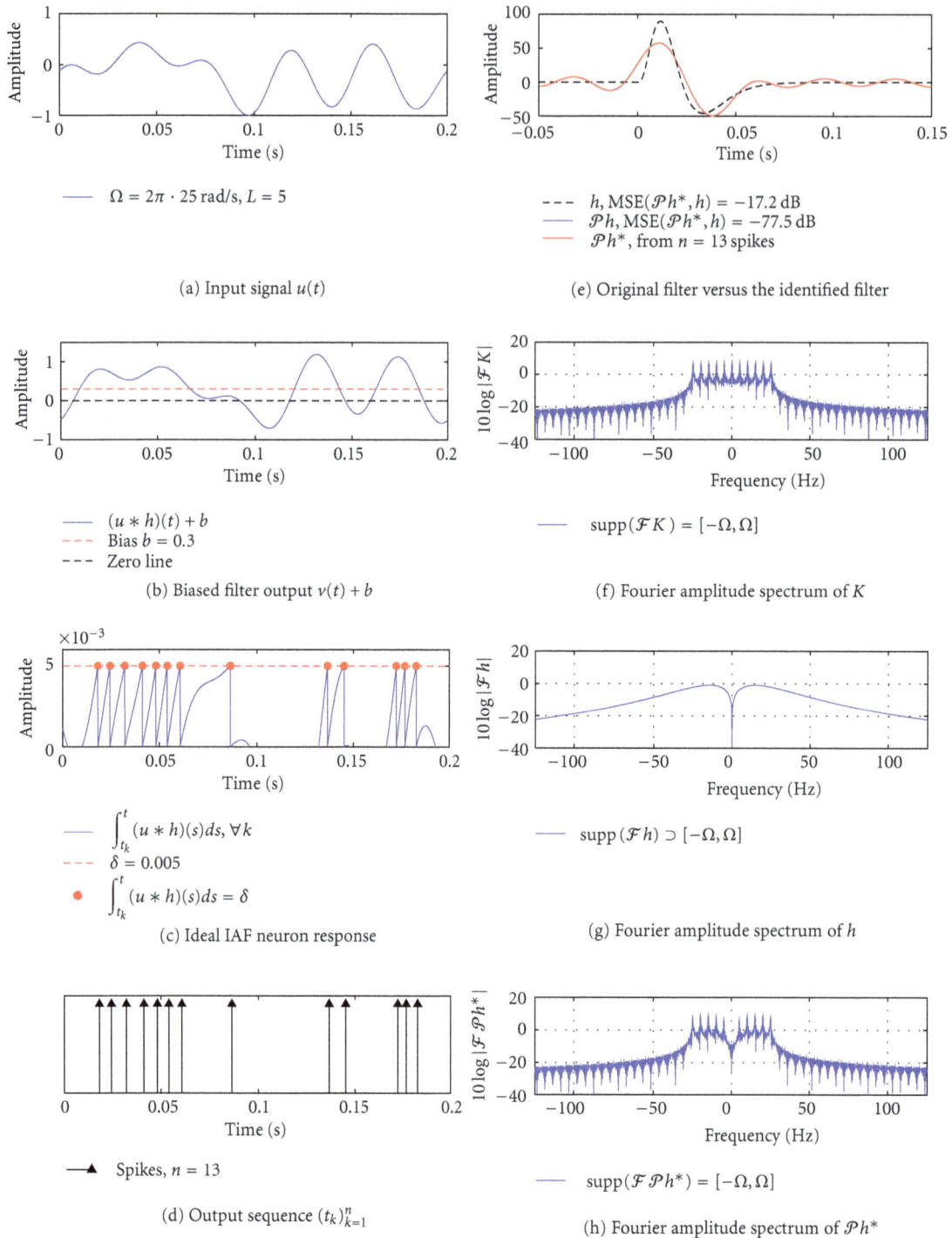

(a) Input signal $u(t)$

(b) Biased filter output $v(t) + b$

(c) Ideal IAF neuron response

(d) Output sequence $(t_k)_{k=1}^n$

(e) Original filter versus the identified filter

(f) Fourier amplitude spectrum of K

(g) Fourier amplitude spectrum of h

(h) Fourier amplitude spectrum of $\mathscr{P}h^*$

FIGURE 5: Channel identification in a SISO [Filter]-[Ideal IAF] circuit using a single I/O pair. (a) An input signal u is bandlimited to 25 Hz. The order of the space is $L = 5$. (b) The corresponding biased output of the filter $v(t) + b$. (c) The filter output in (b) is integrated by the ideal IAF neuron. Whenever the membrane potential reaches a threshold δ, a spike is produced by the neuron and the potential is reset to 0. (d) The neuron generated a total of 13 spikes. (e) The identified impulse response of the filter $\mathscr{P}h^*$ (red) is shown together with the original filter h (dashed black) and its projection $\mathscr{P}h$ (blue). The MSE between $\mathscr{P}h^*$ and $\mathscr{P}h$ is -77.5 dB. (f)–(h) Fourier amplitude spectra of K, h and $\mathscr{P}h^*$. Note that $\mathrm{supp}(\mathscr{F}K) = \mathrm{supp}(\mathscr{F}\mathscr{P}h^*) = [-\Omega, \Omega]$ but $\mathrm{supp}(\mathscr{F}h) \supset [-\Omega, \Omega]$. In other words, $\mathscr{P}h^* \in \mathcal{H}$ but $h \notin \mathcal{H}$.

with $c = 3$ and $\alpha = 200$. The general form of this kernel was suggested in [24] as a plausible approximation to the temporal structure of a visual receptive field. Since the length of the filter support $S = 0.1\,\mathrm{s}$, we will need to use a signal with a period $T \geq 0.1\,\mathrm{s}$. In Figure 5(a), we apply a signal u that is bandlimited to 25 Hz and has a period of $T = 0.2\,\mathrm{s}$, that is, the order of the space $L = T \cdot \Omega/(2\pi) = 5$. The biased output of the filter $v = (u * h) + b$ is then fed into an ideal integrate-and-fire neuron (Figure 5(b)). Here the bias b guarantees that the output of the integrator reaches the threshold value in finite time. Whenever the biased filter output is above zero (Figure 5(b)), the membrane potential is increasing (Figure 5(c)). If the membrane potential $\int_{t_k}^{t}[(u * h)(s) + b]ds$ reaches a threshold δ, a spike is generated by the neuron at a time t_{k+1} and the potential is reset to zero (Figure 5(c)). The resulting spike train $(t_k)_{k=1}^{n}$ at the output of the [Filter]-[Ideal IAF] circuit is shown in Figure 5(d). Note that the circuit generated a total of $n = 13$ spikes in an interval of length $T = 0.2\,\mathrm{s}$. According to Theorem 14, we need at least $n = 2L + 2 = 12$ spikes, corresponding to $2L + 1 = 11$ measurements, in order to identify the projection $\mathcal{P}h$ of the filter h loss-free. Hence, for this particular example, it will suffice to use a single I/O pair $(u, \mathbb{T})$.

In Figure 5(e), we plot the original impulse response of the filter h, the filter projection $\mathcal{P}h$, and the filter $\mathcal{P}h^*$. The latter filter was identified using the algorithm in Theorem 14. Notice that the identified impulse response $\mathcal{P}h^*$ (red) is quite different from h (dashed black). In contrast, and as expected, the blue and red curves corresponding, respectively, to $\mathcal{P}h$ and $\mathcal{P}h^*$ are indistinguishable. The mean squared error (MSE) between $\mathcal{P}h^*$ and $\mathcal{P}h$ amounts to -77.5 dB.

The difference between $\mathcal{P}h$ and h is further evaluated in Figures 5(f)–5(h). By definition of $\mathcal{P}h$ in (6), $\mathcal{P}h = h * \overline{K(\cdot, 0)}$, or $\mathcal{F}(\mathcal{P}h) = \mathcal{F}(h)\mathcal{F}(K(\cdot, 0))$ since $\overline{K} = K$. Hence both the projection $\mathcal{P}h$ and the identified filter $\mathcal{P}h^*$ will contain frequencies that are present in the reproducing kernel K, or equivalently in the input signal u. In Figure 5(f) we show the double-sided Fourier amplitude spectrum of $K(t, 0)$. As expected, we see that the kernel is bandlimited to 25 Hz and contains $2L + 1 = 11$ distinct frequencies. On the other hand, as shown in Figure 5(g), the original filter h is not bandlimited (since it has a finite temporal support). As a result, the input signal u explores h in a limited spectrum of $[-\Omega, \Omega]$ rad/s, effectively projecting h onto the space $\mathcal{H}$ with $\Omega = 2\pi \cdot 25$ rad/s and $L = 5$. The Fourier amplitude spectrum of the identified projection $\mathcal{P}h^*$ is shown in Figure 5(h).

3.3.2. SISO [Filter]-[Ideal IAF] Circuit, Multiple I/O Pairs.
Next, we identify the projection of h onto the space of functions that are bandlimited to 100 Hz and have the same period $T = 0.2\,\mathrm{s}$ as in the first example. This means that the order L of the space of input signals $\mathcal{H}$ is $L = T \cdot \Omega/(2\pi) = 20$. In order to identify the projection $\mathcal{P}h$ loss-free, the neuron has to generate at least $2L + 1 = 41$ measurements. If the neuron produces about 13 spikes

(12 measurements) on an interval of length T, as in the previous example, a single I/O pair will not suffice. However, we can still recover the projection $\mathcal{P}h$ if we use multiple I/O pairs.

In Figure 6 we illustrate identification of the filter using the algorithm in Theorem 11. A total of 48 spikes were produced by the neuron in response to four different signals $u^1, \ldots, u^4$. Since $48 > 2L + N + 1 = 45$, the MSE between the identified filter $\mathcal{P}h^*$ (red) and the projection $\mathcal{P}h$ (blue) is -73.3 dB.

3.3.3. SISO [Filter]-[Ideal IAF] Circuit, $h(t) = \delta(t)$.
Now we consider a special case when the channel does not alter the input signal, that is, when $h(t) = \delta(t)$, $t \in \mathbb{R}$, is the Dirac delta function. As explained in Remark 10, the CIM should identify the projection of $\delta(t)$ onto $\mathcal{H}$, that is, it should identify the kernel $K(t, 0)$. This is indeed the case as shown in Figure 7.

3.3.4. SISO [Filter]-[Nonlinear Oscillator-ZCD] Circuit, Multiple I/O Pairs.
Next we consider a SISO circuit consisting of a channel in cascade with a nonlinear dynamical system that has a stable limit cycle. We assume that the (positive) output of the channel $v(t) + b$ is multiplicatively coupled to the dynamical system (Figure 2(b)) so that the circuit is governed by a set of equations

$$\frac{d\mathbf{y}}{dt} = (v(t) + b)\mathbf{f}(\mathbf{y}). \tag{16}$$

A system (16) followed by a zero-crossing detector is an example of a TEM with multiplicative coupling and has been previously investigated in [6]. It can be shown that such a TEM is input/output equivalent to an IAF neuron with a threshold δ that is equal to the period of the dynamical system on a stable limit cycle [6].

As an example, we consider a [Filter]-[van der Pol - ZCD] TEM with the van der Pol oscillator described by a set of equations

$$\frac{dy_1}{dt} = (u * h + b)\left[\mu\left(y_1 - \frac{1}{3}y_1^3\right) - y_2\right],$$
$$\frac{dy_2}{dt} = (u * h + b)y_1, \tag{17}$$

where μ is the damping coefficient [6]. We assume that y_1 is the only observable state of the oscillator and without loss of generality we choose the zero phase of the limit cycle to be the peak of y_1.

In Figure 8, we show the results of a simulation in which a SISO CIM was used to identify the channel. Input signals (Figure 8(a)) were bandlimited to 50 Hz and had a period $T = 0.5\,\mathrm{s}$, that is, $L = 25$. In the absence of an input, that is, when $u = 0$, a constant bias $b = 1$ (Figure 8(b)) resulted a in period of 34.7 ms on a stable limit cycle (Figure 8(e)). As seen in Figures 8(b) and 8(c), downward/upward deviations of $v^1(t) + b$ in response to u^1 resulted in the slowing-down/speeding-up of the oscillator. In order to identify the

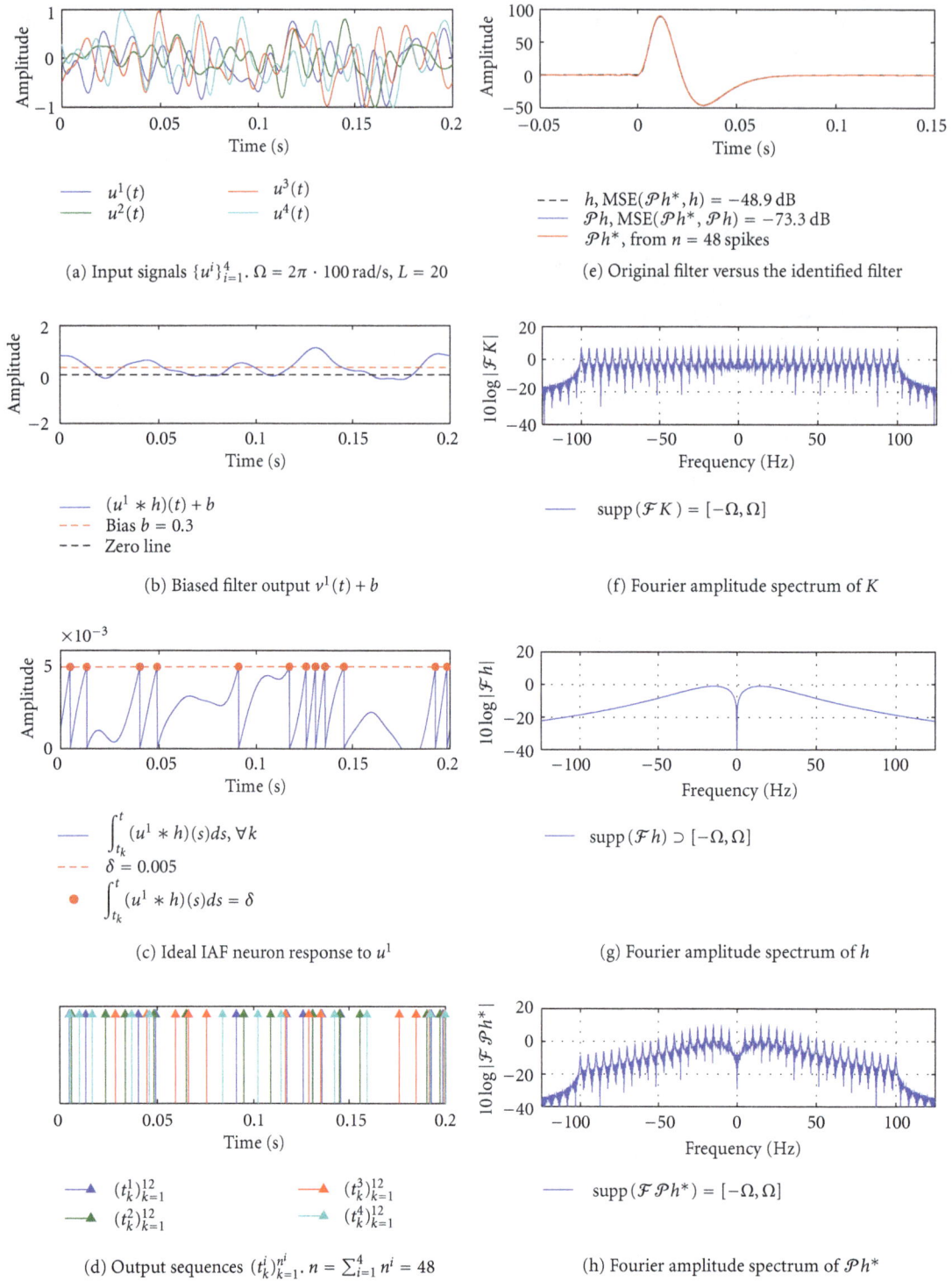

(a) Input signals $\{u^i\}_{i=1}^4$. $\Omega = 2\pi \cdot 100\,\text{rad/s}$, $L = 20$

(b) Biased filter output $v^1(t) + b$

(c) Ideal IAF neuron response to u^1

(d) Output sequences $(t_k^i)_{k=1}^{n^i}$. $n = \sum_{i=1}^4 n^i = 48$

(e) Original filter versus the identified filter

(f) Fourier amplitude spectrum of K

(g) Fourier amplitude spectrum of h

(h) Fourier amplitude spectrum of $\mathscr{P}h^*$

FIGURE 6: Channel identification in a SISO [Filter]-[Ideal IAF] circuit using multiple I/O pairs. (a) Input signals $u^1, \ldots, u^4$ are bandlimited to 100 Hz. The order of the space $L = 20$. (b) Biased output of the filter $v^1(t) + b$ in response to the stimulus u^1. (c) The filter output in (b) is integrated by an ideal IAF neuron. (d) The neuron generated a total of 48 spikes in response to all 4 input signals. (e) The identified impulse response $\mathscr{P}h^*$ (red) is shown together with the original filter h (dashed black) and its projection $\mathscr{P}h$ (blue). The MSE between $\mathscr{P}h^*$ and $\mathscr{P}h$ is -73.3 dB. (f)–(h) Fourier amplitude spectra of K, h and $\mathscr{P}h^*$. Note that $\text{supp}(\mathscr{F}K) = [-\Omega, \Omega] = \text{supp}(\mathscr{F}\mathscr{P}h^*)$ but $\text{supp}(\mathscr{F}h) \supset [-\Omega, \Omega]$. In other words, $\mathscr{P}h^* \in \mathcal{H}$ but $h \notin \mathcal{H}$.

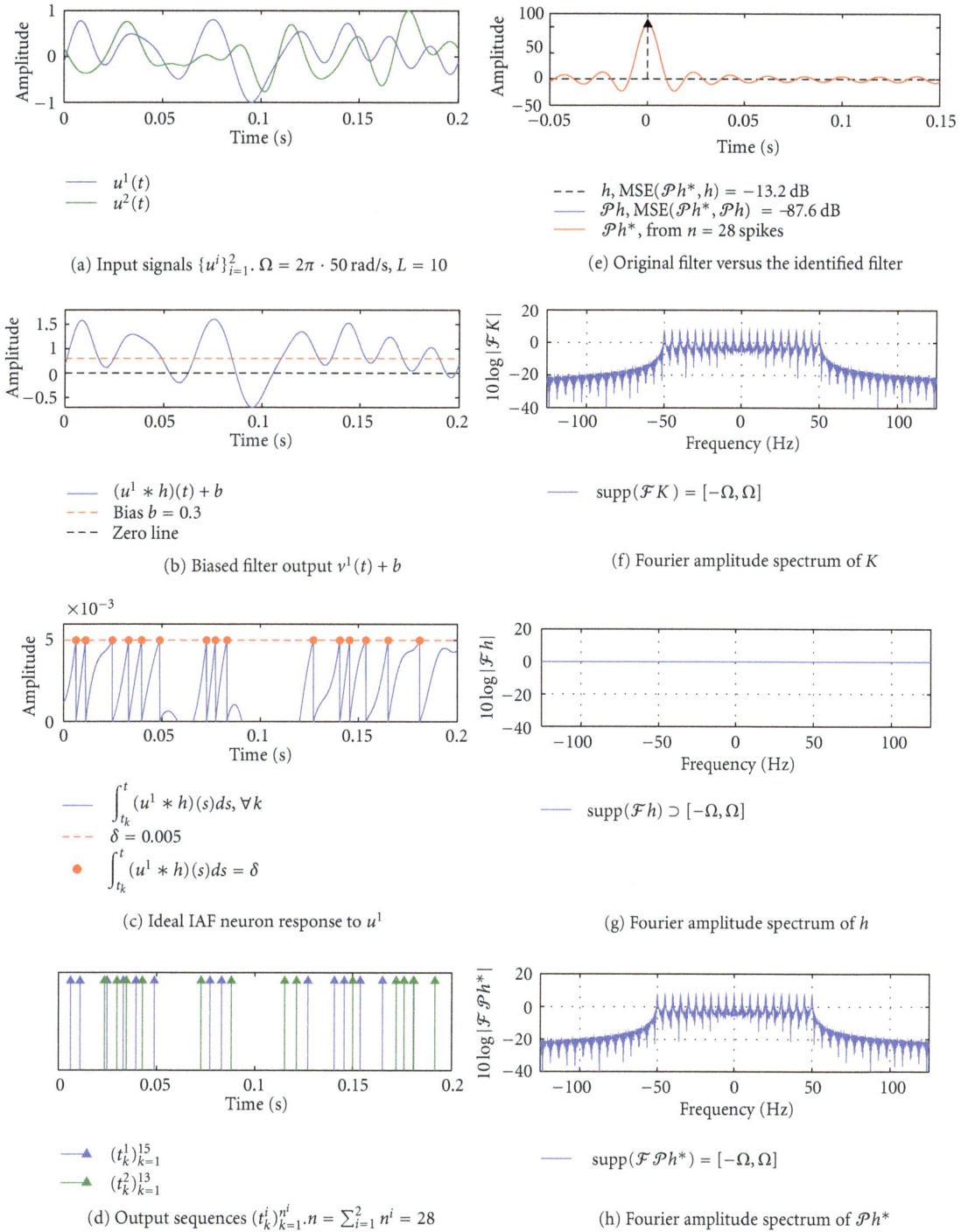

(a) Input signals $\{u^i\}_{i=1}^2$. $\Omega = 2\pi \cdot 50$ rad/s, $L = 10$

(b) Biased filter output $v^1(t) + b$

(c) Ideal IAF neuron response to u^1

(d) Output sequences $(t_k^i)_{k=1}^{n^i}$. $n = \sum_{i=1}^2 n^i = 28$

(e) Original filter versus the identified filter

(f) Fourier amplitude spectrum of K

(g) Fourier amplitude spectrum of h

(h) Fourier amplitude spectrum of $\mathscr{P}h^*$

FIGURE 7: Channel identification for $h(t) = \delta(t)$. (a) Input signals u^1, u^2 are bandlimited to 50 Hz. The order of the space $L = 10$. (b) Biased output of the filter $v^1(t) + b$ in response to the stimulus u^1. (c) The filter output in (b) is integrated by an ideal IAF neuron. (d) The neuron generated a total of 28 spikes in response to 2 input signals. (e) The identified filter $\mathscr{P}h^*$ (red) is exactly the kernel $K(t, 0)$ for $\mathcal{H}_{\Omega,L}^1$ with $\Omega = 2\pi \cdot 10$ rad/s and $L = 10$. Also shown is the original filter $h = \delta$ (dashed black) and its projection $\mathscr{P}h = \delta * \overline{K(\cdot, 0)} = K(\cdot, 0)$ (blue). The MSE between $\mathscr{P}h^*$ and $\mathscr{P}h$ is -87.6 dB. (f)–(h) Fourier amplitude spectra of K, h and $\mathscr{P}h^*$. As before, $\mathscr{P}h^* \in \mathcal{H}$ but $h \notin \mathcal{H}$.

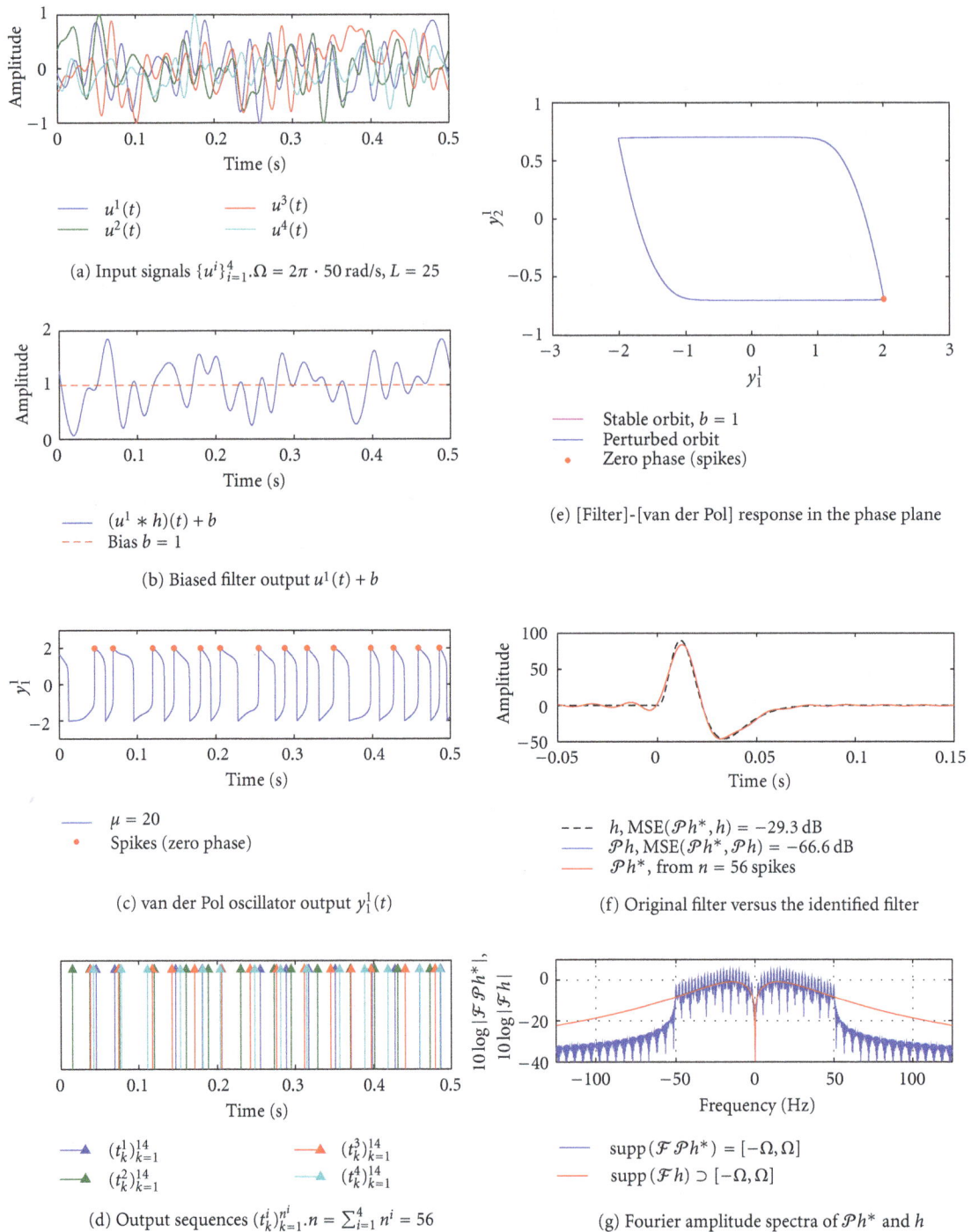

(a) Input signals $\{u^i\}_{i=1}^4$. $\Omega = 2\pi \cdot 50$ rad/s, $L = 25$

(b) Biased filter output $u^1(t) + b$

(c) van der Pol oscillator output $y_1^1(t)$

(d) Output sequences $(t_k^i)_{k=1}^{n^i}$. $n = \sum_{i=1}^4 n^i = 56$

(e) [Filter]-[van der Pol] response in the phase plane

(f) Original filter versus the identified filter

(g) Fourier amplitude spectra of $\mathscr{P}h^*$ and h

FIGURE 8: Channel identification in a SISO [Filter]-[van der Pol-ZCD] circuit using multiple I/O pairs. (a) Input signals $u^1, \ldots, u^4$ are bandlimited to 50 Hz. The order of the space $L = 25$. (b) Biased output of the filter $v^1(t) + b$ in response to the stimulus u^1. (c) Downward and upward deviations of $v^1(t) + b$ from the bias b cause the oscillator to slow down and to speed up, respectively. The damping coefficient $\mu = 20$. (d) The oscillator produced a total of 56 spikes in response to 4 stimuli. Here spikes correspond to the peaks of the observable state variable y_1^1. (e) A limit cycle of the van der Pol oscillator for $\mu = 20$ is shown in the phase plane. In the absence of channel output, the bias b resulted in a constant period of oscillation $T(b) = 34.7$ ms. The red dot denotes the zero phase (spike) of an oscillation. (f) The identified filter $\mathscr{P}h^*$ (red) is shown together with the original filter h (dashed black) and its projection $\mathscr{P}h$ (blue). The MSE between $\mathscr{P}h^*$ and $\mathscr{P}h$ is -66.6 dB. (g) Fourier amplitude spectra of h and $\mathscr{P}h^*$. As before, $\mathscr{P}h^* \in \mathcal{H}$ but $h \notin \mathcal{H}$.

filter projection onto a space of order $L = 25$ loss-free, we used a total of $n = 56$ zeros at the output of the zero-crossing detector (Figure 8(d)). This is 1 more zero than the rank requirement of $2L + N + 1 = 55$ zeros, or equivalently of $2L + 1 = 51$ measurements. The MSE between the identified filter $\mathcal{P}h^*$ (red) and the projection $\mathcal{P}h$ (blue) is -66.6 dB.

3.4. Convergence of the SISO CIM Estimate. Recall, that the original problem of interest is that of recovering the impulse response of the filter h. The CIM lets us identify the projection $\mathcal{P}h$ of that filter onto the input space. A natural question to ask is whether $\mathcal{P}h$ converges to h and if so how and under what conditions. We formalize this below.

Proposition 12. *If $\int_0^T |h(t)|^2 dt < \infty$, then $\mathcal{P}h \rightarrow h$ in the L^2 norm and almost everywhere on $t \in [0, T]$ with increasing Ω, L and fixed T.*

Proof. Let $T = 2\pi L/\Omega = $ const. Then $K(t, 0) = (1/T) \sum_{l=-L}^{L} e^{(j\Omega l/L)t} = (1/T) \sum_{l=-L}^{L} e^{(j2\pi l/T)t}$ and by definition of $\mathcal{P}h$ in (6), we have

$$(\mathcal{P}h)(t) = \int_0^T \left[\frac{1}{T} \sum_{l=-L}^{L} e^{(j2\pi l/T)(t-s)} \right] h(s)ds$$

$$= \sum_{l=-L}^{L} \left[\frac{1}{T} \int_0^T h(s)e^{-(j2\pi l/T)s}ds \right] e^{(j2\pi l/T)t} \quad (18)$$

$$= \sum_{l=-L}^{L} \hat{h}(l)e^{(j2\pi l/T)t} = S_L^h(t),$$

where S_L^h is the Lth partial sum of the Fourier series of h and $\hat{h}(l)$ is the lth Fourier coefficient. Hence the problem of convergence of $\mathcal{P}h$ to h is the same as that of the convergence of the Fourier series of h. We thus have convergence in the L^2 norm and convergence almost everywhere follows from Carleson's theorem [23]. $\square$

Remark 13. More generally, if $\int_0^T |h(t)|^p dt < \infty$, $p \in (1, \infty)$, then $\mathcal{P}h \rightarrow h$ in the L^p norm and almost everywhere by Hunt's theorem [23].

It follows from Proposition 12 that $\mathcal{P}h$ approximates h arbitrarily closely (in the L^2 norm, or MSE sense), given an appropriate choice of Ω and L. Since the number of measurements needed to identify the projection $\mathcal{P}h$ increases linearly with L, a single channel identification problem leads us to consider a countably infinite number of time encoding problems in order to identify the impulse response of the filter with arbitrary precision. To provide further intuition about the relationship between h and $\mathcal{P}h$, we compare the two in time and frequency domains for multiple values of Ω and L in Figure 9.

4. MISO Channel Identification Machines

In this section we consider the identification of a bank of M filters with impulse responses h^m, $m = 1, 2, \ldots, M$. We present a MISO CIM algorithm in Section 4.1, followed by an example demonstrating its performance in Section 4.2.

4.1. An Identification Algorithm for MISO Channels. Consider now the MISO ASDM-based circuit in Figure 2(c), where the signal $\mathbf{u} = [u^1(t), u^2(t), \ldots, u^M(t)]^T$, $t \in [0, T]$, $M \in \mathbb{N}$, is transformed into the time sequence $(t_k)_{k=1}^n$. This circuit is also an instance of a TEM and (assuming $z(t_1) = b$) its t-transform is given by

$$\int_{t_k}^{t_{k+1}} \sum_{m=1}^{M} (u^m * h^m)(s)ds = \langle v, \phi_k \rangle = q_k, \quad (19)$$

where $v = \sum_m (u^m * h^m)(t)$, $\phi_k \in \mathcal{H}$ with $\phi_k = \sum_l \phi_{l,k} e_l(t)$ and $q_k = (-1)^k [2C\delta - b(t_{k+1} - t_k)]$. One simple way to identify filters h^m, $m = 1, 2, \ldots, M$, is to identify them one by one as in Theorem 11. For instance, this can be achieved by applying signals of the form $\mathbf{u} = [0, \ldots, 0, u^m, 0, \ldots, 0]$ when identifying the filter h^m. In a number of applications, most notably in early olfaction [25], this model of system identification cannot be applied. An alternative procedure that allows to identify all filters at once is given below.

Theorem 14 (MISO channel identification machine). *Let $\{\mathbf{u}^i \mid \mathbf{u}^i \in \mathcal{H}^M\}_{i=1}^N$ be a collection of N linearly-independent vector-valued signals at the input of a MISO [Filter]-[ASDM-ZCD] circuit with filters $h^m \in H$, $m = 1, \ldots, M$. The filter projections $\mathcal{P}h^m$ can be perfectly identified from a collection of I/O pairs $\{(\mathbf{u}^i, \mathbb{T}^i)\}_{i=1}^N$ as*

$$(\mathcal{P}h^m)(t) = \sum_{l=-L}^{L} h_l^m e_l(t), \quad (20)$$

$m = 1, \ldots, M$. Here the coefficients h_l^m are given by $\mathbf{h} = \mathbf{\Phi}^+ \mathbf{q}$ with $\mathbf{q} = [\mathbf{q}^1, \mathbf{q}^2, \ldots, \mathbf{q}^N]^T$, $[\mathbf{q}^i]_k = q_k^i$ and $\mathbf{h} = [h_{-L}^1, \ldots, h_{-L}^M, h_{-L+1}^1, \ldots, h_{-L+1}^M, \ldots, h_L^1, \ldots, h_L^M]^T$, provided that the matrix $\mathbf{\Phi}$ has rank $r(\mathbf{\Phi}) = M(2L + 1)$. The matrix $\mathbf{\Phi}$ is given by

$$\mathbf{\Phi} = \begin{bmatrix} \mathbf{\Phi}^1 & 0 & \cdots & 0 \\ 0 & \mathbf{\Phi}^2 & \cdots & 0 \\ \vdots & \vdots & \ddots & \vdots \\ 0 & 0 & \cdots & \mathbf{\Phi}^N \end{bmatrix} \begin{bmatrix} \mathbf{U}^1 \\ \mathbf{U}^2 \\ \vdots \\ \mathbf{U}^N \end{bmatrix}, \text{ with}$$

$$\mathbf{U}^i = \begin{bmatrix} \mathbf{u}_{-L}^i & 0 & \cdots & 0 \\ 0 & \mathbf{u}_{-L+1}^i & \cdots & 0 \\ \vdots & \vdots & \ddots & \vdots \\ 0 & 0 & \cdots & \mathbf{u}_L^i \end{bmatrix}, \quad (21)$$

where $\mathbf{u}_l^i = [u_l^{i1}, u_l^{i2}, \ldots, u_l^{iM}]$, $i = 1, 2, \ldots, N$. Finally, the elements of matrix $\mathbf{\Phi}^i$ are given by

$$[\mathbf{\Phi}^i]_{kl} = \begin{cases} (t_{k+1}^i - t_k^i), & l = 0, \\ \dfrac{L\sqrt{T} \left(e_l(t_{k+1}^i) - e_l(t_k^i)\right)}{jl\Omega}, & l \neq 0. \end{cases} \quad (22)$$

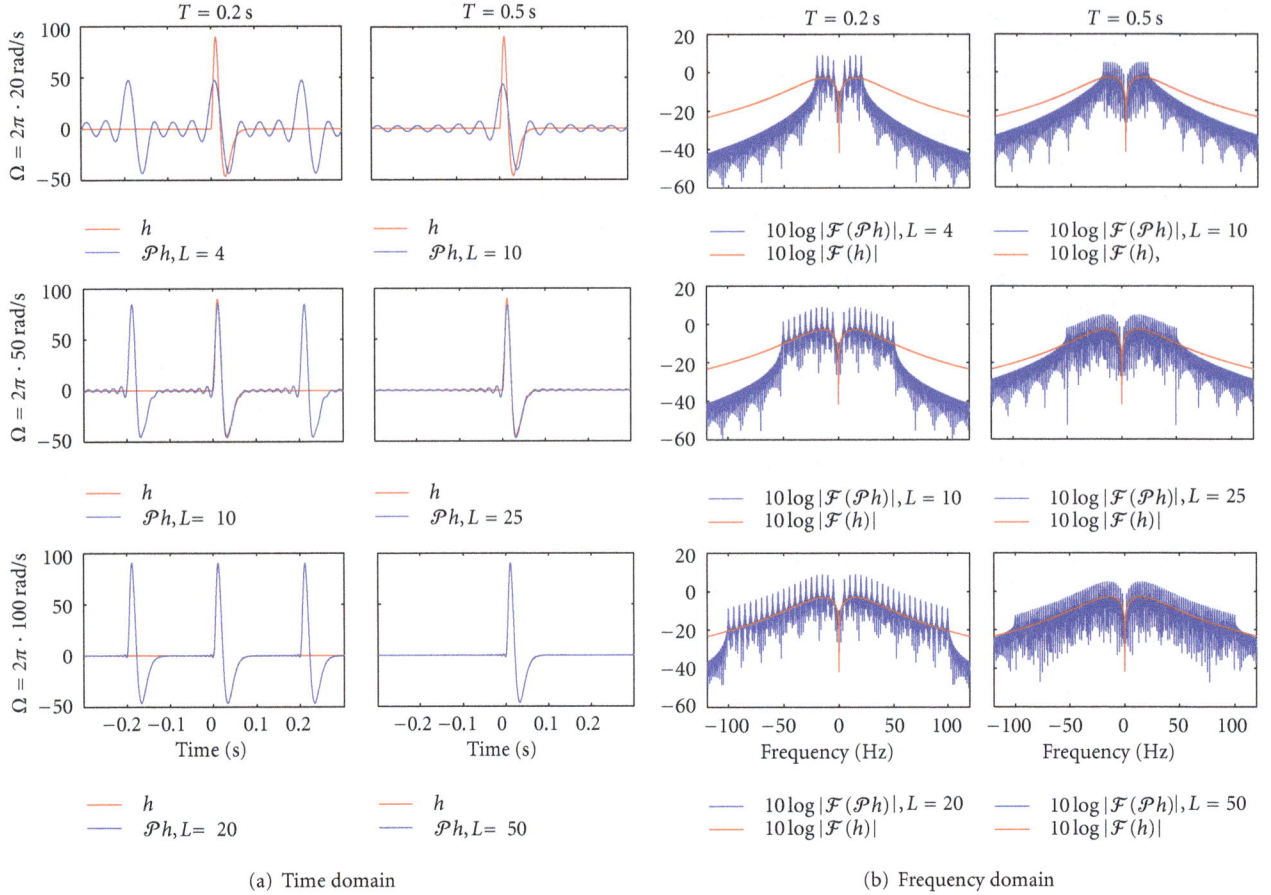

(a) Time domain

(b) Frequency domain

FIGURE 9: Comparison between h and $\mathscr{P}h$ in time and frequency domains. (a) h (red) and its projection $\mathscr{P}h$ (blue) are shown for several values of Ω and L in the time domain. $\Omega = 2\pi \cdot 20$ rad/s, $2\pi \cdot 50$ rad/s and $2\pi \cdot 100$ rad/s in the top, middle and bottom row, respectively. The period T is fixed at $T = 0.2$ s in the left column and $T = 0.5$ s in the right column. (b) Fourier amplitude spectra of h (red) and $\mathscr{P}h$ (blue) for the same values of Ω and L as in (a). Note that the differentiating filter h clearly removes the zero-frequency (dc) coefficient corresponding to $l = 0$ in all cases.

Proof. Since $\mathscr{P}h^m \in \mathcal{H}$ for all $m = 1,\ldots,M$, it can be written as $(\mathscr{P}h^m)(t) = \sum_{l=-L}^{L} h_l^m e_l(t)$. Hence, for the mth component of the stimulus $\mathbf{u}^i$ we get $(u^{im} * h^m)(t) = (u^{im} * \mathscr{P}h^m)(t) = \sqrt{T}\sum_{l=-L}^{L} h_l^m u_l^{im} e_l(t)$ and

$$v^i(t) = \sum_{m=1}^{M} \sqrt{T} \sum_{l=-L}^{L} h_l^m u_l^{im} e_l(t). \qquad (23)$$

Using the definition of $\phi_k^i = \sum_{l=-L}^{L} \phi_{l,k}^i e_l(t)$ and substituting (23) into the t-transform (19), we obtain

$$q_k^i = \left\langle v^i, \phi_k^i \right\rangle = \sum_{m=1}^{M} \sum_{l=-L}^{L} \sqrt{T} h_l^m u_l^{im} \overline{\phi_{l,k}^i}, \qquad (24)$$

or $\mathbf{q}^i = \mathbf{\Phi}^i \mathbf{U}^i \mathbf{h}$ with $[\mathbf{q}^i]_k = q_k^i$, $[\mathbf{\Phi}^i]_{kl} = \sqrt{T} \cdot \overline{\phi_{l,k}^i}$, $\mathbf{U}^i = \mathrm{diag}(\mathbf{u}_{-L}^i,\ldots,\mathbf{u}_L^i)$, $\mathbf{u}_l^i = [u_l^{i1},\ldots,u_l^{iM}]$ and $\mathbf{h} = [h_{-L}^1,\ldots,h_{-L}^M, h_{-L+1}^1,\ldots,h_{-L+1}^M,\ldots,h_L^1,\ldots,h_L^M]^T$. Repeating for all stimuli $\mathbf{u}^i$, $i = 1,\ldots,N$, we obtain $\mathbf{q} = \mathbf{\Phi}\mathbf{h}$ with $\mathbf{\Phi}$ as specified in (21). This system of linear equations can be solved for $\mathbf{h}$, provided that the rank of $\mathbf{\Phi}$ satisfies the condition $r(\mathbf{\Phi}) = M(2L + 1)$. To find the coefficients $\overline{\phi_{l,k}^i}$, we note that $\overline{\phi_{l,k}^i} = \overline{\mathcal{L}_k^i(e_l)}$. Hence, the result follows. $\square$

The MIMO time-encoding interpretation of the channel identification problem for a MISO [Filter]-[ASDM-ZCD] circuit is shown in Figure 10(a). The block diagram of the MISO CIM in Theorem 14 is shown in Figure 10(b).

Remark 15. From (23), we see that $v^i = \sum_{l=-L}^{L} v_l^i e_l(t)$ with $v_l^i = \sqrt{T}\sum_{m=1}^{M} h_l^m u_l^{im}$. Writing this for all $i = 1,\ldots,N$, we obtain $\mathbf{v}_l = \mathbf{U}_l \mathbf{h}_l$, where $[\mathbf{U}_l]_{im} = \sqrt{T} u_l^{im}$, $\mathbf{h}_l = [h_l^1, h_l^2,\ldots, h_l^M]^T$ and $\mathbf{v}_l = [v_l^1, v_l^2,\ldots, v_l^N]^T$. In order to identify the multidimensional channel this system of equations must have a solution for every l. A necessary condition for the latter is that $N \geq M$, that is, the number N of test signals $\mathbf{u}^i$ is greater than the number of signal components M.

Remark 16. The rank condition $r(\mathbf{\Phi}) = M(2L + 1)$ can be satisfied by increasing the number N of input signals $\mathbf{u}^i$. Specifically, if on average the system is providing ν measurements in a time interval $t \in [0, T]$, then the minimum number of test signals is $N = \lceil M(2L + 1)/\nu \rceil$.

4.2. Example: MISO [Filter]-[ASDM-ZCD] Circuit. We now describe simulation results for identifying the channel in a

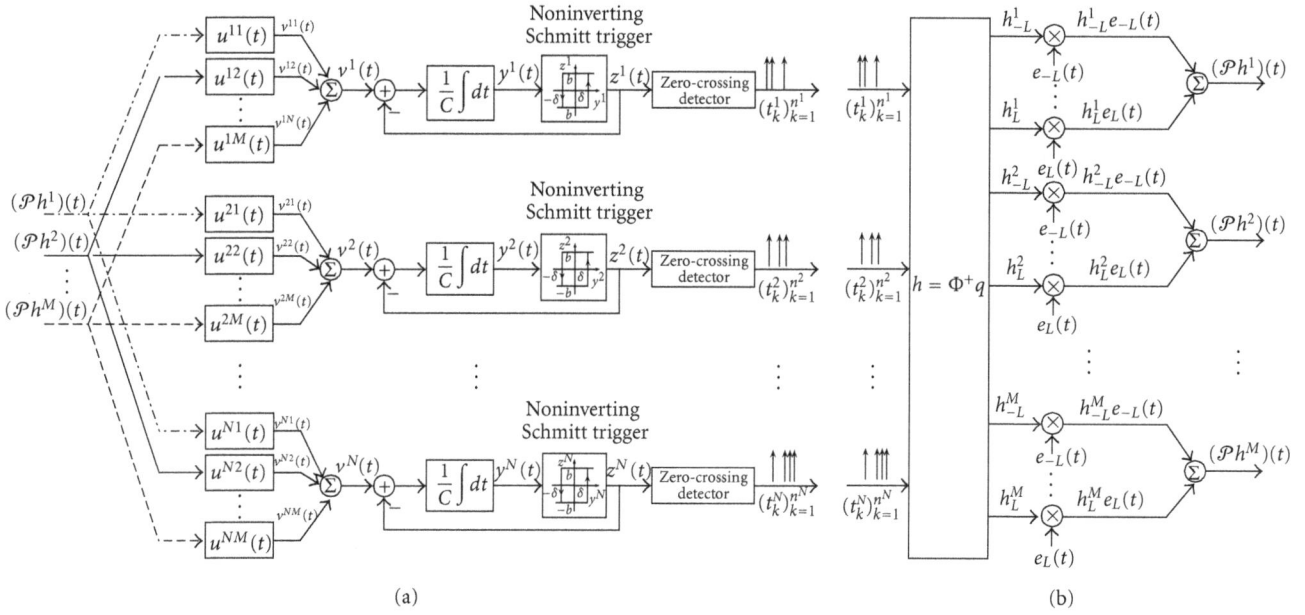

(a)   (b)

FIGURE 10: MISO CIM algorithm for the [Filter]-[ASDM-ZCD] circuit. (a) Time encoding interpretation of the MISO channel identification problem. (b) Block diagram of the MISO channel identification machine.

MISO [Filter]-[ASDM - ZCD] circuit of Figure 2(c). We use three different filters:

$$
\begin{aligned}
h^1(t) &= ce^{-\alpha t}\left[\frac{(\alpha t)^3}{3!} - \frac{(\alpha t)^5}{5!}\right], \\
h^2(t) &= h^1(t - \beta), \\
h^3(t) &= -h^1(t),
\end{aligned}
\tag{25}
$$

with $t \in [0, 0.1]$ s, $c = 3$ and $\alpha = 200$ and $\beta = 20$ ms. All $N = 5$ signals are bandlimited to 100 Hz and have a period of $T = 0.2$ s, that is, the order of the space $L = 20$. According to Theorem 14, the ASDM has to generate a total of at least $M(2L + 1) + N = 128$ trigger times in order to identify the projections $\mathscr{P}h^1$, $\mathscr{P}h^2$ and $\mathscr{P}h^3$ loss-free. We use all five triplets $\mathbf{u}^i = [u^{i1}, u^{i2}, u^{i3}]$, $i = 1, \ldots, 5$, to produce 131 trigger times.

A single such triplet $\mathbf{u}^1$ is shown in Figure 11(a). The corresponding biased aggregate channel output $v^1(t) - z^1(t)$ is shown in Figure 11(b). Since the Schmitt trigger output $z(t)$ switches between $+b$ and $-b$ (Figure 11(d)), the signal $v^1(t) - z^1(t)$ is piece-wise continuous. Figure 11(c) shows the integrator output. Note that when $z(t) = -b$, the channel output is positively biased and the integrator output $\int_{t_k}^t [v^1(s) - z(s)]ds$ is compared against a threshold $+\delta$. As soon as that threshold is reached, the Schmitt trigger output switches to $z(t) = b$ and the negatively-biased channel output is compared to a threshold $-\delta$. Passing the ASDM output $z^1(t)$ through a zero-crossing device (Figure 11(d)), we obtain a corresponding sequence of trigger times $(t_k^1)_{k=1}$. The set of all 131 trigger times is shown in Figure 11(e). Three identified filters $\mathscr{P}h^{1*}$, $\mathscr{P}h^{2*}$ and $\mathscr{P}h^{3*}$ are plotted in Figures 11(f)–11(h). The MSE between filter projections and

filters recovered by the algorithm in Theorem 14 is on the order of -60 dB.

5. Generalizations

We shall briefly generalize the results presented in previous sections in two important directions. First, we consider a general class of signal spaces for test signals in Section 5.1. Then we discus channel models with noisy observations in Section 5.2.

5.1. Hilbert Spaces and RKHSs for Input Signals. Until now we have presented channel identification results for a particular space of input signals, namely the space of trigonometric polynomials. The finite-dimensionality of this space and the simplicity of the associated inner product makes it an attractive space to work with when implementing a SISO or a MISO CIM algorithm. However, fundamentally the identification methodology relied on the the geometry of the Hilbert space of test signals [5, 26]; computational tractability was based on kernel representations in an RKHS.

Theorem 17. *Let $\{u^i \mid u^i \in \mathcal{H}(I)\}_{i=1}^N$ be a collection of N linearly independent and bounded stimuli at the input of a [Filter]-[Asynchronous Sampler] circuit with a linear processing filter $h \in H$ and the t-transform*

$$
\mathcal{L}_k^i(\mathscr{P}h) = q_k^i,
\tag{26}
$$

where $\mathcal{L}_k^i : \mathcal{H} \to \mathbb{R}$ is a bounded linear functional mapping $\mathscr{P}h$ into a measurement q_k^i. Then there is a set of sampling functions $\{(\phi_k^i)_{k\in\mathbb{Z}}\}_{i=1}^N$, in $\mathcal{H}$ such that

$$
q_k^i = \left\langle \mathscr{P}h, \phi_k^i \right\rangle,
\tag{27}
$$

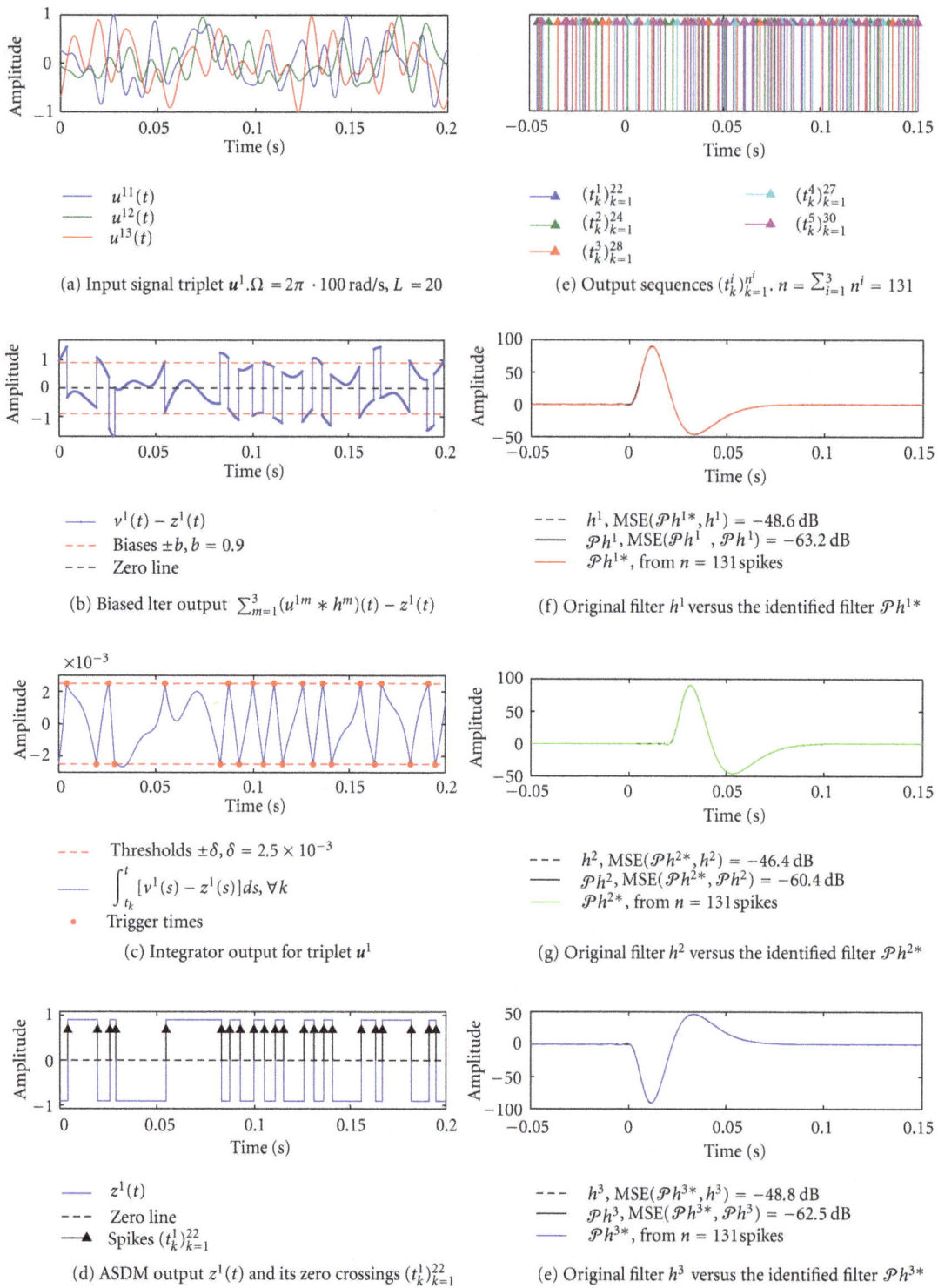

(a) Input signal triplet $\boldsymbol{u}^1$. $\Omega = 2\pi \cdot 100$ rad/s, $L = 20$

(e) Output sequences $(t_k^i)_{k=1}^{n^i}$. $n = \sum_{i=1}^{3} n^i = 131$

(b) Biased lter output $\sum_{m=1}^{3}(u^{1m} * h^m)(t) - z^1(t)$

(f) Original filter h^1 versus the identified filter $\mathscr{P}h^{1*}$

(c) Integrator output for triplet $\boldsymbol{u}^1$

(g) Original filter h^2 versus the identified filter $\mathscr{P}h^{2*}$

(d) ASDM output $z^1(t)$ and its zero crossings $(t_k^1)_{k=1}^{22}$

(e) Original filter h^3 versus the identified filter $\mathscr{P}h^{3*}$

FIGURE 11: Channel identification in a MISO [FIlter]-[ASDM] circuit using multiple I/O pairs. (a) An input triplet signal $\mathbf{u}^1 = [u^{11}, u^{12}, u^{13}]$ is bandlimited to 100 Hz. The order of the space $L = 20$. (b) Biased aggregate output of the channel $v^1(t) - z^1(t)$ in response to the triplet $\mathbf{u}^1$. (c) Integrator output $\int_{t_k}^{t}[v^1(s) - z^1(s)]ds$ (blue) is compared against two thresholds $+\delta$ and $-\delta$ (dashed red). Trigger times of the noninverting Schmitt trigger are indicated by red dots. (d) The ASDM output $z^1(t)$ (blue) is passed through a zero-crossing detector to produce a sequence of trigger times $(t_k^1)_{k=1}^{22}$. (e) A total of 131 trigger times were generated by the ASDM in response to five input triplets. (f)–(h) Identified filters $\mathscr{P}h^{1*}$ (red), $\mathscr{P}h^{2*}$ (green) and $\mathscr{P}h^{3*}$ (blue) are shown together with the original filters h^1, h^2, h^3 (dashed black) and their projections $\mathscr{P}h^1$, $\mathscr{P}h^2$ and $\mathscr{P}h^3$ (black). The MSE achieved by the identification algorithm is less than -60 dB.

for all $k \in \mathbb{Z}$, $i = 1, 2, \ldots, N$. Furthermore, if $\mathcal{H}$ is an RKHS with a kernel $K(s, t)$, $s, t \in I$, then $\phi_k^i(t) = \overline{\mathcal{L}_k^i(K(\cdot, t))}$. Let the set of representation functions $\{(\psi_k^i)_{k \in \mathbb{Z}}\}_{i=1}^N$, span the Hilbert space $\mathcal{H}$. Then

$$(\mathcal{P}h)(t) = \sum_{i=1}^N \sum_{k \in \mathbb{Z}} h_k^i \psi_k^i(t). \tag{28}$$

Finally, if $\{(\phi_k^i)_{k \in \mathbb{Z}}\}_{i=1}^N$ and $\{(\psi_k^i)_{k \in \mathbb{Z}}\}_{i=1}^N$ are orthogonal basis or frames for $\mathcal{H}$, then the filter coefficients amount to $\mathbf{h} = \mathbf{\Phi}^+ \mathbf{q}$, where $\mathbf{h} = [\mathbf{h}^1, \mathbf{h}^2, \ldots, \mathbf{h}^N]^T$ with $[\mathbf{h}^i]_k = h_k^i$, $[\mathbf{\Phi}^{ij}]_{lk} = \langle \phi_l^i, \psi_k^j \rangle$ and $\mathbf{q} = [\mathbf{q}^1, \mathbf{q}^2, \ldots, \mathbf{q}^N]^T$ with $[\mathbf{q}^i]_l = q_k^i$ for all $i, j = 1, 2, \ldots, N$, and $k, l \in \mathbb{Z}$.

Proof. By the Riesz representation theorem, since the linear functional $\mathcal{L}_k^i : \mathcal{H} \to \mathbb{R}$ is bounded, there is a set of sampling functions $\{(\phi_k^i)_{k \in \mathbb{Z}}\}_{i=1}^N$ in $\mathcal{H}$ such that $\mathcal{L}_k^i(\mathcal{P}h) = \langle \mathcal{P}h, \phi_k^i \rangle$. If $\mathcal{H}$ is an RKHS, a sampling function ϕ_k^i can be computed using the reproducing property of the kernel K as in

$$\phi_k^i(t) = \langle \phi_k^i, K(\cdot, t) \rangle \equiv \overline{\langle K(\cdot, t), \phi_k^i \rangle} = \overline{\mathcal{L}_k^i(K(\cdot, t))}. \tag{29}$$

Finally, writing all inner products $\langle \phi_k^i, \mathcal{P}h \rangle = q_k^i$ yields, with the notation above, a system of linear equations $\mathbf{\Phi}\mathbf{h} = \mathbf{q}$ and the filter coefficients amount to $\mathbf{h} = \mathbf{\Phi}^+ \mathbf{q}$. $\qquad \square$

5.1.1. Example: Paley-Wiener Space.

As an example, we consider the Paley-Wiener space which is closely related to the space of trigonometric polynomials. Specifically, the finite-dimensional space $\mathcal{H}$ can be thought of as a discretized version of the infinite-dimensional Paley-Wiener space

$$\Xi = \{ u \in \mathbb{L}^2(\mathbb{R}) \mid \text{supp}(\mathcal{F}u) \subseteq [-\Omega, \Omega] \} \tag{30}$$

in the frequency domain. An element $u \in \mathcal{H}$ has a line spectrum at frequencies $l\Omega/L$, $l = -L, -L+1, \ldots, L$. This spectrum becomes dense in $[-\Omega, \Omega]$ as $L \to \infty$. The space Ξ with the inner product $\langle \cdot, \cdot \rangle : \Xi \times \Xi \to \mathbb{R}$ given by

$$\langle u, w \rangle = \int_{\mathbb{R}} u(t)w(t)dt \tag{31}$$

is also an RKHS with an RK [22]

$$K(s, t) = \frac{\sin(\Omega(t - s))}{\pi(t - s)}, \tag{32}$$

with $t, s \in \mathbb{R}$. Defining the projection of the filter h onto Ξ as $(\mathcal{P}h)(t) = \int_{\mathbb{R}} h(s)\overline{K(s, t)}ds$, we find that Lemma 8 still holds with $\phi_k \in \Xi$ and we can extend Theorem 11 to the following.

Proposition 18. *Let $\{u^i \mid \text{supp}(\mathcal{F}u^i) = [-\Omega, \Omega]\}_{i=1}^N$ be a collection of N linearly independent and bounded stimuli at the input of a [Filter]-[Ideal IAF] neural circuit with a dendritic processing filter $h \in H$. If $\sum_{j=1}^N (b/C\delta) > \Omega/\pi$, then $(\mathcal{P}h)(t)$ can be perfectly identified from the collection of I/O pairs $\{(u^i, \mathbb{T}^i)\}_{i=1}^N$ as*

$$(\mathcal{P}h)(t) = \sum_{i=1}^N \sum_{k \in \mathbb{Z}} h_k^i \psi_k^i(t), \tag{33}$$

where $\psi_k^i(t) = K(t, t_k^i)$, $i = 1, 2, \ldots, N$, and $k \in \mathbb{Z}$. Finally, $\mathbf{h} = \mathbf{\Phi}^+ \mathbf{q}$, where $\mathbf{h} = [\mathbf{h}^1, \mathbf{h}^2, \ldots, \mathbf{h}^N]^T$ with $[\mathbf{h}^i]_k = h_k^i$, $[\mathbf{\Phi}^{ij}]_{lk} = \int_{t_l^i}^{t_{l+1}^i} u^i(s - t_k^j)ds$ and $\mathbf{q} = [\mathbf{q}^1, \mathbf{q}^2, \ldots, \mathbf{q}^N]^T$ with $[\mathbf{q}^i]_l = C\delta - b(t_{l+1}^i - t_l^i)$ for all $i, j = 1, 2, \ldots, N$, and $k, l \in \mathbb{Z}$.

Proof. As before, the spikes $(t_k^i)_{k \in \mathbb{Z}}$ in response to each test signal u^i, $i = 1, 2, \ldots, N$, represent distinct measurements $q_k^i = \langle \phi_k^i, \mathcal{P}h \rangle$ of $(\mathcal{P}h)(t)$. Thus we can think of the $\{(q_k^i)_{k \in \mathbb{Z}}\}_{i=1}^N$'s as projections of $\mathcal{P}h$ onto $\{(\phi_k^i)_{k \in \mathbb{Z}}\}_{i=1}^N$, where $\phi_k^i(t) = \mathcal{L}_k^i(K(\cdot, t)) = \int_{t_k^i}^{t_{k+1}^i} \int_{\mathbb{R}} u^i(z)K(s - z, t)dz\,ds = \int_{t_k^i}^{t_{k+1}^i} u^i(s - t)ds$. Since the signals are linearly independent and bounded [5], it follows that, if $\sum_{i=1}^N (b/C\delta) > \Omega/\pi$ or equivalently if the number of test signals $N > \Omega C\delta/\pi b$, the set of functions $\{(\psi_k^i)_{k \in \mathbb{Z}}\}_{i=1}^N$ with $\psi_k^i(t) = K(t, t_k^i)$, is a frame for Ξ [5, 26]. Hence

$$(\mathcal{P}h)(t) = \sum_{i=1}^N \sum_{k \in \mathbb{Z}} h_k^i \psi_k^i(t). \tag{34}$$

If the set of functions $\{(\phi_k^i)_{k \in \mathbb{Z}}\}_{i=1}^N$ forms a frame for Ξ, we can find the coefficients h_k^i, $k \in \mathbb{Z}$, $i = 1, 2, \ldots, N$, by taking the inner product of (34) with each element of $\{\phi_l^i(t)\}_{i=1}^N$: $\langle \phi_l^i, \mathcal{P}h \rangle = \sum_{k \in \mathbb{Z}} h_k^1 \langle \phi_l^i, \psi_k^1 \rangle + \sum_{k \in \mathbb{Z}} h_k^2 \langle \phi_l^i, \psi_k^2 \rangle + \cdots + \sum_{k \in \mathbb{Z}} h_k^N \langle \phi_l^i, \psi_k^N \rangle \equiv q_l^i$, for $i = 1, 2, \ldots, N, l \in \mathbb{Z}$. Letting $[\mathbf{\Phi}^{ij}]_{lk} = \langle \phi_l^i, \psi_k^j \rangle$, we obtain

$$\begin{aligned} q_l^i = &\sum_{k \in \mathbb{Z}} \left[\mathbf{\Phi}^{i1}\right]_{lk} h_k^1 \\ &+ \sum_{k \in \mathbb{Z}} \left[\mathbf{\Phi}^{i2}\right]_{lk} h_k^2 + \cdots + \sum_{k \in \mathbb{Z}} \left[\mathbf{\Phi}^{iN}\right]_{lk} h_k^N, \end{aligned} \tag{35}$$

for $i = 1, 2, \ldots, N, l \in \mathbb{Z}$. Writing (35) in matrix form, we have $\mathbf{q} = \mathbf{\Phi}\mathbf{h}$ with $[\mathbf{\Phi}^{ij}]_{lk} = \langle \phi_l^i, \psi_k^j \rangle = \langle \phi_l^i(\cdot), K(\cdot, t_k^j) \rangle = \phi_l^i(t_k^j) = \int_{t_l^i}^{t_{l+1}^i} u^i(s - t_k^j)ds$. Finally, the coefficients h_k^i, $i = 1, 2, \ldots, N$ and $k \in \mathbb{Z}$, amount to $\mathbf{h} = \mathbf{\Phi}^+ \mathbf{q}$. $\qquad \square$

Simulation results of a SISO CIM for a Paley-Wiener space of test signals is shown in Figure 12 Input signals $u^1, \ldots, u^5$ were bandlimited to 100 Hz and the circuit generated a total of 38 spikes. The MSE between the identified filter $\mathcal{P}h^*$ (red) and the projection $\mathcal{P}h$ (blue) is -71.1 dB.

5.2. Channels with Noisy Observations.

In the derivations above we implicitly assumed that the I/O system was noiseless. In practice, noise is introduced either by the channel or the sampler itself. Here we revisit the t-transform in (5) and show that the analysis/methodology employed in the previous sections can be extended within an appropriate mathematical setting to I/O systems with noisy measurements.

Recall the t-transform of an ideal IAF neuron is given by $\int_{t_k}^{t_{k+1}} (u * h)(t)dt = \langle \mathcal{P}h, \phi_k \rangle = q_k$, $k = 1, 2, \ldots, n-1$, where n is the number of spikes generated by the neuron in an interval of length T. The measurements q_k were obtained by applying a piece-wise linear operator on the channel output $v = u * h$.

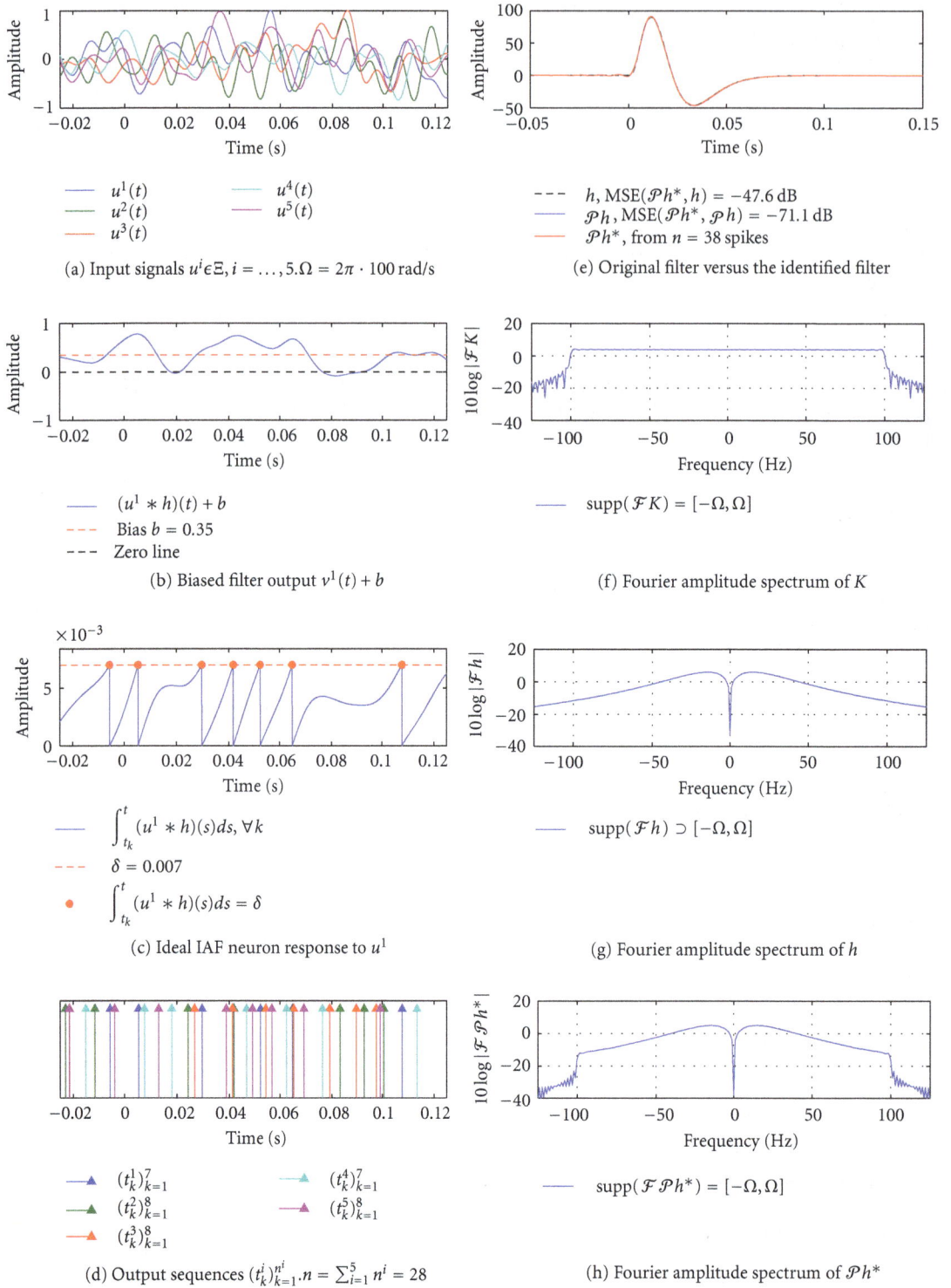

(a) Input signals $u^i \epsilon \Xi, i = \dots, 5. \Omega = 2\pi \cdot 100$ rad/s

(b) Biased filter output $v^1(t) + b$

(c) Ideal IAF neuron response to u^1

(d) Output sequences $(t_k^i)_{k=1}^{n^i} . n = \sum_{i=1}^{5} n^i = 28$

(e) Original filter versus the identified filter

(f) Fourier amplitude spectrum of K

(g) Fourier amplitude spectrum of h

(h) Fourier amplitude spectrum of $\mathcal{P}h^*$

FIGURE 12: Channel identification in a SISO [Filter]-[Ideal IAF] circuit using signals from the Paley-Wiener space Ξ. (a) In contrast to Figure 6, input signals $u^i \in \Xi$, $i = 1, \dots, 5$. (b) Biased output of the filter $v^1(t) + b$ in response to the stimulus u^1. (c) The filter output in (b) is integrated by an ideal IAF neuron. (d) The neuron generated a total of 38 spikes in response to all 5 input signals. (e) The identified impulse response of the filter $\mathcal{P}h^*$ (red) is shown together with the original filter h (dashed black) and its projection $\mathcal{P}h$ (blue). The MSE between $\mathcal{P}h^*$ and $\mathcal{P}h$ is -71.1 dB. (f)–(h) Fourier amplitude spectra of K, h, and $\mathcal{P}h^*$. In contrast to Figure 6, K and $\mathcal{P}h^*$ do not exhibit a discrete (line) spectrum. Again, $\mathcal{P}h^* \in \Xi$ but $h \notin \Xi$.

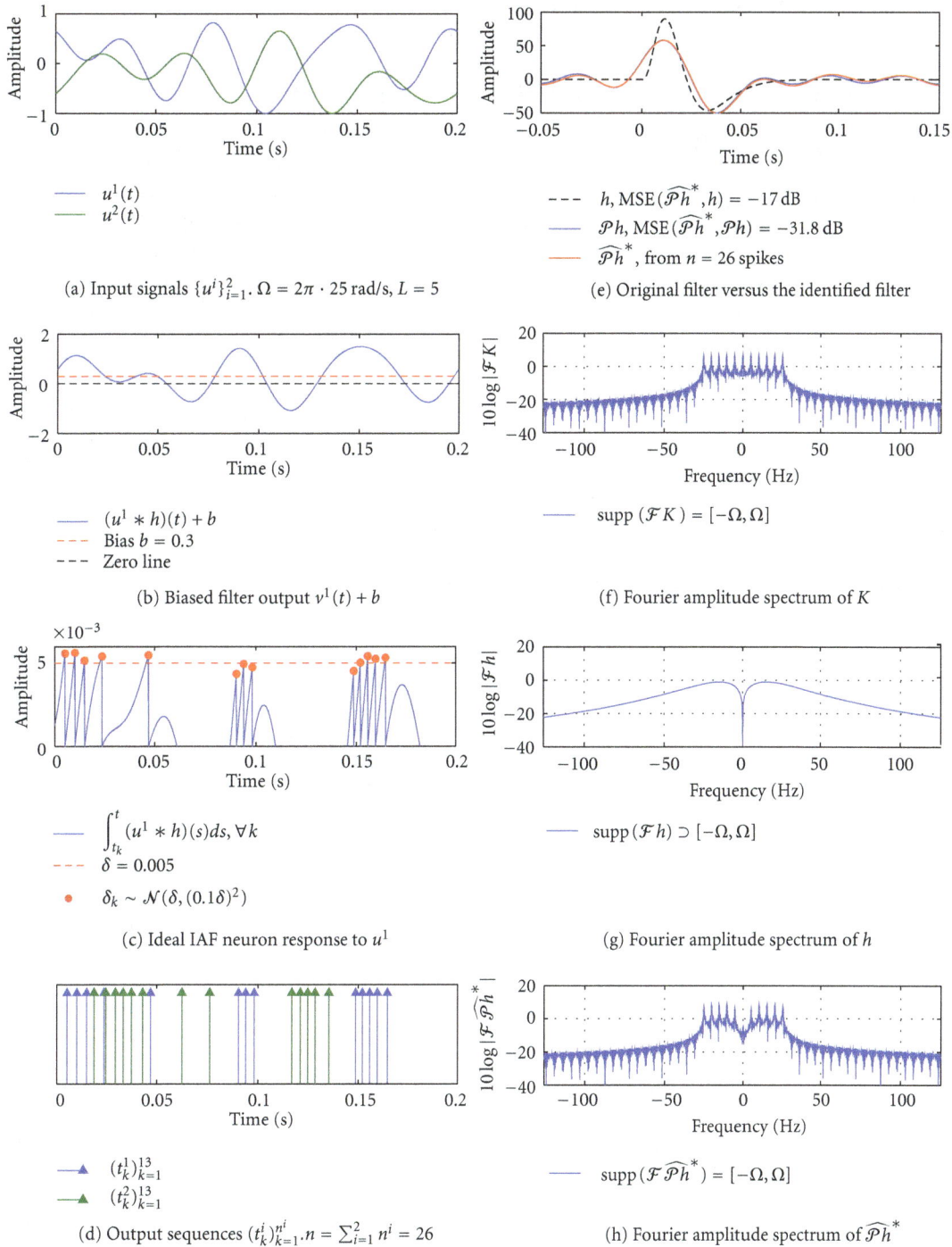

(a) Input signals $\{u^i\}_{i=1}^2$. $\Omega = 2\pi \cdot 25$ rad/s, $L = 5$

(b) Biased filter output $v^1(t) + b$

(c) Ideal IAF neuron response to u^1

(d) Output sequences $(t_k^i)_{k=1}^{n^i}$. $n = \sum_{i=1}^2 n^i = 26$

(e) Original filter versus the identified filter

(f) Fourier amplitude spectrum of K

(g) Fourier amplitude spectrum of h

(h) Fourier amplitude spectrum of $\widehat{\mathcal{P}h}^*$

FIGURE 13: Noisy channel identification in a SISO [Filter]-[Ideal IAF] circuit using multiple I/O pairs. (a) Input signals u^1, u^2 are bandlimited to 25 Hz. The order of the space $L = 5$. (b) Biased output of the filter $v^1(t) + b$ in response to the stimulus u^1. (c) Thresholds are random with $\delta_k \sim \mathcal{N}(\delta, (0.1\delta)^2)$. (d) The neuron produced a total of 26 spikes in response to 2 stimuli. (e) The optimal estimate $\widehat{\mathcal{P}h}^*$ (red) is shown together with the original filter h (dashed black) and its projection $\mathcal{P}h$ (blue). Note that the MSE between $\widehat{\mathcal{P}h}^*$ and $\mathcal{P}h$ is -31.8 dB. (f)–(h) Fourier amplitude spectra of K, h and $\widehat{\mathcal{P}h}^*$. As before, $\text{supp}(\mathcal{F}K) = [-\Omega, \Omega] = \text{supp}(\mathcal{F}\widehat{\mathcal{P}h}^*)$ but $\text{supp}(\mathcal{F}h) \supset [-\Omega, \Omega]$. In other words, $\widehat{\mathcal{P}h}^* \in \mathcal{H}$ but $h \notin \mathcal{H}$.

If either the channel or the sampler introduce an error, we can model it by adding a noise term ε_k to the t-transform [9]:

$$\langle \mathscr{P}h, \phi_k \rangle = q_k + \varepsilon_k. \tag{36}$$

Here we will assume that $\varepsilon_k \sim \mathscr{N}(0, \sigma^2)$, $k = 1, 2, \ldots, n-1$, are i.i.d.

In the presence of noise it is not possible to identify the projection $\mathscr{P}h$ loss-free. However, we can still identify an estimate $\widehat{\mathscr{P}h}$ of $\mathscr{P}h$ that is optimal for an appropriately defined cost function. For example, we can formulate a bi-criterion Tikhonov regularization problem

$$\min_{\widehat{\mathscr{P}h} \in \mathscr{H}} \sum_{i=1}^{N} \sum_{k=1}^{n-1} \left(\langle \widehat{\mathscr{P}h}, \phi_k^i \rangle - q_k^i \right)^2 + \lambda \left\| \widehat{\mathscr{P}h} \right\|_{\mathscr{H}}^2, \tag{37}$$

where the scalar $\lambda > 0$ provides a trade-off between the faithfulness of the identified filter projection $\widehat{\mathscr{P}h}$ to measurements $(q_k)_{k=1}^{n-1}$ and its norm $\|\widehat{\mathscr{P}h}\|_{\mathscr{H}}$.

Theorem 19. *Problem (37) can be solved explicitly in analytical form. The optimal solution is achieved by*

$$\left(\widehat{\mathscr{P}h} \right)(t) = \sum_{l=-L}^{L} h_l e_l(t), \tag{38}$$

with $\mathbf{h} = (\mathbf{\Phi}^H \mathbf{\Phi} + \lambda \mathbf{I})^{-1} \mathbf{\Phi}^H \mathbf{q}$, $\mathbf{\Phi} = [\mathbf{\Phi}^1; \mathbf{\Phi}^2; \ldots; \mathbf{\Phi}^N]$ *and* $\mathbf{\Phi}^i$, $i = 1, 2, \ldots, N$, *as defined in* (13).

Proof. Since the minimizer $\widehat{\mathscr{P}h}$ is in $\mathscr{H}$, it is of the form given in (38). Substituting this into (37), we obtain

$$\min_{\mathbf{h} \in \mathbb{C}^{2L+1}} \|\mathbf{\Phi}\mathbf{h} - \mathbf{q}\|_{\mathbb{R}^{n-1}}^2 + \lambda \|\mathbf{h}\|_{\mathbb{C}^{2L+1}}^2, \tag{39}$$

where $\mathbf{\Phi} = [\mathbf{\Phi}^1; \mathbf{\Phi}^2; \ldots; \mathbf{\Phi}^N]$ with $\mathbf{\Phi}^i$, $i = 1, 2, \ldots, N$, as defined in (13). This quadratic optimization problem can be solved analytically by expressing the objective as a convex quadratic function $J(\mathbf{h}) = \mathbf{h}^H \mathbf{\Phi}^H \mathbf{\Phi}\mathbf{h} - 2\mathbf{q}^H \mathbf{\Phi}\mathbf{h} + \mathbf{q}^H \mathbf{q} + \lambda \mathbf{h}^H \mathbf{h}$ with H denoting the conjugate transpose. A vector $\mathbf{h}$ minimizes J if and only if $\nabla J = 2(\mathbf{\Phi}^H \mathbf{\Phi} + \lambda \mathbf{I})\mathbf{h} - 2\mathbf{\Phi}^H \mathbf{q} = 0$, that is, $\mathbf{h} = (\mathbf{\Phi}^H \mathbf{\Phi} + \lambda \mathbf{I})^{-1} \mathbf{\Phi}^H \mathbf{q}$. $\square$

Remark 20. In Section 3.2, identification of the projection $(\mathscr{P}h)(t) = \sum_{l=-L}^{L} h_l e_l(t)$ amounted to finding $\mathscr{P}h \in \mathscr{H}$ such that the sum of the residuals $(\langle \mathscr{P}h, \phi_k \rangle - q_k)^2$ was minimized [9]. In other words, we were solving an unconstrained convex optimization problem of the form

$$\min_{\mathscr{P}h \in \mathscr{H}} \sum_{i=1}^{N} \sum_{k=1}^{n-1} \left(\langle \mathscr{P}h, \phi_k^i \rangle - q_k^i \right)^2 \iff \min_{\mathbf{h} \in \mathbb{C}^{2L+1}} \|\mathbf{\Phi}\mathbf{h} - \mathbf{q}\|_{\mathbb{R}^{n-1}}^2, \tag{40}$$

where $\mathbf{h} = [h_{-L}, \ldots, h_L]$ and $\mathbf{\Phi} = [\mathbf{\Phi}^1; \mathbf{\Phi}^2; \ldots; \mathbf{\Phi}^N]$ with $\mathbf{\Phi}^i$, $i = 1, 2, \ldots, N$, as defined in (13).

5.2.1. Example: Noisy SISO [Filter]-[Ideal IAF] Circuit. In the following example, we assume that noise is added to the measurements $(q_k^i)_{k=1}^{n-1}$, $i = 1, 2$, by the neuron and we model that noise by introducing random thresholds that are normally distributed with a mean δ and a standard deviation 0.1δ, that is, $\delta_k \sim \mathscr{N}(\delta, (0.1\delta)^2)$: $\int_{t_k^i}^{t_{k+1}^i} (u^i * h)(t)dt = C\delta_k - b(t_{k+1}^i - t_k^i) = [C\delta - b(t_{k+1}^i - t_k^i)] + C(\delta_k - \delta) = q_k^i + \varepsilon_k^i$, where $\varepsilon_k^i \sim \mathscr{N}(0, (0.1\delta)^2)$. Thus random thresholds result in additive noise $\varepsilon_k^i \sim \mathscr{N}(0, (0.1C\delta)^2)$, $i = 1, 2$.

In Figure 13(a) we show two stimuli that were used to probe the [Filter]-[Ideal IAF] circuit. Both stimuli are bandlimited to 25 Hz and have a period of $T = 0.2$ s, that is, the order of the space is $L = 5$. The response of the neuron to a biased filter output $v^1(t) + b$ (Figure 13(b)) is shown in Figure 13(c). Note the significant deviations in thresholds δ_k around the mean value of $\delta = 0.05$. Although a significant amount of noise is introduced into the system, we can identify an optimal estimate $\widehat{\mathscr{P}h}^*$ that is still quite close to the true projection $\mathscr{P}h$. The MSE of identification is -31.8 dB.

6. Conclusion

In this paper we presented a class of channel identification problems arising in the context of communication channels in [Filter]-[Asynchronous Sampler] circuits. Our results are based on a key structural conditional duality result between time decoding and channel identification. The conditional duality result shows that given a class of test signals, the projection of the filter onto the space of input signals can be recovered loss-free. Moreover, the channel identification problem can be converted into a time decoding problem. We considered a number of channel identification problems that arise both in communications and in neuroscience. We presented CIM algorithms that allow one to recover projections of both one-dimensional and multi-dimensional filters in such problems and demonstrated their performance through numerical simulations. Furthermore, we showed that under natural conditions on the impulse response of the filter, the filter projection converges to the original filter almost everywhere and in the mean-squared sense (L^2 norm), with increasing bandwidth and order of the space. Thus in order to identify the impulse response of the filter with arbitrary precision, we are lead to consider a countably infinite number of time encoding problems. Finally, we generalized our results to a large class of test signal spaces and to channel models with noisy observations.

Acknowledgments

This work was supported in part by the NIH under the grant no. R01 DC008701-05 and in part by AFOSR under grant no. FA9550-12-1-0232. The authors' names are alphabetically ordered.

References

[1] L. Tong, B. M. Sadler, and M. Dong, "Pilot-assisted wireless transmissions: general model, design criteria, and signal

processing," *IEEE Signal Processing Magazine*, vol. 21, no. 6, pp. 12–25, 2004.

[2] M. Unser, "Sampling—50 years after Shannon," *Proceedings of the IEEE*, vol. 88, no. 4, pp. 569–587, 2000.

[3] J. J. Benedetto and P. J. S. G. Ferreira, Eds., *Modern Sampling Theory, Mathematics and Applications*, Birkhäuser, 2001.

[4] A. A. Lazar and L. T. Tóth, "Perfect recovery and sensitivity analysis of time encoded bandlimited signals," *IEEE Transactions on Circuits and Systems I*, vol. 51, no. 10, pp. 2060–2073, 2004.

[5] A. A. Lazar and E. A. Pnevmatikakis, "Faithful representation of stimuli with a population of integrate-and-fire neurons," *Neural Computation*, vol. 20, no. 11, pp. 2715–2744, 2008.

[6] A. A. Lazar, "Population encoding with Hodgkin-Huxley neurons," *IEEE Transactions on Information Theory*, vol. 56, no. 2, pp. 821–837, 2010.

[7] H. G. Feichtinger and Gröchenig K., "Theory and practice of irregular sampling," in *Wavelets: Mathematics and Applications*, Studies in Advanced Mathematics, pp. 305–363, CRC Press, 1994.

[8] S. Yan Ng, *A continuous-time asynchronous Sigma Delta analog to digital converter for broadband wireless receiver with adaptive digital calibration technique [Ph.D. thesis]*, Department of Electrical and Computer Engineering, Ohio State University, 2009.

[9] A. A. Lazar, E. A. Pnevmatikakis, and Y. Zhou, "Encoding natural scenes with neural circuits with random thresholds," *Vision Research*, vol. 50, no. 22, pp. 2200–2212, 2010, Special Issue on Mathematical Models of Visual Coding.

[10] M. C.-K. Wu, S. V. David, and J. L. Gallant, "Complete functional characterization of sensory neurons by system identification," *Annual Review of Neuroscience*, vol. 29, pp. 477–505, 2006.

[11] U. Friederich, D. Coca, S. Billings, and M. Juusola, "Data modelling for analysis of adaptive changes in fly photoreceptors," *Neural Information Processing*, vol. 5863, no. 1, pp. 34–48, 2009.

[12] T. W. Berger, D. Song, R. H. M. Chan, and V. Z. Marmarelis, "The neurobiological basis of cognition: identification by multi-input, multioutput nonlinear dynamic modeling," *Proceedings of the IEEE*, vol. 98, no. 3, pp. 356–374, 2010.

[13] T. W. Berger, D. Song, R. H. M. Chan et al., "A hippocampal cognitive prosthesis: multi-input, multi-output nonlinear modeling and VLSI implementation," *IEEE Transactions on Neural Systems and Rehabilitation Engineering*, vol. 20, no. 2, pp. 198–211, 2012.

[14] Z. Song, M. Postma, S. A. Billings, D. Coca, R. C. Hardie, and M. Juusola, "Stochastic, adaptive sampling of information by microvilli in fly photoreceptors," *Current Biology*, vol. 22, pp. 1–10, 2012.

[15] F. J. Doyle III, R. K. Pearson, and B. A. Ogunnaike, *Identification and Control Using Volterra Models*, Springer, 2002.

[16] L. Ljung, "Perspectives on system identification," *Annual Reviews in Control*, vol. 34, no. 1, pp. 1–12, 2010.

[17] F. E. Theunissen, S. V. David, N. C. Singh, A. Hsu, W. E. Vinje, and J. L. Gallant, "Estimating spatio-temporal receptive fields of auditory and visual neurons from their responses to natural stimuli," *Network*, vol. 12, no. 3, pp. 289–316, 2001.

[18] R. de Boer and P. Kuyper, "Triggered correlation," *IEEE Transactions on Biomedical Engineering*, vol. 15, no. 3, pp. 169–179, 1968.

[19] O. Schwartz, E. J. Chichilnisky, and E. P. Simoncelli, "Characterizing neural gain control using spike-triggered covariance,"

[20] A. A. Lazar and Y. B. Slutskiy, "Identifying dendritic processing," *Advances in Neural Information Processing Systems*, vol. 23, pp. 1261–1269, 2010.

[21] W. P. Torres, A. V. Oppenheim, and R. R. Rosales, "Generalized frequency modulation," *IEEE Transactions on Circuits and Systems I*, vol. 48, no. 12, pp. 1405–1412, 2001.

[22] A. Berlinet and C. Thomas-Agnan, *Reproducing Kernel Hilbert Spaces in Probability and Statistics*, Kluwer Academic Publishers, 2004.

[23] L. Grafakos, *Modern Fourier Analysis*, vol. 250 of *Graduate Texts in Mathematics*, Springer, 2008.

[24] E. H. Adelson and J. R. Bergen, "Spatiotemporal energy models for the perception of motion," *Journal of the Optical Society of America A*, vol. 2, no. 2, pp. 284–299, 1985.

[25] A. J. Kim, A. A. Lazar, and Y. B. Slutskiy, "System identification of *Drosophila* olfactory sensory neurons," *Journal of Computational Neuroscience*, vol. 30, no. 1, pp. 143–161, 2011.

[26] O. Christensen, *An Introduction to Frames and Riesz Bases. Applied and Numerical Harmonic Analysis*, Birkhäuser, 2003.

Advances in Neural Information Processing Systems, vol. 14, pp. 269–276, 2002.

9

Color Image Quantization Algorithm Based on Self-Adaptive Differential Evolution

Qinghua Su[1] and Zhongbo Hu[1,2]

[1] School of Mathematics and Statistic, Hubei Engineering University, Xiaogan, Hubei 432000, China
[2] School of Sciences, Wuhan University of Technology, Wuhan, Hubei 430070, China

Correspondence should be addressed to Qinghua Su; suqhdd@126.com

Academic Editor: Daoqiang Zhang

Differential evolution algorithm (DE) is one of the novel stochastic optimization methods. It has a better performance in the problem of the color image quantization, but it is difficult to set the parameters of DE for users. This paper proposes a color image quantization algorithm based on self-adaptive DE. In the proposed algorithm, a self-adaptive mechanic is used to automatically adjust the parameters of DE during the evolution, and a mixed mechanic of DE and K-means is applied to strengthen the local search. The numerical experimental results, on a set of commonly used test images, show that the proposed algorithm is a practicable quantization method and is more competitive than K-means and particle swarm algorithm (PSO) for the color image quantization.

1. Introduction

Color image quantization, one of the common image processing techniques, is the process of reducing the number of colors presented in a color image with less distortion [1]. The main purpose of color quantization is reducing the use of storage media and accelerating image sending time [2]. Color image quantization consists of two essential phases. The first one is to design a colormap with a smaller number of colors (typically 8–256 colors [3]) than that of a color image. The second one is to map each pixel in the color image to one color in the colormap. Most of the color quantization methods focus on creating an optimal colormap. For being an NP-hard problem, it is not feasible to find the optimal colormap without a prohibitive amount of time [4]. To address this problem, researchers have applied several stochastic optimization methods, such as GA and PSO. In particular, the literature [5–8] has compared the color image quantization algorithm using PSO (PSO-CIQ) and several other well-known color image quantization methods. The experimental results show that PSO-CIQ has higher performance.

Differential evolution algorithm (DE) [9–11] is a population-based heuristic search approach. DE has been applied to the classification for gray images [12–14]. In the literature

[12–14], DE and PSO show similar performance. However, due to simple operation, litter parameters, and fast convergence, DE is the better choice to use than PSO [12]. However, few researches have been done for using DE to solve the color image quantization. This paper applies DE to solve the color image quantization. However, the performance of DE is decided by two important parameters, the scaling factor F and the crossover rate CR. In practice, it is difficult to set the two parameters. For this difficulty, this paper proposes a color image quantization algorithm based on self-adaptive DE (SaDE-CIQ). In SaDE-CIQ, the self-adaptive mechanics in the literature [15, 16] are used to automatically adjust the parameters of DE during the evolution, and K-means is mixed into DE with a little probability for strengthening the local search. SaDE-CIQ starts with an initialized population, in which each individual represents a candidate colormap. A small number of candidate colormaps are adjusted by K-means. Then the adjusted candidate colormaps and the rest ones are repeatedly updated by DE operations, in which the parameters are automatically adjusted. The optimal solution is the optimal colormap, by which the quantized image is generated. By some commonly used color images, the performance of SaDE-CIQ in the color image quantization is compared with that of K-means and PSO.

This paper is organized as follows. Section 2 introduces the classical DE briefly. In Section 3, a self-adaptive mechanic of DE parameters and a mixed mechanic of DE and K-means are introduced. In Section 4, SaDE-CIQ is proposed. In Section 5, numerical experiments are performed to compare the color image quantization qualities of SaDE-CIQ, K-means, and PSO. Section 6 concludes this paper.

2. Classical Differential Evolution

DE is a simple and powerful stochastic global optimization algorithm, and several DE variants or strategies have been presented. In this paper, we focus on the classical DE, which applies the simple arithmetic operations: mutation, crossover, and selection to evolve the population. Before the introduction of the classical DE, the following symbols used throughout this paper are defined:

   (i) $g(x)$: objective function or fitness function,

   (ii) D: the dimension of an optimization problem,

   (iii) NP: population size,

   (iv) $X = \{x^1, x^2, \ldots, x^{NP}\}$: population,

   (v) $x^j = (x_1^j, x_2^j, \ldots, x_D^j)$: the jth individual in the population X, $j = 1, 2, \ldots, NP$,

   (vi) F: scaling factor,

   (vii) CR: crossover rate.

Consider the following optimization problem:

$$\min g(x), \quad x = (x_1, x_2, \ldots, x_D) \in \prod_{i=1}^{D}[L_i, U_i], \quad (1)$$

$$i = 1, 2, \ldots, D,$$

where L_i and U_i are the lower bound and upper bound of variable x_i and $\prod_{i=1}^{D}[L_i, U_i]$ is the feasible domain of this problem.

An initial population $X = \{x^1, x^2, \ldots, x^{NP}\}$ including NP individuals is generated randomly, where each individual $x^j = (x_1^j, x_2^j, \ldots, x_D^j)$, $j = 1, 2, \ldots, NP$.

The following mutation operation is performed to generate a donor vector for each individual x^j:

$$u^j = \left(u_1^j, u_2^j, \ldots, u_D^j\right) = x^{r_1} + F \cdot \left(x^{r_2} - x^{r_3}\right),$$

$$j = 1, 2, \ldots, NP,$$

$$y_i^j = \begin{cases} L_i, & \text{if } \left(u_i^j < L_i\right), \\ U_i, & \text{if } \left(u_i^j > U_i\right), \quad i = 1, 2, \ldots, D, \\ & \qquad\qquad\quad j = 1, 2, \ldots, NP, \\ u_i^j, & \text{otherwise,} \end{cases} \quad (2)$$

$$y^j = \left(y_1^j, y_2^j, \ldots, y_D^j\right), \quad j = 1, 2, \ldots, NP,$$

where y^j is donor vector, r_1, r_2, and r_3 are three uniformly different integers on $[1, NP]$, and the scaling factor F is a parameter on $[0, 1]$.

Then, the following crossover operation is performed to obtain a trial vector for each individual x^j:

$$z^j = \left(z_1^j, z_2^j, \ldots, z_D^j\right), \quad j = 1, 2, \ldots, NP,$$

$$z_i^j = \begin{cases} y_i^j, & \text{if } \left(\text{rand}_i \le CR \text{ or } i = \text{rnbr}_j\right) \\ x_i^j, & \text{if } \left(\text{rand}_i > CR \text{ and } i \ne \text{rnbr}_j\right), \\ & \qquad\qquad\quad i = 1, 2, \ldots, D, \end{cases} \quad (3)$$

where z^j is trial vector, the crossover rate CR is a parameter on $[0, 1]$, rand_i is a uniformly random value on $[0, 1]$, and rnbr_j is a uniformly random integer on $[1, D]$ for each different j to assure that at least one component of z^j is taken from the donor vector.

Finally, according to the fitness values of the fitness function, the population is updated by the following selection operation:

$$x^{j'} = \begin{cases} x^j, & \text{if } \left(g\left(x^j\right) \le g\left(z^j\right)\right) \\ z^j, & \text{if } \left(g\left(x^j\right) > g\left(z^j\right)\right), \\ & \qquad\quad j = 1, 2, \ldots, NP, \end{cases} \quad (4)$$

where $x^{j'}$ is the updated individual for the next generation population $X' = \{x^{1'}, x^{2'}, \ldots, x^{NP'}\}$.

As stated previously, for obtaining the best solution of the fitness function $g(x)$, DE starts with a randomly generated initial population and repeatedly updates the population with the mutation, crossover, and selection operations until the stopping condition is satisfied.

3. A Self-Adaptive Mechanic and a Mixed Mechanic of DE

The scaling factor F and the crossover rate CR can influence the convergence and stability of DE. In practice, it is more difficult to set right F and CR for user. One of the effective methods to solve this difficult problem is to self-adaptively control the parameters in DE during evolution. In the following SaDE-CIQ, F and CR are automatically updated for each individual in each generation by the self-adaptive mechanics of the literature [15, 16]:

$$F^j = \begin{cases} 0.1 + \text{rand} * 0.9, & \text{if rand} < 0.1, \\ F^j, & \text{otherwise,} \end{cases} \quad (5)$$

$$CR^j = \begin{cases} \text{rand}, & \text{if rand} < 0.1, \\ CR^j, & \text{otherwise.} \end{cases} \quad (6)$$

Generally, self-adaptively adjusting parameters maybe have negative effect on the performance of DE. For improving the color quantization quality of self-adaptive DE, a mixed mechanic is used in the following SaDE-CIQ. K-means is a quickly cluster algorithm with better local search ability. In the mixed mechanic, a small number of individuals from a population are selected by a little probability p. Before being updated by DE, the selected individuals are adjusted by K-means quickly. The mixed operation can simplify the search space of DE and improve its convergence speed.

```
Input color image I = {z}
Set parameters NP, t_max, p
x_i^{j,0} = rand(0, 1) · 255, i = 1, 2, ..., D
x^{j,0} = (x_1^{j,0}, x_2^{j,0}, ..., x_D^{j,0}), j = 1, 2, ..., NP          //population initialization
F = 0.5, CR = 0.6
for t = 0 to t_max
    Select a small number of individuals from current population according to p, and adjust them by K-means.
    for j = 1, 2, ..., NP
        rnbr_j = rand(1, D)
        if x^{r2,t} − x^{r3,t} ≤ 100
            u^{j,t} = x^{r1,t} + F · (x^{r2,t} − x^{r3,t})
        else u^{j,t} = x^{r1,t} + F · [(x^{r2,t} − x^{r3,t}) %100]
        Update F by formulas (5)
        for i = 1, 2, ..., D
            if u_i^{j,t} < 0 then y_i^{j,t} = 0
            else if u_i^{j,t} > 255 then y_i^{j,t} = 255
                else y_i^{j,t} = u_i^{j,t}
                end if
            end if                                                        // mutation
            rand_i = rand(0, 1)
            if rand_i ≤ CR or i = rnbr_j then z_i^{j,t} = y_i^{j,t}
            else z_i^{j,t} = x_i^{j,t}                                    //crossover
        z^{j,t} = (z_1^{j,t}, z_2^{j,t}, ..., z_D^{j,t})
        Update CR by formulas (6)
        Calculate g(x^{j,t}) and g(z^{j,t})
        if g(x^{j,t}) > g(z^{j,t}) then x^{j,t+1} = z^{j,t}
        else x^{j,t+1} = x^{j,t}                                         //selection
    Find the global optimal solution x^best = (x_1^best, x_2^best, ..., x_D^best)
    Output the optimal colormap {c_1, c_2, ..., c_K}, c_k = (x_{1+3(k−1)}^best, x_{2+3(k−1)}^best, x_{3+3(k−1)}^best), k = 1, 2, ..., K
    Construct the quantized image I'
```

PSEUDOCODE 1: The pseudocode of the SaDE-CIQ.

4. Color Image Quantization Algorithm Based on Self-Adaptive DE

In RGB color space, each color pixel of a color image is a combination of red, green, and blue (RGB). For color images, the data space is $[0, 255]^3$. For a given color image I, the color number of the image I is set to be N, and the set of all colors belonging to I is set to be S. S', called a colormap, is a subset with K colors in $[0, 255]^3$, where $K < N$. The color image quantization is to design a colormap S' and to create a map $f : S \rightarrow S'$, by which each color pixel in S is replaced by one of the colors in S'. Thus, a new color image I', called the quantized image of I, with the K colors in S' is constructed. The objective to quantize the color image I is to minimize the color error between the color image I and its quantized image I'.

The color image quantization consists of two major phases:

(i) design a better colormap with a reduced number of colors (typically 8–256);

(ii) create the mapping relationship between the color image and the colormap by which a quantized image is obtained.

In the process of color image quantization, it is most important to design an optimal colormap. To address this problem, researchers have applied some heuristic techniques for color image quantization. These techniques can be mainly categorized into preclustering approaches and postclustering approaches. Although being time consuming, the postclustering approaches are superior to the preclustering approaches in the quantization quality. Postclustering approaches perform clustering of the color space [17]. A postclustering algorithm starts with an initial colormap concluding K colors. Each color pixel of the image I is mapped to the color in the colormap with the minimal color distance from the color pixel. Thus, all the color pixels in image I are clustered into K clusters whose centers are separately the K colors in the colormap. Then the colormap and the K clusters are iteratively modified to improve the optimum.

This section describes a new postclustering color image quantization approach using self-adaptive DE, called the color image quantization algorithm based on self-adaptive DE (SaDE-CIQ). A measure is given to quantify the quality of the resultant quantized image, after which the SaDE-CIQ is introduced.

Peppers

(a)

Baboon

(b)

Lena

(c)

Airplane

(d)

FIGURE 1: Test images.

4.1. Measure of Quality.

The mean square error (MSE) is the most general measure of quality of a quantized image [4]. It represents the color error between image I and its quantized image I'. In the following SaDE-CIQ, the MSE is set to be the fitness function, which is defined as follows:

$$\text{MSE} = \frac{1}{N_p} \left\{ \sum_{r=1}^{N_p} \left[\min_{k=1}^{K} d\left(p_r, c_k \right) \right] \right\}, \qquad (7)$$

where the symbols used in (7), which will be used in the remaining parts of this paper, are explained as follows:

(i) N_p: the number of image pixels,

(ii) K: the color number in the colormap,

(iii) $p_r = (p_{r1}, p_{r2}, p_{r3})$: the rth pixel of the color image I, $r = 1, 2, \ldots, N_p$,

(iv) c_k: the kth color triple in the colormap, $k = 1, 2, \ldots, K$.

4.2. Description of the SaDE-CIQ.

In the SaDE-CIQ, the classical DE with the previously mentioned self-adaptive mechanic and mixed mechanic is used for the color image quantization. A population $X = \{x^1, x^2, \ldots, x^{NP}\}$ represents a set of candidate colormaps. Each individual represents a candidate colormap with K color triples in the RGB color space $[0, 255]^3$. The jth individual is denoted by

$$x^j = \left(c_1^j, c_2^j, \ldots, c_K^j \right)$$

$$= \left(x_1^j, x_2^j, x_3^j, x_4^j, x_5^j, x_6^j, \ldots, x_{3K-2}^j, x_{3K-1}^j, x_{3K}^j \right), \qquad (8)$$

$$j = 1, 2, \ldots, NP,$$

where $c_k^j = (x_{1+3(k-1)}^j, x_{2+3(k-1)}^j, x_{3+3(k-1)}^j)$, $k = 1, 2, \ldots, K$. Thus, the dimension of each individual is $D = 3 \times K$, and the

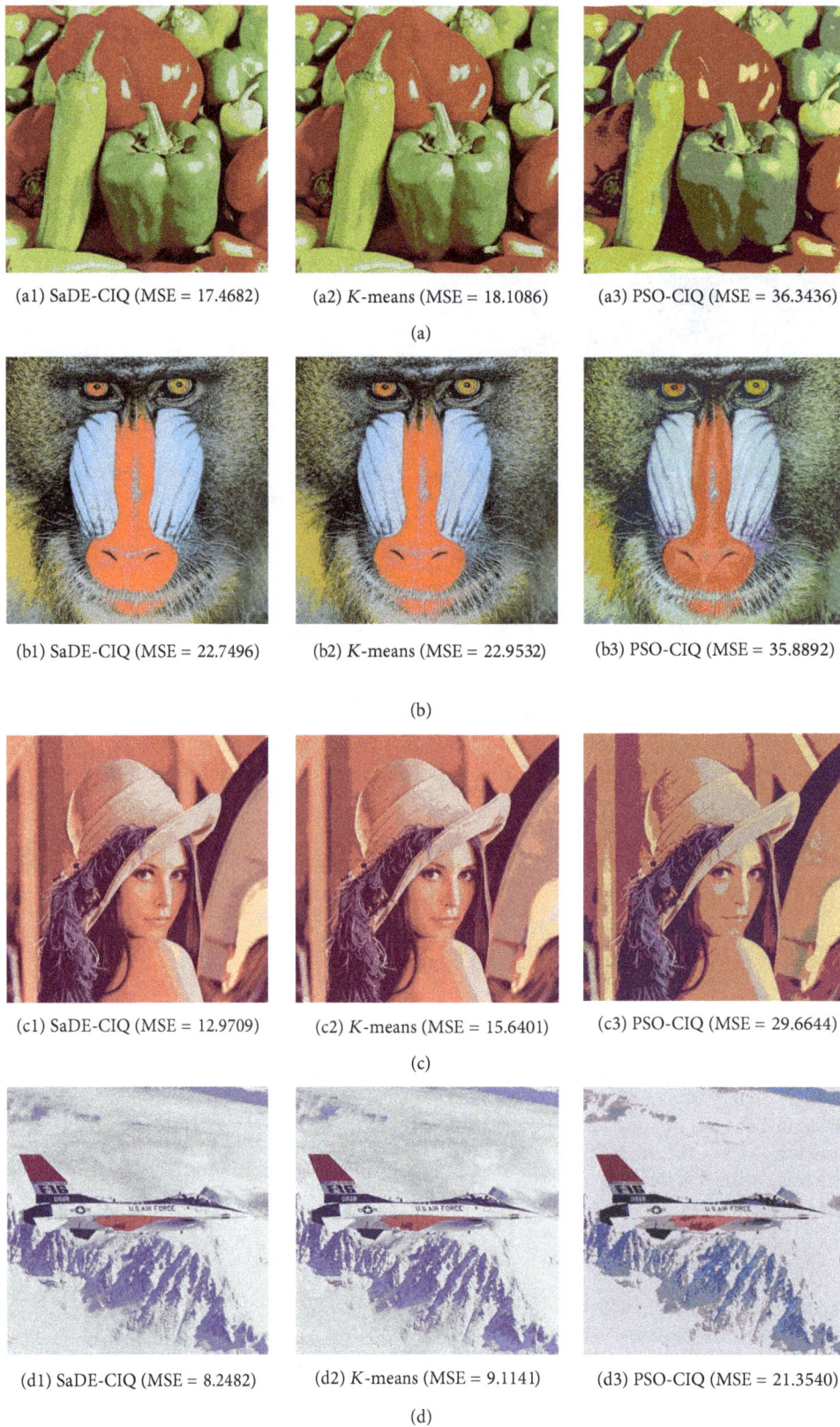

(a1) SaDE-CIQ (MSE = 17.4682)   (a2) *K*-means (MSE = 18.1086)   (a3) PSO-CIQ (MSE = 36.3436)

(a)

(b1) SaDE-CIQ (MSE = 22.7496)   (b2) *K*-means (MSE = 22.9532)   (b3) PSO-CIQ (MSE = 35.8892)

(b)

(c1) SaDE-CIQ (MSE = 12.9709)   (c2) *K*-means (MSE = 15.6401)   (c3) PSO-CIQ (MSE = 29.6644)

(c)

(d1) SaDE-CIQ (MSE = 8.2482)   (d2) *K*-means (MSE = 9.1141)   (d3) PSO-CIQ (MSE = 21.3540)

(d)

FIGURE 2: The quantized images obtained by SaDE-CIQ, *K*-means, and PSO-CIQ.

TABLE 1: The MSEs resulting from SaDE-CIQ and PSO-CIQ.

Alg.	Peppers		Baboon		Lena		Airplane	
	min	max	min	max	min	max	min	max
SaDE-CIQ	17.4682	18.7266	22.7496	23.3382	12.9709	13.8055	8.2482	8.9740
K-means	18.1086	21.2676	22.9532	24.9563	15.6401	19.1314	9.1141	10.4430
PSO-CIQ	36.3436	40.9532	35.8892	41.9940	29.6644	34.5867	21.3540	24.3200

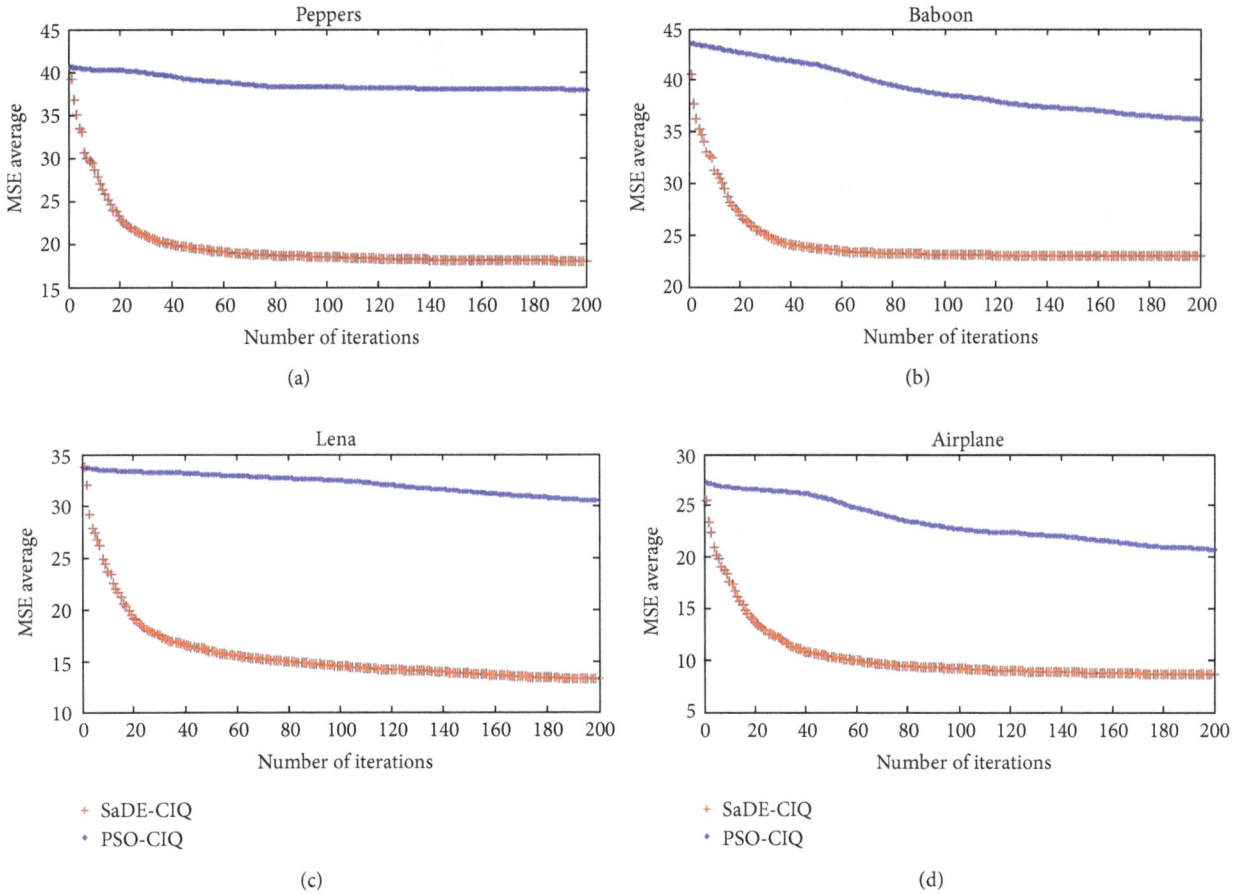

FIGURE 3: The average MSE variations with the number of iterations of SaDE-CIQ and PSO-CIQ.

feasible domain is $[0, 255]^{3 \times K}$. The quality of each individual is measured by the MSE in (7):

$$
\begin{aligned}
g\left(x^j\right) = \text{MSE}\left(x^j\right) &= \frac{1}{N_p} \left\{ \sum_{r=1}^{N_p} \left[\min_{k=1}^{K} d\left(p_r, c_k^j\right) \right] \right\} \\
&= \frac{1}{N_p} \left\{ \sum_{r=1}^{N_p} \left[\min_{k=1}^{K} \sqrt{\sum_{q=1}^{3} \left(p_{rq} - x_{q+3(k-1)}^j\right)^2} \right] \right\},
\end{aligned}
$$

$$
j = 1, 2, \ldots, NP.
$$

(9)

The stopping condition of the algorithm is to reach a specified maximal number of iterations $t_{\max}$.

In the first phase of the SaDE-CIQ, an optimal colormap is designed. A set of NP candidate colormaps are initialized. Each colormap consists of K randomly selected color triples in the color space $[0, 255]^3$. K-means is applied to adjust a small number of colormaps randomly selected from all the candidate colormaps by a little probability p. Then the adjusted colormaps and the rest ones are repeatedly updated by mutation and crossover operations, where F and CR with the initial values 0.5 and 0.6 are updated by formulas (5) and (6). The K-means and DE operations are performed until the stopping condition is satisfied. The last optimal solution is the optimal colormap. In the second phase of the SaDE-CIQ, the mapping relationship is created according to the minimal color distance principle. By replacing each pixel in the color image I with its corresponding color in the optimal colormap, I is to be reconstructed to obtain the quantized image I'.

See Pseudocode 1 of the SaDE-CIQ.

5. Numerical Experiments

In this section, the SaDE-CIQ is tested on a set of four commonly used test images in the quantization literature. In addition, the performance of the SaDE-CIQ is compared with that of K-means and the color image quantization algorithm using PSO (PSO-CIQ) presented in literature the [5].

5.1. Images and Parameters Set. The set of test images include Lena, Peppers, Baboon, and Airplane, which have the same size 512×512 pixels. They are shown in Figure 1.

The parameters in the SaDE-CIQ are set as the population size $NP = 100$, the maximal number of iterations $t_{max} = 200$, and the mixed probability $p = 0.05$.

The PSO-CIQ has more parameters than the SaDE-CIQ. They are set as the swarm size $NP = 100$, the inertia weight $\omega = 0.72$, the acceleration constants $c_1 = c_2 = 1.49$, the maximum velocity $V_{max} = 0.4$, and the maximal number of iterations $t_{max} = 200$. These parameters except for the last one are as same as those in the literature [5].

5.2. Experimental Results. For each algorithm, the test images are quantized into 16 colors. The colors quantized images with the smallest MSEs over 10 simulations are shown in Figure 2. The smallest MSEs and the largest MSEs over 10 simulations are listed in Table 1, and for SaDE-CIQ and PSO-CIQ, the MSE variations with the number of iterations are exhibited in Figure 3.

5.3. Analysis of Experimental Results. As shown in Figure 2, the SaDE-CIQ outperforms K-means and PSO-CIQ in the visual quality of the quantized images for all test images. The quantized images a-1, b-1, c-1, and d-1 have richer layers and more details than the other quantized images.

As illustrated in Table 1, the SaDE-CIQ generates a smaller MSEs than K-means and PSO-CIQ for each test image.

Shown in Figure 3, the SaDE-CIQ has a smaller average MSE than the PSO-CIQ at each same number of iterations. Moreover, the average MSE resulting from the SaDE-CIQ decreases more quickly than that resulting from the PSO-CIQ with the increasing number of iterations.

The above experimental results can be summarized as follows:

(i) the SaDE-CIQ is an effective color image quantization method;

(ii) the SaDE-CIQ has better quantization quality than K-means and PSO-CIQ;

(iii) the SaDE-CIQ converges more quickly than the PSO-CIQ.

6. Conclusions

This paper presents a color image quantization algorithm based on self-adaptive DE (SaDE-CIQ). Numerical experiments are implemented to investigate the performance of the SaDE-CIQ and to compare it against K-means and PSO-CIQ

presented in the literature [5]. For a set of commonly used test images, the experimental results demonstrate the feasibility of the SaDE-CIQ and its superiority to K-means and PSO-CIQ in the quantization quality. In addition, the SaDE-CIQ has simpler operation, litter parameters, and faster convergence than the PSO-CIQ.

Acknowledgments

This work was supported in part by the National Natural Science Foundation of China under Grant no. 61070009, the Fundamental Research Funds for the Central Universities under Grant no. 2012-YB-19, and the Nature Science Foundation of Hubei, China, under Grant no. 2011CDC161.

References

[1] J.-P. Braquelaire and L. Brun, "Comparison and optimization of methods of color image quantization," *IEEE Transactions on Image Processing*, vol. 6, no. 7, pp. 1048–1052, 1997.

[2] F. Alamdar, Z. Bahmani, and S. Haratizadeh, "Color quantization with clustering by F-PSO-GA," in *Proceedings of the IEEE International Conference on Intelligent Computing and Intelligent Systems (ICIS '10)*, vol. 3, pp. 233–238, October 2010.

[3] P. Scheunders, "A genetic c-means clustering algorithm applied to color image quantization," *Pattern Recognition*, vol. 30, no. 6, pp. 859–866, 1997.

[4] M. G. Omran, A. P. Engelbrecht, and A. Salman, "A color image quantization algorithm based on Particle Swarm Optimization," *Soft Computing in Multimedia Processing*, vol. 29, no. 3, pp. 261–269, 2005.

[5] Q. Sa, X. Liu, X. He, and D. Yan, "Color image quantization using particle swarm optimization," *Journal of Image and Graphics*, vol. 12, no. 9, pp. 1544–1548, 2007 (Chinese).

[6] X. Zhou, Q. Shen, and J. Wang, "Color quantization algorithm based on particle swarm optimization," *Microelectronics and Computer*, vol. 25, no. 3, pp. 51–54, 2009 (Chinese).

[7] Y. Xu and Z. Jiang, "A K-mean color image quantization method based on particle swarm optimization," *Journal of Northwest University*, vol. 42, no. 3, 2012 (Chinese).

[8] F. Alamdar, Z. Bahmani, and S. Haratizadeh, "Color quantization with clustering by F-PSO-GA," in *Proceedings of the IEEE International Conference on Intelligent Computing and Intelligent Systems (ICIS '10)*, pp. 233–238, October 2010.

[9] R. Storn and K. Price, "Differential evolution—a simple and efficient heuristic for global optimization over continuous spaces," *Journal of Global Optimization*, vol. 11, no. 4, pp. 341–359, 1997.

[10] Y. Shen, M. Li, and H. Yin, "A novel differential evolution for numerical optimization," *International Journal of Advancements in Computing Technology*, vol. 4, no. 4, pp. 24–31, 2012.

[11] H. Li and J. Tang, "Improved differential evolution with local search," *Journal of Convergence Information Technology*, vol. 7, no. 4, pp. 197–204, 2012.

[12] M. G. H. Omran, A. P. Engelbrecht, and A. Salman, "Differential evolution methods for unsupervised image classification," in *Proceedings of the IEEE Congress on Evolutionary Computation (IEEE CEC '05)*, pp. 966–973, September 2005.

[13] S. Das and A. Konar, "Automatic image pixel clustering with an improved differential evolution," *Applied Soft Computing Journal*, vol. 9, no. 1, pp. 226–236, 2009.

[14] Q. Su, Z. Huang, and Z. Hu, "Binarization algorithm based on differential evolution algorithm for gray images," *ICNC-FSKD*, vol. 6, pp. 2624–2628, 2012.

[15] J. Brest, S. B. Bošković, V. Žume, and M. S. Maučec, "Performance comparison of self-adaptive and adaptive differential evolution algorithms," *Soft Computing*, vol. 11, no. 7, pp. 617–629, 2007.

[16] J. Brest, V. Žumer, and M. S. Maučec, "Self-adaptive differential evolution algorithm in constrained real-parameter optimization," in *Proceedings of the IEEE Congress on Evolutionary Computation (CEC '06)*, pp. 215–222, July 2006.

[17] S.-C. Cheng and C.-K. Yang, "A fast and novel technique for color quantization using reduction of color space dimensionality," *Pattern Recognition Letters*, vol. 22, no. 8, pp. 845–856, 2001.

10

Quantitative Tools for Examining the Vocalizations of Juvenile Songbirds

Cameron D. Wellock and George N. Reeke

Laboratory of Biological Modeling, The Rockefeller University, 1230 York Avenue, New York, NY 10065, USA

Correspondence should be addressed to Cameron D. Wellock, cwellock@rockefeller.edu

Academic Editor: Francois Benoit Vialatte

The singing of juvenile songbirds is highly variable and not well stereotyped, a feature that makes it difficult to analyze with existing computational techniques. We present here a method suitable for analyzing such vocalizations, windowed spectral pattern recognition (WSPR). Rather than performing pairwise sample comparisons, WSPR measures the typicality of a sample against a large sample set. We also illustrate how WSPR can be used to perform a variety of tasks, such as sample classification, song ontogeny measurement, and song variability measurement. Finally, we present a novel measure, based on WSPR, for quantifying the apparent complexity of a bird's singing.

1. Introduction

A bird's song can be a powerful marker of identity, used by other birds—and humans—to identify the singer's species or even to identify a single individual. In many species this song is innate, but for the Oscine songbirds, every bird must acquire its own song [1, 2]. With one such bird, the zebra finch (*Taeniopygia guttata*), it is the males that sing, and juvenile males learn their song from nearby adults such as their father [3]. The learning process has two overlapping but distinct parts: in the first, the animal hears the songs of other birds and somehow commits to memory a model of the song it will sing; in the second, the animal learns how to produce a version of this memorised song through practice [1].

As adults, zebra finches sing in bouts during which they perform their single song motif a variable number of times. The song motif of a zebra finch is on the order of one second long and is composed of multiple syllables, elements separated by silence or a sharp drop in amplitude. Syllables can often be broken down further into notes, segments of distinct sound quality. These notes may demonstrate pronounced frequency modulation and complex harmonics. Adult zebra finches typically exhibit a very high degree of stereotypy in their song, with one performance of the song's motif being very similar to any other. Two typical examples are shown in Figure 1.

In the early stages of a juvenile's song production, vocalizations tend to sound very little like the song of an adult, instead sounding more like a kind of babbling [4]. This earliest stage is called "subsong" [1]. From this, the juvenile progresses to a style of vocalization, "plastic song" [1], which is low in stereotypy but in which the precursors of adult-like sounds can be identified. Eventually, at approximately 80 days posthatch [5], the juvenile learns to produce its song with a high degree of stereotypy and its song-learning process is complete. For the zebra finch, this song will remain largely unchanged for the rest of the animal's life.

Another class of vocalization is the "call," which can serve multiple purposes [2]. Calls are typically short (200–500 ms) continuous sounds that might be described as "honk-like." Zebra finches of both sexes, including juveniles, produce calls. Examples of juvenile song and calls are shown in Figure 2.

In the course of our research, we have at times wanted a tool to identify and compare juvenile vocalizations, primarily to assist in the sorting of large numbers of recorded samples. Although a number of tools exist to compare the songs of adult birds, we have found that, due to the low stereotypy of

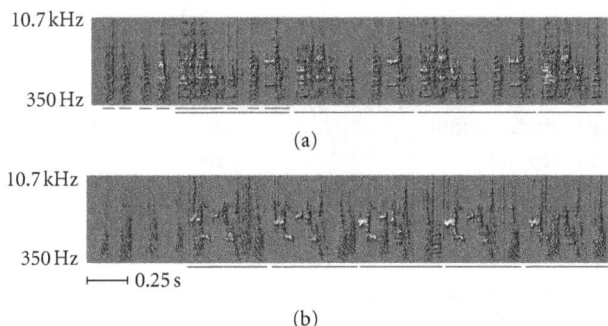

FIGURE 1: (a) Spectrogram of a bout of singing from an adult zebra finch. Noted in the figure are the following song parts: introductory notes, underlined in red; syllables, underlined in green; the silent interval between syllables, underlined in yellow. The blue lines mark the repetitions of the bird's motif. Note that each performance of the motif appears much like the others, except for the truncated final motif. (b) Spectrogram of a bout of singing from a different zebra finch. Although its song is also highly stereotyped, it is visibly different from the song of the bird featured in (a). For convenience, blue lines once again mark repetitions of the bird's motif.

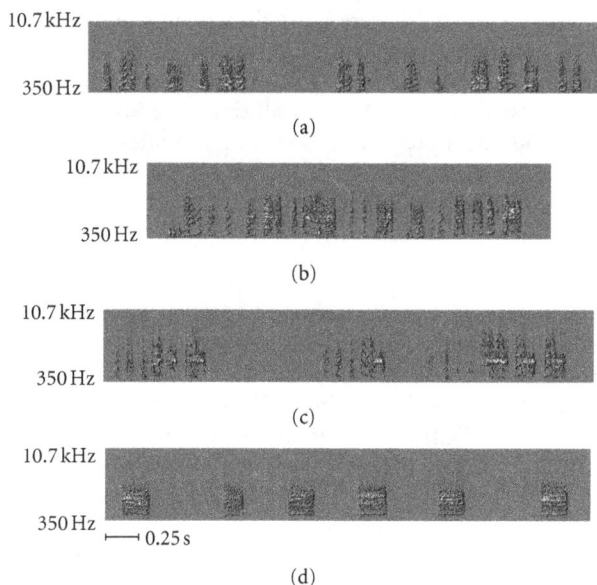

FIGURE 2: (a) Spectrogram of a juvenile's vocalizations produced during the babbling phase, at approximately 36 days posthatch. Note the general lack of stereotypy. (b) Spectrogram of a juvenile's vocalizations produced during the early plastic song phase, at 41 days posthatch. (c) Spectrogram of a juvenile's vocalizations produced during the plastic song phase, at 47 days posthatch. Although the sounds are more adult-like in terms of spectral profile, they still lack the stereotypy of adult birds. (d) Composite spectrogram of a series of calls from a juvenile zebra finch (40 days posthatch). By eye and by ear, these are easily differentiated from adult song.

juvenile singing, these tools do not perform well on samples from juveniles.

The simplest method of comparing song samples is to calculate some measures of correlation between samples, either on their waveforms or spectrograms. This method is employed by several popular tools [6, 7] but works adequately only if the sounds being compared are very similar in timing, ordering, and tone. A related technique is dynamic time warping (DTW) [8], which can compensate for differences in timing but not ordering. DTW-based analyses can be performed on spectrograms or spectrogram-derived measures, such as cepstra [9]. Another strategy, used by at least one popular tool [10], might be described as heuristic feature analysis. A set of measures (e.g., peak frequency, frequency modulation, and spectral entropy) is used to characterise a sample, and these measures are used to compare two samples according to some set of criteria. Although these tools typically do not require the samples being compared to be highly similar, it has been our experience that, with juvenile vocalizations, these methods can produce similarity scores that vary greatly between pairs of samples that, to a human observer, appear more or less equally similar.

The key feature that all these existing methods have in common is that they are designed to compare one single sample against another single sample. For highly stereotyped adult birdsong, this approach makes perfect sense, but, for juveniles, it may not be appropriate: the high variability of juvenile song means that two samples from the same bird, taken seconds apart, may not be "similar" in any reasonable sense, and yet both are representative of that animal. With a large enough sample set, however, we should be able to identify all the characteristic sounds produced by a bird and be able to describe new samples in terms of how typical they are, even if the new sample does not seem particularly similar to any other sample.

Other methods for song analysis exist, such as the spectrotemporal modulation analysis used by Ranjard et al. [11], the rhythm analysis of Saar and Mitra [12], or the PCA-based feature analyses of Feher et al. [13]; however these methods as presented are unaware of syllable sequencing [11] or are very highly specialized [12] and are not suitable for general-purpose use.

In this paper, we present a new method for comparing a sample of juvenile birdsong against a model built from a set of training samples. We call this method windowed spectral pattern recognition (WSPR). This method provides a measure of typicality for comparing test samples to the training samples. We show that WSPR is effective as a classifier and may be better suited to this task than another popular tool. We also show that WSPR is relatively robust to changes in a key parameter. Lastly, we demonstrate that the models produced by WSPR can be used to provide measures of song ontogeny, stereotypy, and complexity

2. Methods

2.1. Housing and Care of Juvenile Zebra Finches. Audio recordings from three juvenile male zebra finches provided the data used in this paper. From hatching until 25 days posthatch, the juveniles were housed with their mothers, fathers, and clutch mates in a family setting. From 25 days to

35 days, the juveniles were housed in small cohorts of 2–4 individuals along with an adult tutor. From 35 days to between 50 and 60 days, the juveniles were housed singly in auditory isolation chambers. At all times the juveniles were given food and water *ad libitum*. The juveniles were cared for in accordance with the standards set by the American Association of Laboratory Animal Care and Rockefeller University's Animal Care and Use Committee.

2.2. Recording of Juvenile Birds and Manual Identification of Samples. Continuous recordings were made of three isolated juvenile male zebra finches from 35 days posthatch to 60 days posthatch with Behringer ECM-8000 measurement microphones (Behringer International GmbH, Willich, Germany) and Rolls MP13 preamplifiers (Rolls Corporation, Murray, UT). A MCC PCI-DAS6013 digital acquisition card (Measurement Computing Corporation, Norton, MA) was used to digitise the audio inputs. Recordings were made at 44.1 kHz, 16 bits/sample, and stored as lossless FLAC [14] files.

We examined recordings with Audacity sound editing software [15] and manually identified vocalization bouts as being calls, song, or neither. Vocalization bouts identified as calls or song were eliminated if they contained excessive levels of spurious noise—flapping of wings, footfalls on metal bars, and the like—or if they were less than one second long. 2026 samples were taken from the three birds. Each bird's samples were assigned to one of four different sample sets: song training, song testing, call training, and call testing.

2.3. Building a Model and Scoring Using WSPR. A test model was built using the WSPR command-line tool from a combination of both training sample sets, using the parameters given in Table 1. During the clustering phase of the WSPR algorithm, the set of spectra that was clustered as well as their cluster assignments were extracted and silhouette statistics [16] were computed using the "cluster" package [17] for the *R* statistical computing environment [18]. For comparison, a random dataset was also generated and clustered. A set of 7 500 vectors, each the same length as the WSPR spectra, was produced, with every value in each vector being a randomly generated number from a $[0, 1]$ uniform distribution. This random dataset was clustered using *R*'s "*k* means" function, and silhouette statistics were computed as for the clustered spectra.

2.4. Binary Classification of Juvenile Vocalization Samples. For each bird, a binary classifier was constructed using the WSPR algorithm for classifiers described in the appendix. The classifier contained one model for song, built from the song training samples, and one for calls, built from the call training samples. The parameters used in the construction of these models are found in Table 1.

All testing samples were presented to the classifier. Samples were assigned to a group by the classifier, and the Matthews correlation coefficient (MCC) [19] was used to assess the accuracy of the assignments. The number of

TABLE 1: Parameters used in all examples, unless specified otherwise.

Parameter	Value
STFT window width	500 samples (11.6 msec)
STFT step size	100 samples (2.9 msec)
STFT bandpass cutoffs	500 Hz–7500 Hz
Model window width	11 symbols (34.0 msec)
Number of power spectra clustered	7 500
Number of prototypes generated	120
Silence cutoff level	0.01 (arbitrary units)

samples used as training and testing data for each bird, as well as mean sample lengths, is given in Table 2.

For comparison, Sound Analysis Pro+ (SA+) [10] was also used to classify samples from the first bird. From the 669 original samples, two hundred were randomly chosen, with fifty from each of the four sample sets (song training, song testing, call training, and call testing). The samples were loaded into the SA+ software and run in a series of pairwise comparisons using SA+'s "batch similarity" tool, so that each test sample was compared against one training sample from the "call" set and one training sample from the "song" set. SA+'s volume threshold was reduced, but otherwise was run with all settings at their default values. The calculated similarity scores were then exported from SA+ for statistical analysis.

When used as a classifier, the same classification method described in the appendix (Classification Using Multiple Models) was used on the SA+ scores, with the exception that the SA+-generated scores were used in place of WSPR's raw scores.

2.5. Measuring Song Ontogeny and Stereotypy. Recordings from the juveniles examined previously were taken, and, for each bird, two models were made: an early model, consisting of the earliest 100 song samples; a late model, consisting of the latest 100 song samples. The remaining samples from each bird were grouped by day and scored against the models.

For one juvenile, all samples were grouped into blocks of five consecutive days each, and models were generated for each group, and the standard error of the nonstandardized scores samples used to build the model against the model was calculated.

2.6. Testing the Effects of Parameter Selection on Score Distributions. Fifteen models were built with varying numbers of prototypes: $10, 20, \ldots, 150$. All models were built using the same set of song training data for the first bird as described previously. Except as noted in the results, the WSPR parameters are found in Table 1. Each sample from the first bird's song test data was scored against all 15 models. Means and standard deviations were calculated for the scores from each model.

2.7. Estimating the Stereotypy and Complexity of Sample Sets. Additional recordings were made of an adult zebra finch,

TABLE 2: Summary of sample set sizes used to build and test models.

	Bird 1	Bird 2	Bird 3
Total samples	466	569	991
Manually classified as song	166	150	500
Manually classified as calls	300	419	491
Used as song training data	50	100	150
Used as call training data	100	150	150
Used as song testing data	116	50	350
Used as call testing data	200	269	341
Average sample length, song	4.5 seconds	1.1 seconds	0.4 seconds
Average sample length, call	1.8 seconds	1.1 seconds	0.8 seconds

over 100 days old, with equipment and conditions identical to those used for the juvenile recordings, except that the DAQ digitiser was bypassed and the computer's built-in audio input was used instead. One hundred samples of adult song were manually identified and extracted from the recordings. For each of the three juvenile birds, the WSPR algorithm was used to generate separate models for song and calls on all available samples, including both training and testing data from the earlier experiments. For the adult bird, a model was generated for its song on the 100 collected samples. For all models, all samples were concatenated and the combined samples were truncated to a length of exactly two million audio samplings (approximately 45 seconds); each model was built from its corresponding concatenated sample. All samples were scored against the models they were used to train, and the standard deviations of all scores against each model were calculated. The models were generated using the parameters found in Table 1, with the following exceptions: STFT window width, 4096 samples; STFT step size, 1024 samples; model prototypes, 50; model window width, 25. The WSPR complexity of each model was also calculated according to the algorithm found in the appendix (Calculating the Complexity of a Model).

3. Results

3.1. Building a Model Using the WSPR Algorithm. Model building is composed of two discrete steps: creating an encoding and producing tables of observed frequencies of patterns. To create an encoding, a set of 100 samples of juvenile plastic song was taken from a single individual. Samples were converted from digitized waveforms into a frequency-versus-time representation (a spectrogram) using a discrete-time Fourier transform (DTFT), as illustrated in Figure 3(a).

From the set of all training samples' spectrograms, 7 500 spectra were chosen without replacement. These were clustered using a k-means clustering algorithm [20] into 120 clusters. The k-means clustering algorithm works to divide the 7 500 spectra into k clusters, with all the items in each cluster more similar to each other than to the members of any other cluster. Each cluster represents a single kind of "sound" that the bird makes: clusters may represent single notes, harmonic stacks, staccato bursts, or other types of sound.

The members of each cluster were averaged to produce a set of prototypical sounds, one prototype per cluster, and each prototype was assigned a unique index number (its "symbol"); these prototypes formed the basis for the encoding. Sample prototypes can be found in Figure 3(b).

The silhouette statistic [16] was used to characterize how well the clusters divided the underlying spectra. The silhouette statistic is a unitless value between −1 and +1; a silhouette value of 1 implies that an item is ideally clustered, a value of −1 implies that an item should be assigned to another cluster, and a value of 0 implies that an item could just as easily be assigned to another cluster as to its current cluster. For the clusters used to produce the prototypes, the mean silhouette value was 0.264. In contrast, the mean silhouette value for a randomized dataset was 0.0252. This suggests that many clusters are only moderately separated from their neighbors, which is reasonable given the large number of clusters and the high variability of the underlying bird vocalizations.

Sounds were encoded by first converting from waveform to frequency-versus-time representation, as before. Each discrete frequency spectrum was compared to the full set of prototypes, and the spectrum was encoded as the index number of the prototype it was most similar to (determined by root mean square deviation). Each sample was thus converted from a waveform, to sequence of frequency spectra, to a sequence of symbols. With this, the encoding step of building a model was completed. The average sample was 2.42 seconds long; once encoded, the average sample was 1063 symbols long.

The second part of the model-building process is the one in which patterns in the bird's song are identified. A window width (w) of 11 was set, and an anchor position (a) of 6 was calculated. An array of dimension $120 \times 11 \times 120$ was created; all values in the array were set to zero. A count was tallied of the number of times symbol y was seen at position z, given that symbol x was seen at position a, for all x, y, and z, by scanning each sample and tallying the observed symbols.

3.2. Scoring a Sample. A single test sample was first encoded using the same encoding method described for model building. After encoding, the sample was scanned over in a manner very similar to how the frequency array was built; however, instead of modifying the array, the values in the

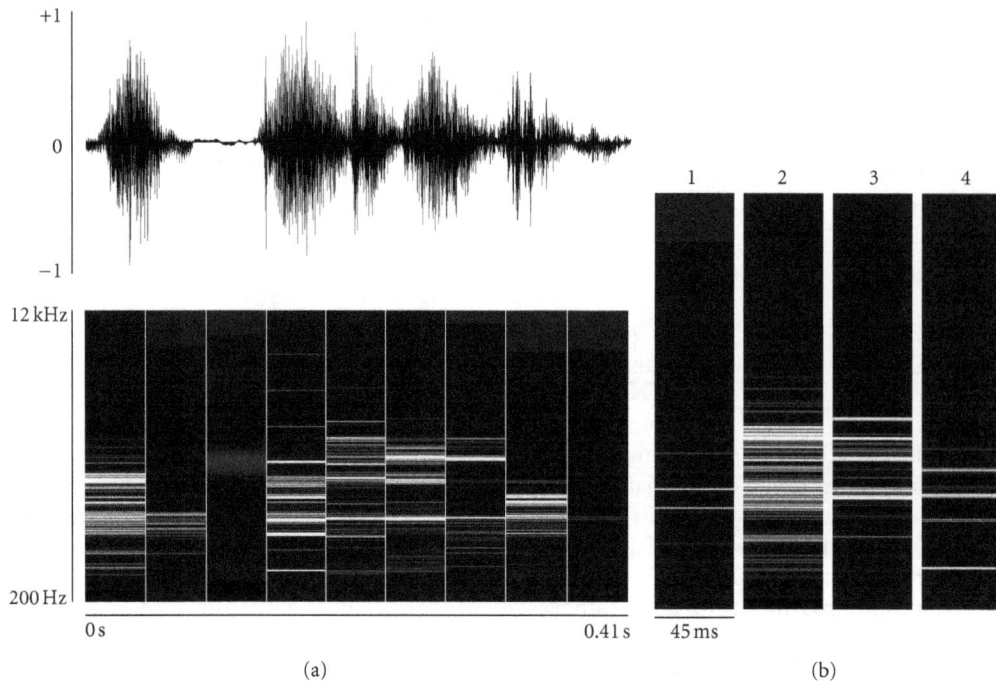

FIGURE 3: (a) Using a discrete Fourier transform, sounds are converted from waveform (top pane) to a sequence of frequency spectra (bottom pane)—in essence, a spectrogram. Note that each discrete frequency spectrum accounts for a period of time much larger than the sampling rate; the effect is exaggerated here to make this clear. (b) Examples of possible prototypes. In the WSPR algorithm, every segment of sound will be matched to a similar prototype and coded as the prototype's index number (1, 2, 3, 4, etc.)

array were incorporated into a score, so that more common sequences of symbols will score higher than less common ones. The exact formula used is described in the appendix. Once the nonnormalized (raw) score was generated, it was standardized (as a z-score) in order to make the scores easier to interpret. The standardization procedure is also described in the appendix. A single arbitrary sample produced a raw score of 0.34, a z-score of 0.45, and a P value of 0.32, implying that the sample was fairly typical of the model's training data, which in this case was to be expected, as the test sample was identified by the authors as being qualitatively "of a kind" with the training data.

3.3. Binary Classification.
The motivation for developing this method was to quickly classify very large sets of recorded samples, so it seemed fitting to examine its fitness for this purpose.

In the authors' recording setup, juveniles were recorded continuously, twenty-four hours a day. It was not possible to listen to all of this audio—indeed, for months, recordings were being accumulated much faster than a single person could listen to, even if that person listened to them every minute of every day.

A simple amplitude threshold check was able to eliminate most of the recordings; however, this still left tens of thousands of audio events—samples—that needed to be examined. One of the primary goals in developing the WSPR tool was to create a reasonably robust tool that could sort through such large sample sets in minutes or hours, rather

than days, and further reduce the amount of work that would need to be done manually.

A timing test was able to show that the WSPR algorithm was indeed suitable for use with such large datasets. A model was built from 200 samples; on a reasonably fast machine (Intel Core i7, 2.67 GHz clock speed), the model-building process took 12.1 seconds. Scoring 200 samples against that model took only 1.6 seconds. Assuming those 200 samples are representative of a larger set, it would take about 15 minutes to score 100 000 samples. By contrast, 200 pairwise comparisons were done using the SA+ program with the same sample set. These 200 comparisons took 593 minutes to complete on the same machine. Scaling up, comparing 100 000 sample pairs would take about 200 days to complete.

While speed is important, it is of little use if the results are inaccurate. To test WSPR's accuracy, a WSPR classifier was built comprised of two models, one of "call" samples, and one of "song" samples. Figure 4 shows the raw scores against both "call" and "song" models for the test data in scatter plots. The MCCs for the classifications of each bird's samples were 0.93, 0.75, and 0.65, with a cumulative MCC of 0.78.

It is also worth comparing the accuracy of WSPR classifications to SA+ scores. Figure 5 shows the raw scores produced by the SA+ program. The MCC for the SA+-based classifier was 0.57, somewhat less than that for the WSPR classifier. On this task, WSPR made about 1/3 as many classification errors as SA+, although both produced fairly good results.

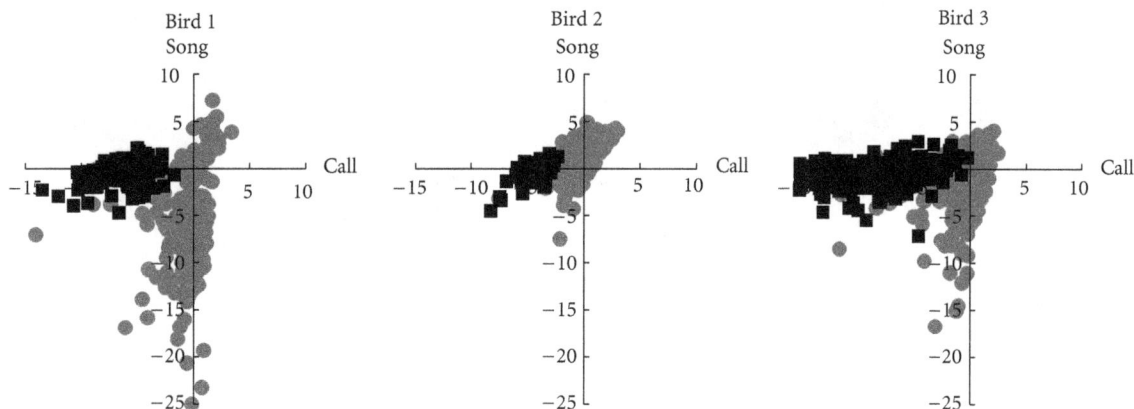

FIGURE 4: Performance on a classification task using standardized z-scores. Each test sample was scored against both "song" and "call" models. Gray points were manually assigned to the "call" class, while black points were manually assigned to the "song" class. For all birds, the two classes of sounds are well separated.

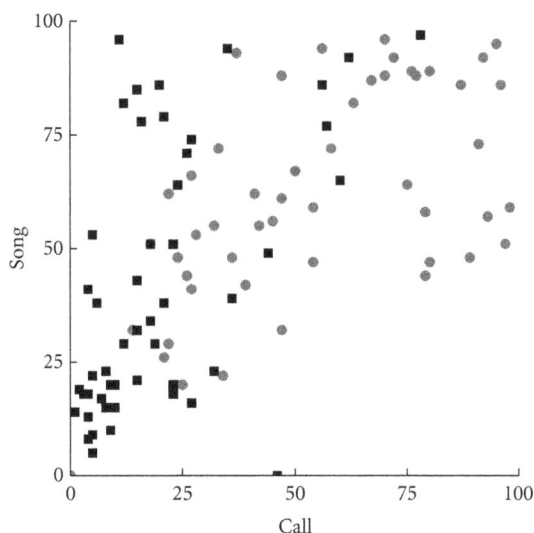

FIGURE 5: SA+ scores, performance on a classification task. Black points were manually assigned to the "song" class, and gray points were manually assigned to the "call" class.

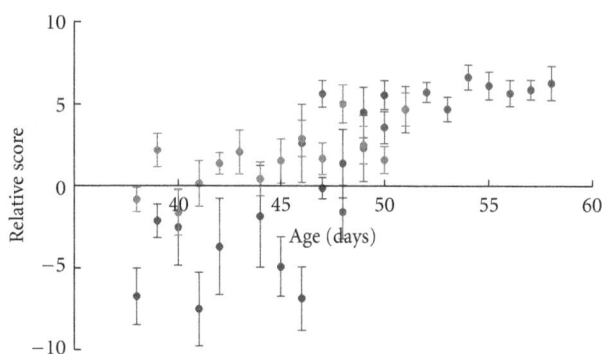

FIGURE 6: Measuring progress in song development. Each colour represents a different bird. For each bird, an early model and late model were built of the first and last 100 samples available; all other samples were compared against both models and their difference calculated, so that negative scores suggest a sample was more typical of the early model, and positive scores suggest a sample was more typical of the late model. Each point is the mean of all samples for that day, and error bars indicate standard error. All birds progress from being essentially early-like to being late-like, but unevenly and at different rates.

3.4. Song Onteny.

3.4. Song Ontogeny. In addition to its use as a classifier, the WSPR tool may also be useful for more analytical tasks. To that end, a test was devised in which WSPR was used to track the ontological development of three juvenile zebra finches.

To do this, once again two models were created for each finch, one from a set of early samples, near day 35, and one from a set of later samples, near day 50. Sets of intermediate samples were then taken, organized by day, and scored against each model. The difference between these two scores, specifically the late-model score minus the early-model score, indicates the extent to which the test sample was more typical of the late model than the early model.

Figure 6 shows that the bird's songs do progress over time towards similarity with each bird's late model. According to the scores, the birds' songs develop unevenly at times and at

different rates, an observation in accord with the authors' personal experiences.

WSPR might also be used to measure stereotypy, a task it seems well suited for given its emphasis on large sample sets. One simple and intuitive measure of stereotypy using WSPR would be the standard deviation or standard error of a sample set against a model; this is the measure used here. This measure is essentially one of variability: the lower the standard error, the greater the stereotypy.

The samples from one bird were grouped into five-day periods, and a model was built for each period. The samples were then scored against their models, and the standard error of the scores was used as a measure of apparent stereotypy. Figure 7 shows the change in standard error for scores as a bird's song develops. As one would expect, variability

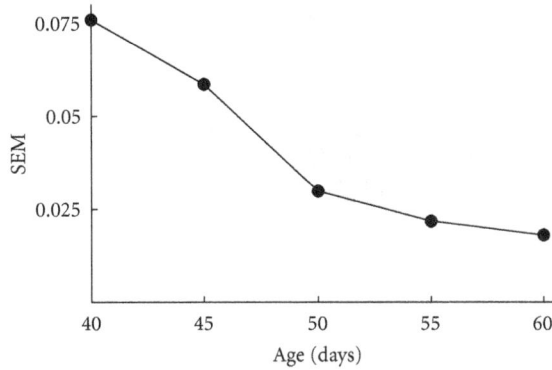

FIGURE 7: Nonstandardized score standard error as a bird develops its song. As the bird matures, the variability in its singing decreases.

decreases and stereotypy increases as the bird ages. There is a noticeable decrease in the rate at which variability declines around day 50.

3.5. Effect of Parameter Selection on Scores. It is important to know how sensitive the WSPR algorithm is to changes in parameters. There is a possibility that small changes in parameters might lead to large changes in scoring accuracy, a situation that would pose a practical problem for the use of the algorithm. There are three key parameters in the model that can be manipulated: the width of the STFT window, the number of prototypes, and the width of the model window.

Two of these parameters, the width of the STFT window and the width of the model window, are determined by the data being analyzed: for the width of the STFT window, the expected maximum length of time over which a sound would be approximately constant; for the width of the model window, the expected length over which patterns would be identifiable. As such, one would expect the scores to vary considerably as these parameters are changed; furthermore, the structure of the data should suggest ranges for these parameters.

There is therefore only one major parameter remaining that must be set in an ad hoc fashion: the number of prototypes. There are potential problems with having either too few or too many prototypes. If there are too few prototypes, sounds with qualitatively different spectral profiles will be assigned to the same prototype, the specificity of the encodings will fall, and the model may produce additional false positives. If there are too many prototypes, sounds that are qualitatively similar will be assigned to different prototypes and the number of false negatives produced will rise.

Figure 8 shows how the mean score and standard deviation of scores change in relation to the number of prototypes used to build the model. It can be seen that both the score means, and to a lesser extent, the standard deviations, level off when more than roughly 100 prototypes are used. This suggests that the method is insensitive to this parameter as long as sufficiently large set of prototypes is used.

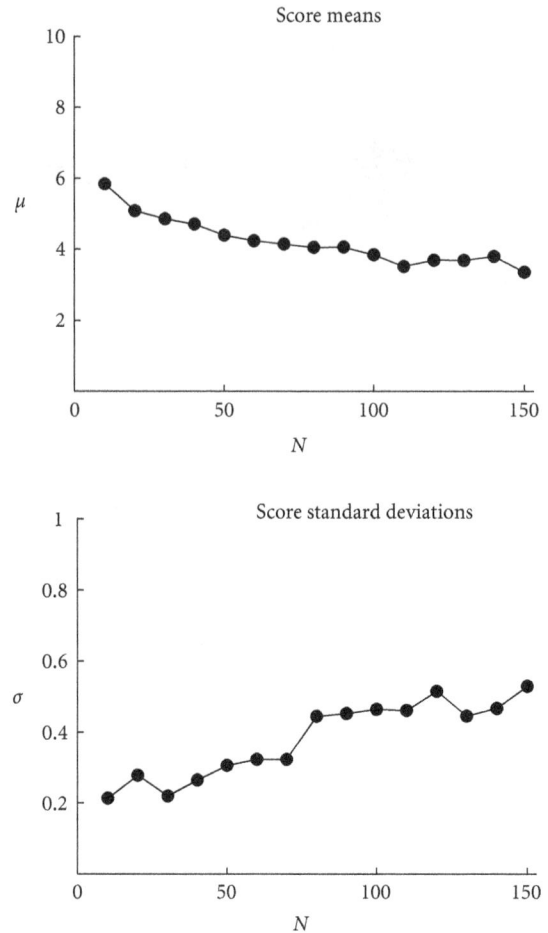

FIGURE 8: Scores and standard deviations as a function of the number of prototypes used (N). Scores were raw (nonstandardized). Neither scores nor standard deviations change abruptly in the face of small changes to the number of prototypes.

3.6. Stereotypy and Complexity. Finally, there is the possibility of using the WSPR algorithm as a basis for measuring the complexity of a bird's song. Exactly what is meant by "complexity" in the context of birdsong is open to debate, but most researchers would probably agree it involves the number of distinct sounds an animal makes, as well as the patterns of those sounds. For many years, people have used informal measures of song complexity, such as the number of distinct notes or syllables in a song [21]. In addition to a high degree of subjectivity, these measures can be difficult to apply to birds with variable songs, such as juveniles or species that improvise when singing. As an alternative, WSPR can be used to generate a measure of complexity based on ideas from statistical complexity theory and information theory.

A WSPR model contains a considerable amount of information about the sounds in a bird's repertoire and the likely sequences of sound it will produce—exactly the kind of information needed to measure complexity. Our method mines a WSPR model to produce a measure of the model's complexity, which is in turn a reflection of the complexity of the sample set used to build the model (see the appendix).

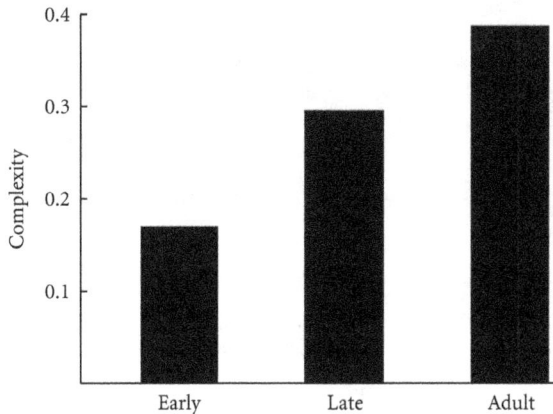

FIGURE 9: Complexity measured for early juvenile, late juvenile, and adult. As birds age, the apparent complexity of their song increases.

This method was tested against vocalizations from a single bird as it learned to produce its song.

Figure 9 shows complexity scores for models built on early juvenile, late juvenile, and adult models. There is a marked rise in complexity as the bird's song develops, which coincides with intuitive expectations.

4. Discussion

4.1. About the Method. The WSPR algorithm attempts to identify recurring patterns in the vocalization samples submitted as training data. It does not need a high degree of top-level similarity, nor does it look for any particular features. At its heart, the algorithm is built upon a simple expression of conditional probability: given that at this moment the sample sounds like x, what are the probabilities of the sounds heard in the preceding and following moments?

WSPR breaks the sample into short segments and, using spectral analysis techniques, identifies the significant frequency components of each segment. It then estimates how probable such a frequency profile might be and how probable its neighbouring profiles are. The probability estimates are based on the distributions observed in the training data.

4.2. On the Use of k-Means Clustering to Divide Data. How well does k-means clustering divide the data? The mean silhouette statistic calculated for our initial model was 0.264, suggesting that the underlying data clusters only moderately well. The motivation however for using k-means clustering here was not to identify clusters but to discretize the data in a reasonably natural way, to the extent that natural partitions may exist within the data. If natural partitions do not exist, the data will simply be divided into neighboring discrete regions. Other partitioning strategies could be employed: a simple partition of the data range into a large number of evenly sized hypercubes, for example, could also be used to discretize the data, although it runs the risk of having most or all of the data fall into a single hypercube. In practice, we have found k-means clustering to produce a reliable discretization of the data sets we have used.

4.3. On the Meaning of Scores. Scores are best thought of as a quantitative measure of how typical each segment of the sample is and how typical its surrounding segments are as compared to the training data. Raw scores exist on a scale that is unique to the model that produces them, and so raw scores cannot be compared across models. As a result, it will generally be preferable to use standardized z-scores. These are standardized against the distribution of scores from the training data: a z-score of 0.0 means that a sample scored as highly as the average sample from the training data; a z-score of 1.0 means that a sample scored one standard deviation higher than the average sample from the training data; a z-score of -1.0 means that a sample scored one standard deviation lower than the average sample from the training data. It is important to note that, although scores are related to probabilities, they are not and cannot be used as expressions of probability. Estimated P values, however, can also be calculated. These P values estimates are for the two-sided hypothesis that a random score for a sample from a pool like the training data would be more extreme than the current score, assuming a normal distribution of scores: users should verify that the scores produced by a model are approximately normally distributed before accepting the P values estimates.

4.4. Performance of WSPR Compared to SA+ as a Classifier. On the data sets used in this paper, the algorithm categorises samples as song or call correctly about 92% of the time. Compared to SA+, the algorithm makes about 1/3 as many assignment errors, a substantial improvement.

There are some caveats in the use of SA+ as a classifier, and some details that must be discussed regarding how it was used in this paper. SA+ was used to compare one single test sample against two single training samples, one from each category. In contrast, the algorithm described here compares a test sample against a digest of dozens or hundreds of training samples. A fairer comparison would involve using SA+ to compare a test sample against a large set of training samples and then using some averaging function to generate a score against each category. This is not typically how SA+ is used however, and SA+'s computationally intense method makes this infeasible: on a fast computer (2.6 GHz), comparing a one-second sample against 100 would take about 40 minutes. To classify a large group of samples in this manner, say 10 000, would take an unreasonable amount of time; hence we consider that the way in which SA+ was used here as a classifier is a fair reflection of how it would be used in practice.

4.5. Using Models to Estimate Stereotypy and Complexity. Aside from classification and general scoring tasks, the models produced by the WSPR algorithm can also be used to provide two measurements about the training data that may be of interest: stereotypy and complexity.

We propose that stereotypy can be thought of as a low degree of variance between samples. A low standard deviation of the scores of a sample set against a model provides the most direct measure based on this idea.

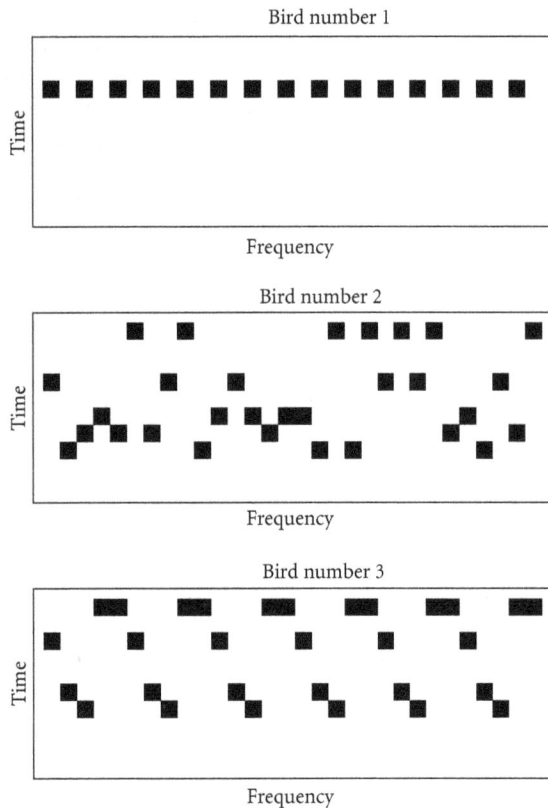

Bird number 1

Bird number 2

Bird number 3

FIGURE 10: Hypothetical spectrograms for three birds. Bird number 1 has a song with a single note, bird number 2 has five notes but no pattern, and bird number 3 has five notes and a clear recurrent pattern

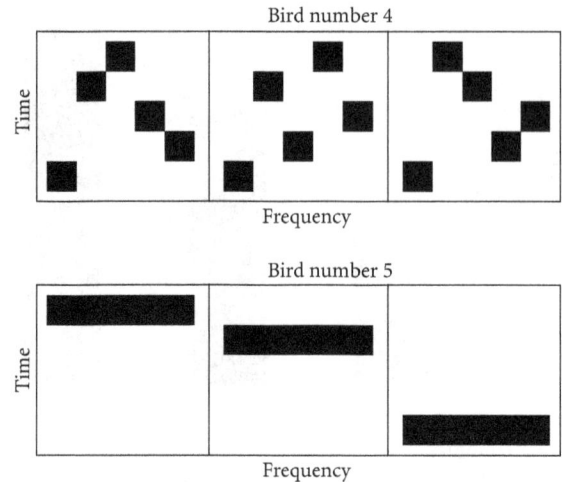

Bird number 4

Bird number 5

FIGURE 11: An illustration of why mutual information may not adequately capture intuitive notions of song complexity. Although the authors believe most people would agree that bird number 4 has a more complex song than bird number 5, the songs of both birds can have equal mutual information.

We can also define a measure of song repertoire complexity, by looking to the field of statistical complexity for inspiration. Measures of statistical complexity attempt to quantify the "structuredness" of a system or process. There is no consensus as to what exactly this means or how it could be best measured, but most proposals generally consider the number of parts in a system as well as the relationships between those parts.

Let us consider this idea of complexity using several examples involving birdsong, illustrated with artificial examples in Figure 10. Bird number 1 has a one-note repertoire, and his song consists of repetitions of this note. He has a highly regular song, but it is very simple. His "system" has only a single "part" and we would suggest that his song is not complex.

Bird number 2 has a five-note repertoire; he sings randomly and each note is sung about 20% of the time. Although the song of bird number 2 has many parts, there are no relationships between these parts—each part is independent of the others, and no patterns emerge from his song beyond the individual notes. We would argue that, although the five notes make this bird's song more complex than bird number 1, he also has a fundamentally simple song structure.

Bird number 3 also has a five-note repertoire, but he sings with sequences of notes that appear regularly. Here, there are

meaningful relationships between parts: some notes follow others at rates much higher than random chance. We would argue that this bird has what most observers would agree to be a more complex song structure.

Is there an existing measure that could be used for the purpose of measuring song complexity? Several measures of statistical complexity exist, for various problem domains [22–24], but the one that seems most relevant to birdsong is the measure of predictive information described by Bialek et al. [25]. To paraphrase, predictive information is how much more you know about the future states of a system upon learning about its past states. If combined with a measure of the number of different states (to prevent bird number 1 from receiving a high complexity score due to the high predictability of his song), predictive information is in accord with our intuitive ideas about birdsong complexity, wherein birds with regular patterns of notes make it possible to predict the note sequence, and the more extensive the patterns are, the more that can be predicted. This measure would also be in accordance with our intuitive notions about the complexity of the songs of the three birds discussed above.

While predictive information seems like a good fit, Bialek et al. [25] use mutual information [26] as their underlying measure, and, under some circumstances, this may lead to counterintuitive results. For example, mutual information would consider a bird with five songs, each containing a different order of five different notes (bird number 4 in Figure 11), just as simple as a bird with five songs of one different note each (bird number 5 in Figure 11): both are equally predictable. For birdsong, a more appropriate underlying measure might be the Kullback-Leibler divergence [27], a measure of the difference between two probability distributions. The Kullback-Leibler divergence (KL divergence) would identify the former bird's repertoire as being more complex than the latter's. By using the KL divergence, we are subtly exchanging the idea of predictive information for a related but different one: the extent to which the past predicts changes in the future. It is our opinion that, of all the measures considered, the KL divergence most closely reflects intuitive ideas about song complexity. The measure we propose uses the data about sound distributions contained in the WSPR models to estimate the structural complexity of a sample set. It is essentially a mean of multiple KL divergences of distributions at different time intervals in the model. The exact formula can be found in the appendix.

An important consideration is that, for any measure, timing and the rate at which song features change are crucially important. On very short timescales, such as microseconds, a bird's song does not change much at all, and the correlation between past and present is total. On very long timescales, such as hours, the correlation between past and present singing behaviour is essentially zero. In between these extremes is a narrow range of timescales that optimally reveal the structure of the bird's song. We have not devised a satisfactory method for automatically identifying these optimal timescales and so can only recommend that care be taken in choosing timing parameters when attempting to measure song complexity using the method we propose.

4.6. Known Issues and Future Directions. One consideration when using WSPR is that background noise in recordings can be problematic: the algorithm does not distinguish background noise from vocalizations or any other noise of interest, and the background noise profile becomes built into the model. In the worst case, models built from samples with significant background noise may assign low scores to test samples simply because the background noise is different. To avoid this problem, noisy recordings should be denoised before either building a model or scoring against an existing model, or recording conditions should be managed to ensure a consistent level of background noise between training samples and test samples. With a sufficiently low level of background noise, WSPR models built from one dataset should be useable with datasets made in different recording environments, enabling the sharing of WSPR models between laboratories.

There are several important points to consider when using the measure of complexity we have provided. The measure is highly sensitive to the size of the sample set, so to make scores comparable across models, we recommend using exactly the same total length of sound to build each model. The measure can also be misled by extended periods of silence, especially if these frequently appear at the beginning or end of samples. To avoid this, we recommend trimming silent intervals from the ends of all samples. The model will also add implicit silence to the beginnings and endings of samples as necessary to make them at least as long as the window size of the model, so we recommend concatenating all samples that are less than twice as long as the window size before building a model.

WSPR bases its analysis on sound spectrograms, in part because this seems the most "natural" interpretation of the data, although many other song analysis tools have found success using higher-order measurements derived from the spectrograms [9–11]. One interesting extension to the work presented here would be to incorporate these higher-order measurements into the WSPR algorithm instead of or in addition to the spectrograms to see if scoring and classification accuracy could be improved further still.

5. Conclusions

In this paper, we have presented a novel method for comparing samples of birdsong against a larger set of samples, WSPR. WSPR is designed to cope with sample sets with low levels of stereotypy, an application that we feel no existing tool adequately addresses. We then extended this method to demonstrate a number of applications: classification problems, the original motivation behind the method's development; tracking song ontogeny; measuring song variability and complexity. We believe that the measure of birdsong complexity presented here represents the first effort of its kind.

Although the methods described in this paper are useable as they are, it is our hope that they may also serve as starting points for further discussion: in general, discussions about analyzing animal vocalizations and algorithms for doing so; in particular, discussions about what complexity means in the context of animal vocalization and how best to measure it.

Appendix

Details of the WSPR Algorithm

The preliminary step in building a model is to produce an encoding scheme for the sounds to be considered by the model. This is a form of vector quantization [28], in which the infinite variety of sounds is reduced to a finite set, or codebook, of representative sounds.

To generate a codebook of size n from a sample set of sounds:

(i) Convert each sample into power spectra via short-time Fourier transform.

(ii) Take $100n$ random samples of power spectrum from all spectrograms.

(iii) Cluster into n groups using k-means clustering.

(iv) For each group:

   (a) Find the geometric mean of each frequency band across all samples in the cluster.

   (b) Normalise so that the mean spectra has a total power of 1.

   (c) The result is the prototypical spectral profile for the group.

   (d) Prepend a "null" spectrum of zeros to the beginning of the prototypes.

All samples must be transformed into a frequency-versus-time representation (a spectrogram) using a discrete-time short-time Fourier transform (STFT) [29]. Variations on the classical STFT are also acceptable; a STFT using a Gaussian window is used in our implementation of the algorithm. Phase information is discarded, as well as parts of the spectra outside a specific band of interest, that is, below 500 Hz or above 7500 Hz.

In cases where one wants to build two or more models from the same underlying encoding, the samples used to generate the codebook should be taken from the joint sample set of the two models.

Encoding Samples. Having constructed a codebook, all samples must now be encoded. To encode a sample:

(i) Convert via STFT to power spectra.

(ii) For each power spectrum s in the spectrogram:

    (a) If the total power in s is below the cutoff threshold, emit a zero (the null spectrum).

    (b) Otherwise, examine the codebook, P, to find the spectrum, p, where the root mean square deviation between s and p is minimized.

    (c) Emit the index number of p, that is, if p is the 3rd spectrum in the codebook, then emit a 3.

    (d) The sequence of emitted numbers is the encoding of the sound.

Constructing the Model. Perform the following steps to construct a model using the encoded samples:

(i) Choose a sliding window length, w. Define a as the middle (anchor) position of the window.

$$a = \left\lfloor \frac{w+1}{2} \right\rfloor. \tag{A.1}$$

(ii) Create an array M of dimension $n \times w \times n$ and a vector T of length n. Initialise all values in M and T to 0. For every encoded sample e, create a set R of all possible subsequences of e of length w.

(iii) For each r in R, perform the following:

$$T_{r_a} \longleftarrow T_{r_a} + 1,$$
$$M_{r_a,j,r_j} \longleftarrow M_{r_a,j,r_j} + 1 \quad \forall j \in \{1,\dots,w\}, \tag{A.2}$$

where r_a is the symbol at the anchor position in r.

(iv) Create an array M^*, same size as M; vector T^*, same size as T.

(v) For i in $1,\dots,n$, perform the following computation:

$$T_i^* = \frac{T_i}{\sum_{j=1}^n T_j}. \tag{A.3}$$

(vi) For j in $1,\dots,w$, k in $1,\dots,n$, perform the following computation:

$$M_{i,j,k}^* = \frac{M_{i,j,k}}{\sum_{l=1}^n M_{i,j,l}}. \tag{A.4}$$

The tuple of $W = (P, T^*, M^*)$ constitutes the constructed model, where r_a is the symbol at the anchor position in r.

Scoring a Sample against the Model. Finally, we must be able to compute a score of a test sample against a model. Perform the following steps to score a test sample against a model W:

(i) Convert the sample to a spectrogram via STFT.

(ii) Encode the sample as described previously, creating encoding e.

(iii) Create a set Q, containing every subsequence of e of length w.

(iv) For each q in Q, calculate the following:

$$\zeta(q, M^*) = T_{q_a} \left(\left(\prod_{j=1}^w M_{q_a,j,q_j}^* \right)^{1/w} \right). \tag{A.5}$$

(i) The score of the sample is

$$Z(Q; W) = \ln \left(\left(\prod_{i=1}^{\text{length}(Q)} \zeta(q_i, M^*) \right)^{1/\text{length}(Q)} \right). \tag{A.6}$$

$Z(Q;W)$ is the non-standardized ("raw") score of Q against the model W.

Standardization of Scores and Estimation of P-Values. After the model is built, every sample in the training set is scored against the model. The mean (μ) and standard deviation (σ) of these raw scores are calculated and stored along with the model. When a test sample is scored against the model, its normalised z-score can be computed as

$$Z^*(Q; W) = \frac{\mu - Z(Q; W)}{\sigma}. \tag{A.7}$$

To calculate an estimated P value, the CDF of a normal distribution with mean μ and standard deviation σ is used to determine the proportion of the distribution that is more extreme than the test score.

Classification Using Multiple Models. Suppose we have x known classes $(1, 2, 3, \dots)$ to which we wish to assign samples and training data sets $\{S_1, S_2, \dots, S_x\}$. We begin by building a joint set of prototypes for all samples from $\{S_1, S_2, \dots, S_x\}$ as described previously. Then, for each set of samples S_i, we build a model W_i using the algorithm described previously.

We calculate the means and standard deviations for each sample set against each model, producing a $x \times x$ table for each statistic:

$$\mu_{i,j} = \text{mean}\big(Z\big(s; W_j\big)\big),$$
$$\sigma_{i,j} = \text{sd}\big(Z\big(s; W_j\big)\big), \quad \forall s \in S_i, \; j \in \{1,\dots,x\}. \tag{A.8}$$

The tuple $(P, \{W_1, W_2, \dots\}, \mu, \sigma)$ constitutes the classifier.

When a sample Q is submitted for classification, we calculate the raw score $Z(Q; W_j)$ for all W_j. Then, we calculate the typicality of $Z(Q; W_j)$ relative to each training set S_i:

$$Y(Q, i, j) = \frac{Z(Q; W_j) - \mu_{i,j}}{\sigma_{i,j}} \quad \forall j \in \{1, \ldots, x\}. \quad (A.9)$$

Finally, we calculate the "atypicality" (or deviation of typicality) for each class as

$$A(Q, j) = \sqrt{\sum_{j=1}^{x} Y(Q, i, j)^2}. \quad (A.10)$$

The i for which $A(Q, i)$ is lowest is the class to which Q is assigned. This method works not by assigning a sample to the model for which it is most typical but by assigning a sample to the class of samples whose scores are most similar across all the models; in practice this seems to provide an improvement in accuracy.

Calculating the Complexity of a Model. Take T^* and M^* from a model W. Recall that T^* is of length n and M^* is of dimension $n \times w \times n$. Then,

$\Gamma(T^*, M^*, k)$

$= \left\{ T_i^* M_{i,j,k}^* \right\}$

$= \{\gamma_1, \gamma_2, \ldots, \gamma_{n \times n}\} \quad \forall i \in \{1, \ldots, n\}, \ j \in \{1, \ldots, n\}.$

$(A.11)$

That is, if $M_{i,j,k}^*$ is the probability of seeing symbol j at position k given that symbol i is at the anchor position, then $\Gamma(T^*, M^*, k)$ for each k is the set of all elements $M_{i,j,k}^*$ for that value of k, each multiplied by the probability of seeing symbol i.

Given the definition of the Kullback-Leibler divergence as

$$D_{\mathrm{KL}}(P, Q) = \sum_i P(i) \log_2 \frac{P(i)}{Q(i)}, \quad (A.12)$$

then our measure of song complexity is calculated as follows:

$C_{\mathrm{song}}(T^*, M^*, x, y)$

$= \frac{1}{y - x + 1} \sum_{i=x}^{y} D_{\mathrm{KL}}(\Gamma(T^*, M^*, a+1), \Gamma(T^*, M^*, a+i+1)),$

$(A.13)$

where M^* and T^* are the components of a model W, a is the anchor position of that model, and x and y are the start and end of a range of positions in the model forward of the anchor position for which the complexity is to be calculated.

Availability of Tools Implementing the WSPR Algorithm

Implementations of the WSPR algorithm in C++ and Mathematica are available for download; a web-based front end has also been developed to facilitate easy access to the WSPR tool. These can all be found at http://wspr.rockefeller.edu/wspr/.

Acknowledgments

This work is supported by the Neurosciences Research Foundation, the Rockefeller University, and the Tri-Institutional Training Program in Computational Biology and Medicine. The authors thank Professor Fernando Nottebohm and members of the Laboratory of Animal Behavior for helpful discussions and the staff of the Rockefeller University Field Research Center for their assistance.

References

[1] P. Marler, "Song learning: the interface between behaviour and neuroethology," *Philosophical Transactions of the Royal Society of London Series B*, vol. 329, no. 1253, pp. 109–114, 1990.

[2] F. Nottebohm, "The origins of vocal learning," *American Naturalist*, vol. 106, pp. 116–140, 1972.

[3] J. Böhner, "Song learning in the zebra finch (*Taeniopygia guttata*): selectivity in the choice of a tutor and accuracy of song copies," *Animal Behaviour*, vol. 31, no. 1, pp. 231–237, 1983.

[4] A. J. Doupe and P. K. Kuhl, "Birdsong and human speech: common themes and mechanisms," *Annual Review of Neuroscience*, vol. 22, pp. 567–631, 1999.

[5] K. Immelmann, N. W. Cayley, and A. H. Chisholm, *Australian Finches in Bush and Aviary*, Angus and Robertson, Sydney, Australia, 1967.

[6] R. Specht, Avisoft-SAS lab pro., 2004.

[7] The Cornell Lab of Ornithology, Raven: Interactive sound analysis software, 2010.

[8] H. Sakoe and S. Chiba, "Dynamic programming algorithm optimization for spoken word recognition," *IEEE Transactions on Acoustics, Speech, and Signal Processing*, vol. 26, no. 1, pp. 43–49, 1978.

[9] J. A. Kogan and D. Margoliash, "Automated recognition of bird song elements from continuous recordings using dynamic time warping and hidden Markov models: a comparative study," *Journal of the Acoustical Society of America*, vol. 103, no. 4, pp. 2185–2196, 1998.

[10] O. Tchernichovski, F. Nottebohm, C. E. Ho, B. Pesaran, and P. P. Mitra, "A procedure for an automated measurement of song similarity," *Animal Behaviour*, vol. 59, no. 6, pp. 1167–1176, 2000.

[11] L. Ranjard, M. G. Anderson, M. J. Rayner et al., "Bioacoustic distances between the begging calls of brood parasites and their host species: a comparison of metrics and techniques," *Behavioral Ecology and Sociobiology*, vol. 64, no. 11, pp. 1915–1926, 2010.

[12] S. Saar and P. P. Mitra, "A technique for characterizing the development of rhythms in bird song," *PLoS One*, vol. 3, no. 1, Article ID e1461, 2008.

[13] O. Feher, H. Wang, S. Saar, P. P. Mitra, and O. Tchernichovski, "De novo establishment of wild-type song culture in the zebra finch," *Nature*, vol. 459, no. 7246, pp. 564–568, 2009.

[14] J. Coalson, FLAC—free lossless audio codec, 2007.

[15] Audacity Development Team, Audacity: free audio editor and recorder, 2010.

[16] P. J. Rousseeuw, "Silhouettes: a graphical aid to the interpretation and validation of cluster analysis," *Journal of Computational and Applied Mathematics*, vol. 20, no. C, pp. 53–65, 1987.

[17] M. Maechler, P. J. Rousseeuw, A. Struyf, and M. Hubert, Cluster analysis basics and extensions, 2005.

[18] Team RDC, R: a language and environment for statistical computing, 2010.

[19] B. W. Matthews, "Comparison of the predicted and observed secondary structure of T4 phage lysozyme," *Biochimica et Biophysica Acta*, vol. 405, no. 2, pp. 442–451, 1975.

[20] J. A. Hartigan and M. A. Wong, "Algorithm AS 136: a K-means clustering algorithm," *Journal of the Royal Statistical Society Series C*, vol. 28, pp. 100–108, 1979.

[21] D. C. Airey and T. J. DeVoogd, "Greater song complexity is associated with augmented song system anatomy in zebra finches," *NeuroReport*, vol. 11, no. 8, pp. 1749–1754, 2000.

[22] G. Tononi, O. Sporns, and G. M. Edelman, "A measure for brain complexity: relating functional segregation and integration in the nervous system," *Proceedings of the National Academy of Sciences of the United States of America*, vol. 91, no. 11, pp. 5033–5037, 1994.

[23] J. P. Crutchfield and K. Young, "Inferring statistical complexity," *Physical Review Letters*, vol. 63, no. 2, pp. 105–108, 1989.

[24] P. Grassberger, "Toward a quantitative theory of self-generated complexity," *International Journal of Theoretical Physics*, vol. 25, no. 9, pp. 907–938, 1986.

[25] W. Bialek, I. Nemenman, and N. Tishby, "Predictability, complexity, and learning," *Neural Computation*, vol. 13, no. 11, pp. 2409–2463, 2001.

[26] T. M. Cover, J. A. Thomas, and J. Wiley, *Elements of Information Theory*, John Wiley & Sons, New York, NY, USA, 1991.

[27] S. Kullback and R. A. Leibler, "On information and sufficiency," *The Annals of Mathematical Statistics*, vol. 22, pp. 79–86, 1951.

[28] R. M. Gray, "Vector quantization," *IEEE ASSP Magazine*, vol. 1, no. 2, pp. 4–29, 1984.

[29] S. H. Nawab and T. F. Quatieri, "Short-time fourier transform," in *Advanced Topics in Signal Processing*, J. Lim and A. Oppenheim, Eds., pp. 289–337, Prentice Hall, Upper Saddle River, NJ, USA, 1987.

11

A Comparative Study of Human Thermal Face Recognition Based on Haar Wavelet Transform and Local Binary Pattern

Debotosh Bhattacharjee, Ayan Seal, Suranjan Ganguly, Mita Nasipuri, and Dipak Kumar Basu

Department of Computer Science and Engineering, Jadavpur University, Kolkata 700032, India

Correspondence should be addressed to Ayan Seal, ayan.seal@gmail.com

Academic Editor: Anton Nijholt

Thermal infrared (IR) images focus on changes of temperature distribution on facial muscles and blood vessels. These temperature changes can be regarded as texture features of images. A comparative study of face two recognition methods working in thermal spectrum is carried out in this paper. In the first approach, the training images and the test images are processed with Haar wavelet transform and the LL band and the average of LH/HL/HH bands subimages are created for each face image. Then a total confidence matrix is formed for each face image by taking a weighted sum of the corresponding pixel values of the LL band and average band. For LBP feature extraction, each of the face images in training and test datasets is divided into 161 numbers of subimages, each of size 8 × 8 pixels. For each such subimages, LBP features are extracted which are concatenated in manner. PCA is performed separately on the individual feature set for dimensionality reduction. Finally, two different classifiers namely multilayer feed forward neural network and minimum distance classifier are used to classify face images. The experiments have been performed on the database created at our own laboratory and Terravic Facial IR Database.

1. Introduction

In the modern society, there is an increasing need to track and recognize persons automatically in various areas such as in the areas of surveillance, closed circuit television (CCTV) control, user authentication, human computer interface (HCI), daily attendance register, airport security checks, and immigration checks [1–3]. Such requirement for reliable personal identification in computerized access control has resulted in an increased interest in biometrics. The key element of biometric technology is its ability to identify a human being and enforce security. Nearly all-biometric systems work in the same manner. First, a person is registered into a database using a specified method. Information about a certain characteristic of the human is captured. This information is usually placed through an algorithm that turns the information into a code that the database stores. When the person needs to be identified, the system will take the information about the person again, translate this new information with the algorithm, and then compare the new code with the stored ones in the database to find out a possible match. Biometrics use physical characteristics or personal traits to identify a person. Physical feature is suitable for identity purpose and generally obtained from living human body. Commonly used physical features are fingerprints, facial features, hand geometry, eye features (iris and retina), and so forth. So biometrics involve using the different parts of the body. Personal trait is sometimes more appropriate for some applications which need direct physical interaction. The most commonly used personal traits are signature and voices and so forth. Among many biometric security systems, face recognition has drawn significant attention of the researchers for the last three decades because of its potential applications in security system. There are a number of reasons to choose face recognition for designing efficient biometric security systems. The most important one is that no physical interaction is needed. This is helpful for the cases where touching is prohibited due to hygienic reasons or religious or cultural traditions. Most of the research works in this area have focused on visible spectrum imaging due to easy availability of low cost visible band optical cameras. But it requires an external source of illumination. Even with a considerable success for automatic face recognition techniques in many practical applications, the task of face recognition based only on the visible spectrum is still a challenging problem under uncontrolled environments. The challenges are even more philosophical when one considers the large variations in the visual stimulus due to illumination conditions, poses [4], facial expressions, aging, and disguises such as facial hair, glasses, or cosmetics. Performance of visual face recognition is sensitive to variations in illumination conditions and usually degrades significantly when the lighting is not bright or when it is not illuminating the face uniformly. The changes caused by illumination on the same individual are often larger than the differences between individuals. Various algorithms (e.g., histogram equalization, eigenfaces, etc.) for compensating such variations have been studied with partial success. These techniques try to reduce the within-class variability introduced by changes in illumination. To overcome this limitation, several solutions have been designed. One solution is using 3D data obtained from 3D vision device. Such systems are less dependent on illumination changes, but they have some disadvantages: the cost of such systems is high, and their processing speed is low. Thermal IR images [5] have been suggested as a possible alternative in handling situations where there is no control over illumination. The wavelength ranges of different infrared spectrums are shown in Table 1.

Thermal IR band is more popular to the researchers working with thermal images. Recently researchers have been using near-IR imaging cameras for face recognition with better results [6], but SWIR and MWIR have not been used significantly till now. Thermal IR images represent the heat patterns emitted from an object, and they do not consider the reflected energy. Objects emit different amounts of IR energy according to their body temperature and characteristics. Previously, thermal IR camera was costly, but recently the cost of such cameras has come down considerably with the development of CCD technology [7]. Thermal images can be captured under different lighting conditions, even under completely dark environment. Using thermal images, the tasks of face detection, localization, and segmentation are comparatively easier and more reliable than those in visible band images [8]. Humans are homoeothermic and hence capable of maintaining constant temperature under different surrounding temperature, and since blood vessels transport warm blood throughout the body, the thermal patterns of faces are derived primarily from the pattern of blood vessels under the skin. The vein and tissue structure of the face is unique for each human being [9], and therefore the IR images are also unique. It is known that even identical twins have different thermal patterns. An infrared camera with good sensitivity can capture images of superficial blood vessels on the human face [10] without any physical interaction. However, it has been indicated by Guyton and Hall [11] that the average diameter of blood vessels is around

TABLE 1: Wavelength ranges for different infrared spectrums.

Spectrum	Wavelength range
Visible spectrum	0.4–0.7 μm (micrometer/micron)
Near infrared (NIR)	0.7–1.0 μm (micrometer/micron)
Short-wave infrared (SWIR)	1–3 μm (micrometer/micron)
Mid-wave infrared (MWIR)	3–5 μm (micrometer/micron)
Thermal infrared (TIR)	8–14 μm (micrometer/micron)

10–15 μm, which is too small to be detected by current IR cameras because of the limitation in spatial resolution. The skin just above a blood vessel is on an average 0.1°C warmer than the adjacent skin, which is beyond the thermal accuracy of current IR cameras. However, the convective heat transfer effect from the flow of "hot" arterial blood in superficial vessels creates characteristic thermal imprints, which are at a gradient with the surrounding tissue. Face recognition based on thermal IR spectrum utilizes the anatomical information [12] of human face as features unique to each individual while sacrificing color recognition. Therefore, the infrared image recognition should focus on thermal distribution patterns on the facial muscles and blood vessels, mainly on cheek, forehead, and nasal tip. These regional thermal distribution patterns can be regarded as the texture pattern unique for a particular face. Wavelet transform can be used to detect the multiscale, multidirectional changes of texture. Local binary patterns (LBPs) are also a well-known texture descriptor and also a successful local descriptor for face recognition under local illumination variations. Therefore, this paper describes that a comparative study of different approach of thermal IR human face recognition system is proposed. The paper is organized as follows: Section 2 presents the outline proposed system. In Section 3, the comparative analyses of these methods in the database created at our own laboratory and Terravic Facial IR Database are presented. Finally, in Section 4, results are discussed and conclusions are given.

2. Outline of the Proposed System

The proposed thermal face recognition system (TFRS) can be subdivided into four main parts, namely, image acquisition, image preprocessing, feature extraction, and classification. The image preprocessing part involves binarization of the acquired thermal face image, extraction of largest component as the face region, finding the centroid of the face region, and finally cropping of the face region in elliptic shape. The two different features extraction techniques have been discussed in this paper. The first one is to find LL band and HL/LH/HH average band images using Haar wavelet transform, and the total confidence matrix is used as a feature vector. The eigenspace projection is performed on feature vector to reduce the dimensionality. This reduced feature vector is fed into a classifier. The second method of features extraction technique is local binary pattern (LBP). As a classifier, a back propagation feed forward neural network or a minimum distance classifier is used in this paper. The block diagram of the proposed system is given in Figure 1. The system starts with acquisition of thermal face image and end with successful classification. The set of image processing and classification techniques which have been used here is discussed in detail in subsequent subsections.

2.1. Thermal Face Image Acquisition. In the present work, unregistered thermal and visible face images are acquired simultaneously with variable expressions, poses, and with/without glasses. Till now 17 individuals have volunteered for this photo shots, and for each individual 34 different templates of RGB color images with different expressions, namely, (Exp1) happy, (Exp2) angry, (Exp3) sad, (Exp4) disgusted, (Exp5) neutral, (Exp6) fearful and (Exp7) surprised are available. Different pose changes about x-axis, y-axis, and z-axis are also available. Resolution of each image is 320×240, and the images are saved in 24-bit JPEG format. Two different cameras are used to capture this database. One is thermal—FLIR 7, and another is Optical—Sony cyber shot. A typical thermal face image is shown in Figure 2(a). This thermal face image depicts interesting thermal information of a facial model.

2.2. Binarization. The binarization of 24-bit colour image is divided into two steps. In the first step, the colour image is converted into an 8-bit grayscale image using

$$I = \left(0.2989 \times \text{red}_{\text{component}}\right) + \left(0.5870 \times \text{green}_{\text{component}}\right)$$
$$+ \left(0.1140 \times \text{blue}_{\text{component}}\right), \tag{1}$$

where "I" is the grayscale image. The grayscale image corresponding to the thermal image of Figure 2(a) is shown in Figure 2(b). Grayscale image is then converted into binary image. For this purpose, mean gray value of grayscale image (say g_{mean}) is computed with the help of

$$g_{\text{mean}} = \frac{\sum_{i=1}^{\text{row}} \sum_{j=1}^{\text{column}} g(i,j)}{(\text{row} \times \text{column})}. \tag{2}$$

If the gray value of any pixel (i,j) (say $g(i,j)$) is greater than or equal to g_{mean}, then the pixel location in the binary image (i,j) is set with 1 (white), else it is set with 0 (black). The binarization process can be mathematically expressed with the help of

$$b(i,j) = \begin{cases} 1 & \text{if } g(i,j) \geq g_{\text{mean}} \\ 0 & \text{otherwise.} \end{cases} \tag{3}$$

In a binary image, black pixels mean background and are represented with "0"s, whereas white pixels mean the face region and are represented with "1"s. The binary image corresponding to the grayscale image of Figure 2(b) is shown in Figure 2(c).

2.3. Extraction of Largest Component. The foreground of a binary image may contain more than one object or components. Say, in Figure 2(c), it has three objects or component.

The large one represents the face region. The others are at the left hand bottom corner and a small dot on the top. Then the largest component has been extracted from binary image using "connected component labeling" algorithm [13]. This algorithm is based on either "4-conneted" neighbours or "8-connected" neighbours method [14]. In "4-connected" neighbours method, a pixel is considered as connected if it has neighbours on the same row or column. This is illustrated in Figure 3(a). Suppose the central pixel of a 3×3 mask "f" is $f(x,y)$, then this method will consider the pixels $f(x+1,y)$, $f(x-1,y)$, $f(x,y+1)$, and $f(x,y-1)$ for checking the connectivity of $f(x,y)$. In "8-connected" method besides the row and columns neighbours, the diagonal neighbours are also checked. That means "4-connected" pixels plus the diagonal pixels are called an "8-connected" neighbour which is illustrated in Figure 3(b). Thus, for a central $f(x,y)$ of a 3×3 mask "f" the "8-connected" neighbour methods will consider $f(x-1,y-1)$, $f(x-1,y)$, $f(x-1,y+1)$, $f(x,y-1)$, $f(x,y+1)$, $f(x+1,y-1)$, $f(x+1,y)$, and $f(x+1,y+1)$ for checking the connectivity of $f(x,y)$.

A pixel to be connected to itself is called reflexive. A pixel and its neighbour are mutually connected is called symmetric. 4-connectivity and 8-connectivity are also transitive: if A is connected to pixel B, and pixel B is connected to pixel C, then there exists a connected path between pixels A and C. A relation (such as connectivity) is called an equivalence relation if it reflexive, symmetric and transitive. All equivalence classes of connected pixels in a binary image, is called connected component labelling. The result of connected component labelling is another image in which everything in one connected region is labeled "1" (for example), everything in another connected region is labeled "2", and so forth. For example, the binary image in Figure 4(a) has three connected components and three labeled connected components is shown in Figure 4(b).

"Connected component labeling" algorithm is given in Algorithm 1.

Using "connected component labeling" algorithm, the largest component of face region is identified from Figure 2(c) which is shown in Figure 5.

2.4. Finding the Centroid [15]. Centroid has been extracted from the largest component of the binary image using

$$X = \frac{\sum m_{f(x,y)} x}{\sum m_{f(x,y)}},$$
$$Y = \frac{\sum m_{f(x,y)} y}{\sum m_{f(x,y)}}, \tag{4}$$

where x, y are the coordinate, of the binary image and m is the intensity value that is $m_{f(x,y)} = f(x,y) = 0$ or 1.

2.5. Cropping of the Face Region in Elliptic Shape. Normally, human face is of ellipse shape. Then, from the above centroid coordinates, human face has been cropped in elliptic shape using "Bresenham ellipse drawing" [16] algorithm. This algorithm takes the distance between the centroid and the right ear as the minor axis of the ellipse and distance

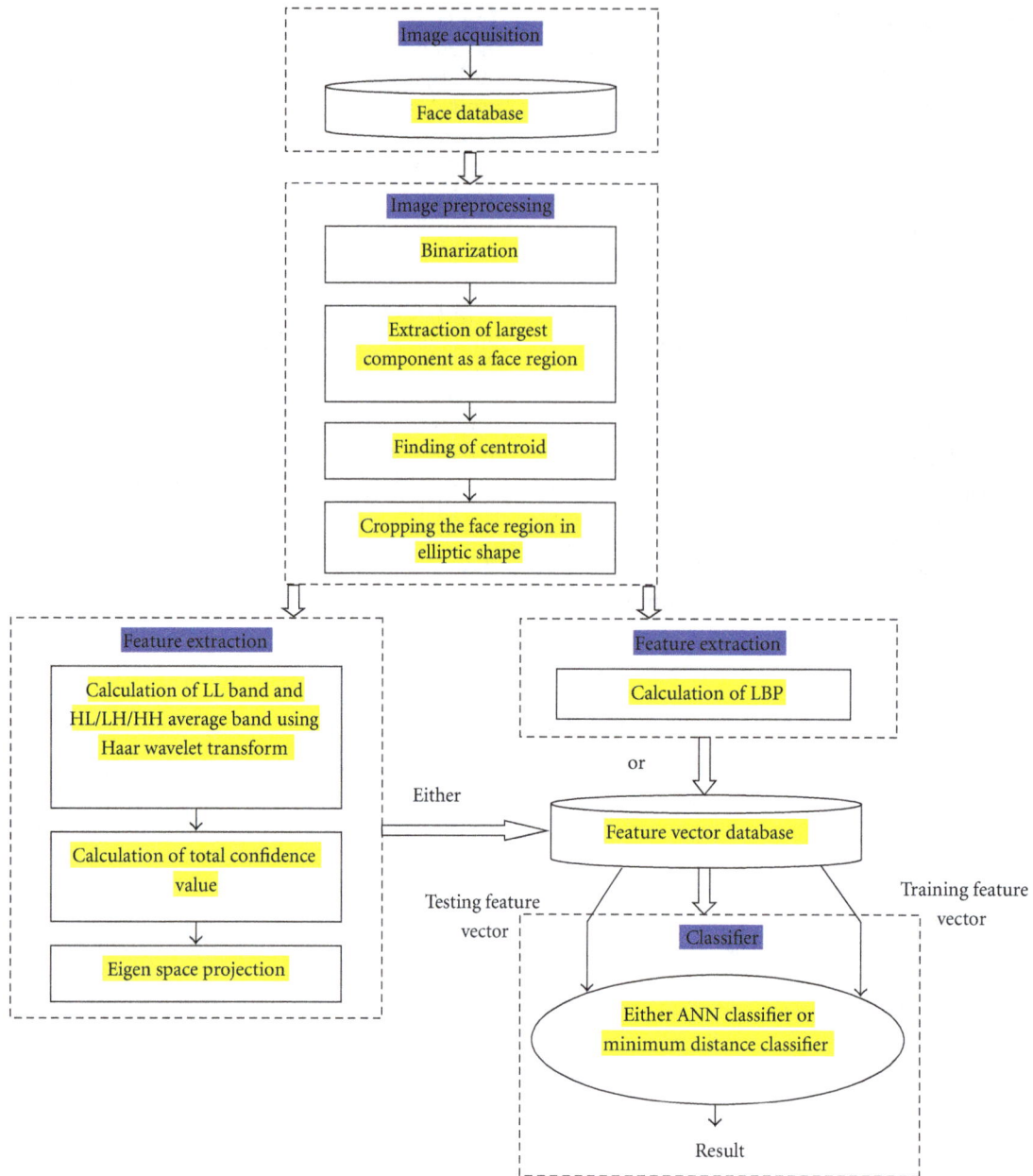

FIGURE 1: Schematic block diagram of the proposed system.

between the centroid and the fore head as major axis of the ellipse. The pixels selected by the ellipse drawing algorithm are mapped onto the gray level image of Figure 2(b), and finally the face region is cropped. This is shown in Figure 6.

2.6. Calculate LL and HL/LH/HH Average Band Using Haar Wavelet Transform. The first method of feature extraction is discrete wavelet transform (DWT). The DWT was invented by the Hungarian mathematician Alfréd Haar in 1909. A key advantage of wavelet transform over Fourier transforms is temporal resolution. Wavelet transform captures both frequency and spatial information. The DWT has a huge

number of applications in science, engineering, computer science, and mathematics. The Haar transformation is used here since it is the simplest wavelet transform of all and can successfully serve our purpose. Wavelet transform has merits of multiresolution, multiscale decomposition and so on. To obtain the standard decomposition [17] of a 2D image, the 1D wavelet transform to each row is applied first. This operation gives an average pixel value along with detail coefficients for each row. These transformed rows are treated as if they were themselves in an image. Now, 1D wavelet transform to each column is applied. The resulting pixel values are all detail coefficients except for a single overall average coefficient. As a result, the elliptical shape facial image is decomposed into four regions

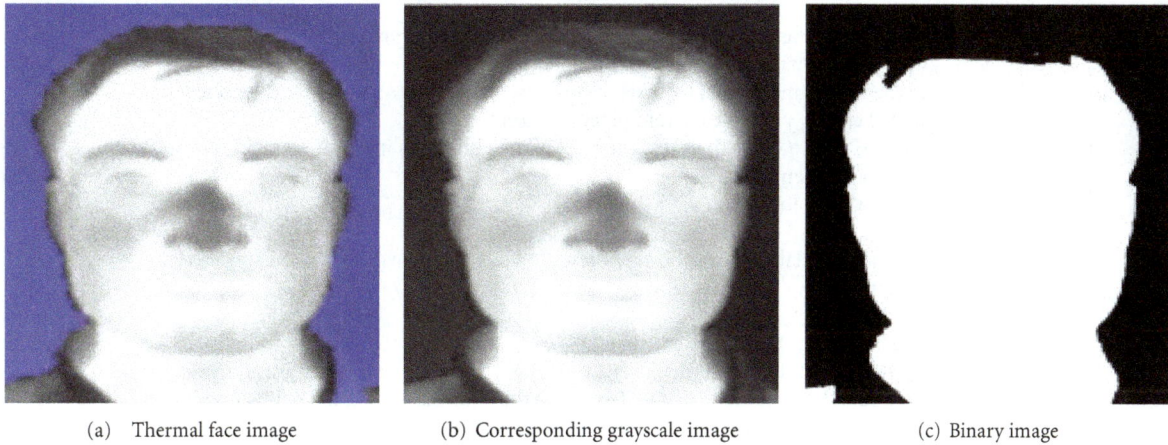

(a)   Thermal face image   (b) Corresponding grayscale image   (c) Binary image

FIGURE 2: Thermal face image and its various preprocessing stages.

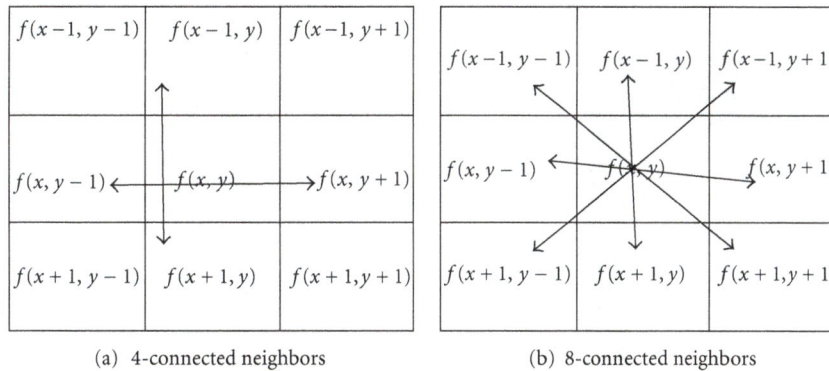

(a)   4-connected neighbors   (b) 8-connected neighbors

FIGURE 3: Different connected neighborhoods.

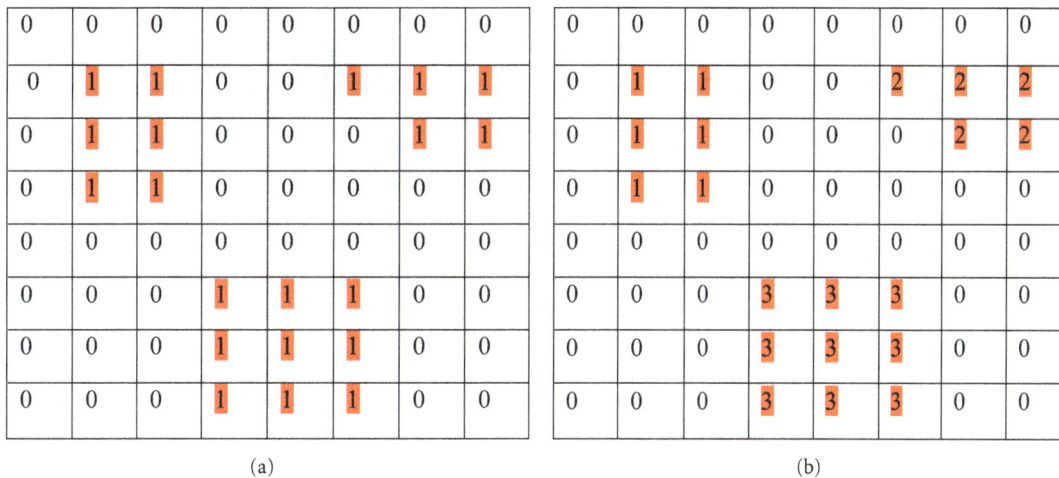

(a)   (b)

FIGURE 4: (a) Connected component, (b) Labeled connected components.

that can be gained. These regions are one low-frequency LL_1 (approximate component), and three high-frequency regions (detailed components), namely, LH_1 (horizontal component), HL_1 (vertical component) and HH_1 (diagonal component), respectively. The low frequency subband LL_1 can be further decomposed into four subbands LL_2, LH_2,

HL_2, and HH_2 at the next coarse scale. LL_i is a reduced resolution corresponding to the low-frequency part of an image. The sketch map of the quadratic wavelet decomposition is shown in Figure 6.

As illustrated in Figure 7, the L denotes low frequency and the H denotes high frequency, and subscripts named

```
//LabelConnectedComponent(im) is a method which takes//one argument that is an image
named if. f(x,y) is the current pixel on xth row and yth column.
(1)  Consider the whole image pixel by pixel in row wise manner in order to get connected
     component. Let f(x, y) be the current pixel position
        if (f(x, y)==1) and f(x, y) does not any labelled neighbour in its 8-connected
        neighbourhood) then
               Create a new label for f(x, y).
    else
        if (f(x, y) has only labelled neighbour) then
               Mark f(x, y) with that label.
    else
        if (f(x, y) has two or more labeled) then
               Choose one of the labels for f(x, y) and memorize that these labels are equivalent.
(2)  Go another pass through the image to determine the equivalencies and labeling each pixel
     with a unique label for its equivalence class.
```

ALGORITHM 1: LabelConnectedComponent(im).

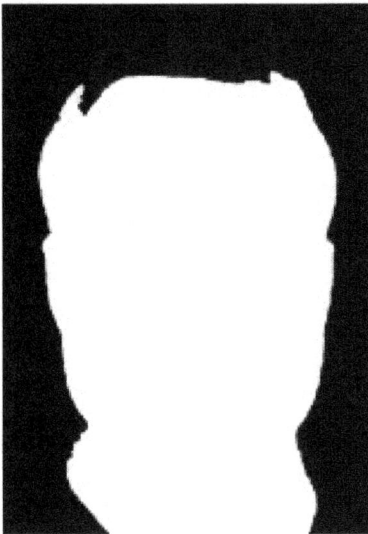

FIGURE 5: The largest component as a face skin region.

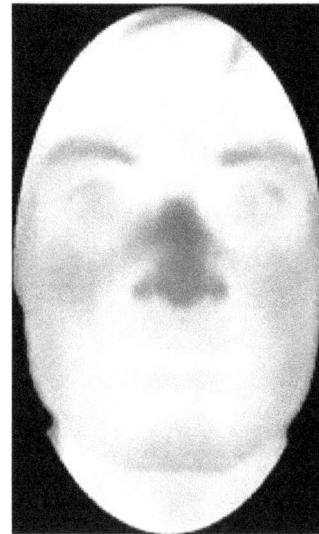

FIGURE 6: Cropped face region in elliptic shape.

from 1 to 2 denote simple, quadratic wavelet decomposition, respectively. The standard decomposition algorithm is given in Algorithm 2.

Let us start with a simple example of 1D wavelet transform [18]. Suppose an image with only one row of four pixels, having intensity values [10 4 9 5]. Now apply the Haar wavelet transform on this image. To do so, first pair up the input intensity values or pixel values, storing the mean in order to get the new lower resolution image with intensity values [7 7]. Obviously, some information might be lost in this averaging process. Some detail coefficients need to be stored to recover the original four intensity values from the two mean values, which capture the missing information. In this example, 3 is the first detail coefficient, since the computed mean is 3 less than 10 and 3 more than 4. This single number is responsible to recover the first two pixels of original four-pixel image. Similarly, the second detail coefficient is 2. Thus, the original image is decomposed

FIGURE 7: Sketch map of the quadratic wavelet decomposition.

into a lower resolution (two-pixel) version and a pair of detail coefficients. Repeating this process recursively on the averages gives the full decomposition, which is shown in Table 2.

```
// Im[1:r,1:c] is an image realized by 2D array, where r is the number of rows and c is the number of column.
        for i = 1 : r
                1D wavelet transform (Im(i,:))
        end
        for j = 1 : c
                1D wavelet transform (Im(:,j))
        end
end
```

ALGORITHM 2: Function StandardDecomposition(Im[1:r,1:c]).

TABLE 2: Resolution, mean, and the detail coefficients of full decomposition.

Resolution	Mean	Detail coefficients
4	[10 4 9 5]	
2	[7 7]	[3 2]
1	[7]	[0]

Thus, the one-dimensional Haar transform of the original four-pixel image is given by [7 0 3 2]. After applying standard decomposition algorithm on Figure 6, the resultant figure is shown in Figure 8.

The pixels of LL_2 image can be rearranged horizontally or vertically. So the image can be treated as a vector (called feature vector).

2.7. Calculation Total Confidence Value.

In the present work, wavelet transform is used on the elliptic shape face region once which divide the whole image into 4 equal sized subimages, namely, low-frequency LL band (approximate component) and three high-frequency bands (detailed components), HL, LH, and HH. Then the pixelwise average of the detail components is computed using

$$D(x, y) = \frac{1}{3}(A(x, y) + B(x, y) + C(x, y)), \quad (5)$$

where $A(x, y)$ is the HL band subimage, $B(x, y)$ is the LH band sub-image and $C(x, y)$ is the HH band subimage. $D(x, y)$ is the average subimage of $A(x, y)$, $B(x, y)$, and $C(x, y)$ band subimages, and x, y are spatial coordinates.

Next, a matrix called total confidence matrix $T(x, y)$ is formed by taking a pixelwise weighted sum of pixel values of LL band and average subimages [19–21] using (6), as given in the following:

$$T(x, y) = (\alpha(x, y) \times L(x, y)) + (\beta(x, y) \times D(x, y)), \quad (6)$$

where $T(x, y)$ is the total confidence value, $L(x, y)$ is the LL band subimages, and $D(x, y)$ is the average of HL/LH/HH band subimages, while $\alpha(x, y)$ and $\beta(x, y)$ denote the weighting factors for pixel values of LL band and HL/LH/HH average band subimages, respectively, which are shown in Figure 9.

After calculating the total confidence matrices for all the images, each matrix is transformed into a horizontal vector, by concatenating the rows of elements in it. This process is repeated for all the images in the database. Let the number of elements in each such horizontal vector be N, where N is the product of the number of rows and columns in LL band or average subimages. By placing the horizontal vectors in row order, a new matrix of size $M \times N$ is formed, where M is the number of images in the database. Thus, $M \times N$ matrix is divided into two parts by the size of $(M/2) \times N$, of which one part will be used for training purpose and the other part for testing purpose only. The first part contains odd number of images like first row, third row, fifth row, and so on from $M \times N$ matrix, and the second part contains even number of images like second row, fourth row, sixth row, and so on from $M \times N$ matrix.

2.8. Eigenface for Recognition.

Principal component analysis (PCA) [22, 23] is performed on training set described above which gives a set of eigenvalues and corresponding eigenvector. Each eigenvector can be shown as sort of ghostly face which is called an eigenface. Each face image in the training set can be represented exactly in terms of a linear combination of these eigenfaces. So the number of rows that is, number of face images in the training set, is equal to the number of a eigenfaces. However, the faces can also be approximated using only the "best" eigenfaces, those that have the largest eigenvalues and which therefore account for the most variance within the set of face images. For this, the eigenvalues are sorted in descending order and eigenvectors corresponding to a few largest eigenvalues are retained. The n-dimensional space that is formed by these eigenvectors or eigenfaces is called eigenspace. The face images in the training set are then projected onto the eigenspace to get the corresponding eigenfaces, which are then used to train a classifier. For the test face images, similar procedure is followed to get their corresponding eigenfaces, which are classified by the trained classifier.

2.9. Local Binary Pattern.

The second method of feature is local binary pattern (LBP). The LBP is a type of feature used for texture classification in computer vision. LBP was first described in 1994 [24, 25]. It has since been found to be a powerful feature for texture classification. As it can be appreciated in Figure 10, the original LBP operator represents each pixel of an image by thresholding its 3×3 neighborhood with the center value and considering the result as a binary

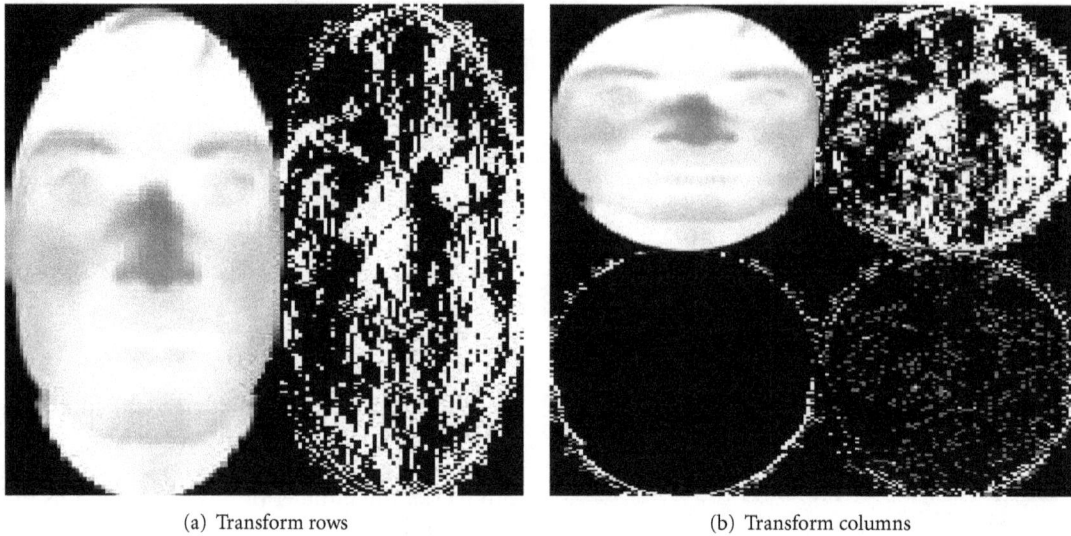

(a) Transform rows   (b) Transform columns

FIGURE 8: Haar wavelet transform.

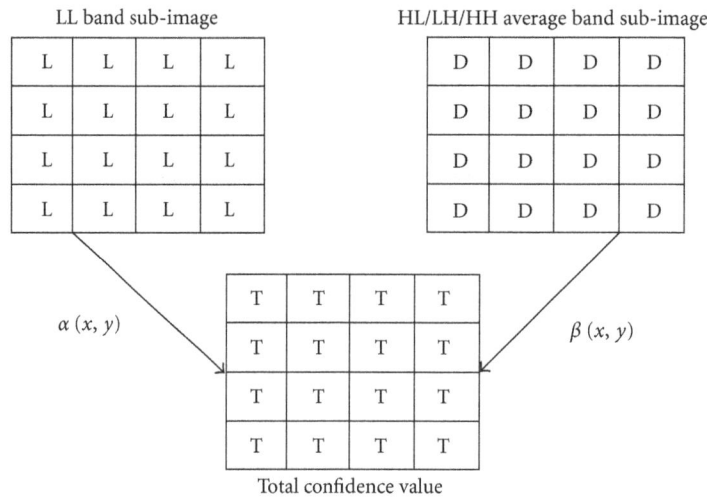

FIGURE 9: Mixing technique.

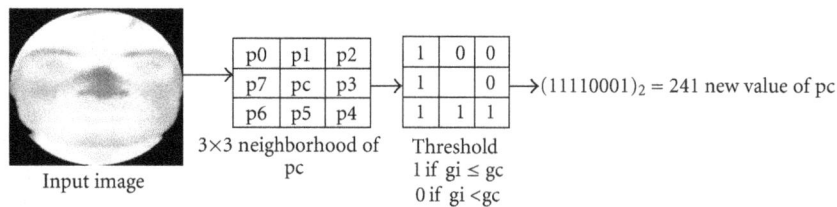

FIGURE 10: Local binary pattern.

number, called the LBP code. In the classification step, the image is usually divided into rectangular regions and histograms of the LBP that codes are calculated over each of them. The histograms of each region are concatenated into a single one, and a dissimilarity measure is used to compare the histograms of two different images.

2.10. Multilayer Feed Forward Neural Network. Artificial neural networks (ANNs) [26, 27] possess extraordinary generalization capability to obtain useful information from complex environment or data. So ANN can be used to extract patterns and detect trends that are too hard to be found by either humans or other computer techniques. A trained ANN can be thought of as an "expert system." The back propagation learning algorithm is one of the most popular neural networks to the scientific and engineering community for modeling and processing of many quantitative phenomena. This learning algorithm is applied to multilayer feed

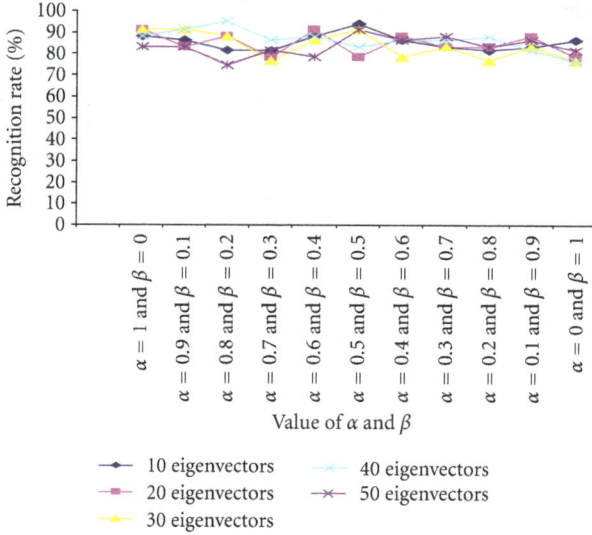

FIGURE 11: Comparative study of recognition performance (own database) with varying numbers of eigenvectors and values of α and β.

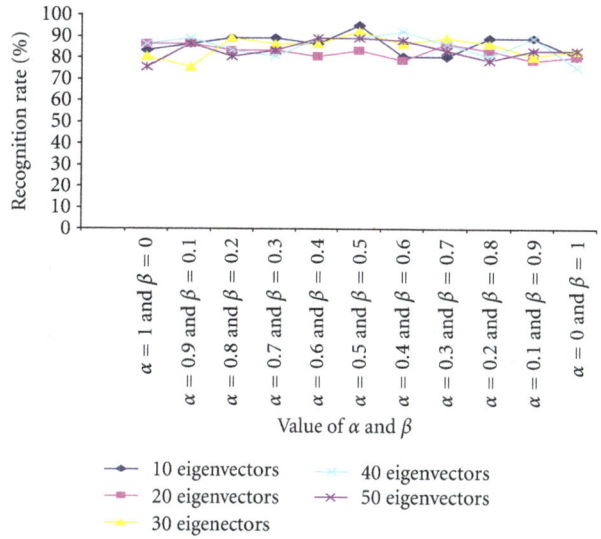

FIGURE 12: Comparative study of recognition rate (performed on Terrivic Facial Thermal Database) with varying numbers of eigenvectors and values of α and β.

forward networks consisting of processing elements with continuous differentiable activation functions. The five layer feed forward back propagation neural network is used here as a classifier. Momentum allows the network to respond to local gradient and to recent trends in the error surface. The momentum is used to back propagation learning algorithm for making weight changes equal to the sum of a fraction of the last weight change and the new change. The magnitude of the effect that the last weight change is allowed is known as momentum constant (mc). The momentum constant may be any number between 0 and 1. The momentum constant zero means, a weight changes according to the gradient and the momentum constant one means, the new weight change is set to equal the last weight change, and the gradient is not considered here. The gradient is computed by summing the gradients calculated at each training example, and the weights and biases are only updated after all training examples have been presented. Tan-sigmoid transfer functions are used to calculate a layer's output from its net, the first input, and the next three hidden layers and the outer most layer gradient descent with momentum training function is used to update weight and bias values.

2.11. Minimum Distance Classifier. Recognition techniques based on matching represent each class by a prototype pattern vector. It places an unknown pattern in the class to which it is closest in terms of a predefined metric. The simplest approach is the minimum distance classifier [15]. It must determine the Euclidean distance between the unknown pattern and each of the prototype vectors. It chooses the smallest distance to take a judgment. The prototype of each pattern class is represented as the mean vector of the patterns of that class which is expressed using

$$m_j = \frac{1}{N_j} \sum_{x \in \omega_j} x_j, \quad j = 1, 2, 3, \ldots, W, \quad (7)$$

where W is the number of pattern classes, ω_j is the set of pattern vectors of class j, and N_j is the number of pattern vectors in ω_j. In order to get the class membership of an unknown pattern vector x, its closest prototype is searched using Euclidean distance measure, which is shown in

$$D_j(x) = \left\| x - m_j \right\|, \quad j = 1, 2, 3, \ldots, W. \quad (8)$$

If $D_j(x)$ is the smallest distance, that is, best match, then assign x to class ω_j.

3. Experiment and Results

Experiments have been performed on our own captured thermal face images at our laboratory and Terravic Facial Infrared database. In our Database, there are $17 \times 34 = 578$ thermal images. The details of our database have been mentioned in Section 2.1. Twelve images are taken in each person for our experiments from two above-mentioned datasets, out of which 6 face images are used to form training set and 6 face images are used to form testing set. We have made all the images of size 112×92. The Terravic Facial Infrared Database contains total number of 20 classes (19 men and 1 woman) of 8-bit gray scale JPEG thermal faces of 320×240. Size of the database is 298 MB, and images with different rotations are left, right, and frontal face images also available with different items like glass and hat [13]. Experimental process can be divided into several ways.

3.1. Harr Wavelet + PCA + ANN. In the first set of experiments, Haar wavelet is used to decompose the cropped face image once which produces 4 subimages as LL, HL, LH, and HH bands. Then the average of HL/LH/HH band subimages is computed using (5). We have used ten different sets of values for (α, β) to generate 10 different confidence

TABLE 3: Recognition performance (own database) with varying numbers of eigenvectors and the values of α and β.

Value of α and β	Recognition rate (%)				
	10 eigenvectors	20 eigenvectors	30 eigenvectors	40 eigenvectors	50 eigenvectors
$\alpha = 1.0$ and $\beta = 0.0$	88.23	91.18	91.18	88.23	83.33
$\alpha = 0.9$ and $\beta = 0.1$	86.27	83.33	91.18	91.18	83.33
$\alpha = 0.8$ and $\beta = 0.2$	81.38	88.23	88.23	**95.09**	74.50
$\alpha = 0.7$ and $\beta = 0.3$	81.38	78.57	76.74	86.27	81.38
$\alpha = 0.6$ and $\beta = 0.4$	88.23	91.18	86.27	88.23	78.57
$\alpha = 0.5$ and $\beta = 0.5$	94.11	78.57	91.18	83.33	91.18
$\alpha = 0.4$ and $\beta = 0.6$	86.27	88.23	78.57	86.27	86.27
$\alpha = 0.3$ and $\beta = 0.7$	83.33	83.33	83.33	86.27	88.23
$\alpha = 0.2$ and $\beta = 0.8$	81.38	83.33	76.74	88.23	83.33
$\alpha = 0.1$ and $\beta = 0.9$	83.33	88.23	83.33	81.38	86.27
$\alpha = 0.0$ and $\beta = 1.0$	86.27	78.57	76.74	76.74	81.38

TABLE 4: Recognition performance (benchmark database) with varying numbers of eigenvectors and the values of α and β.

Value of α and β	Recognition rate (%)				
	10 eigenvectors	20 eigenvectors	30 eigenvectors	40 eigenvectors	50 eigenvectors
$\alpha = 1.0$ and $\beta = 0.0$	83.33	86.27	80.39	86.27	75.49
$\alpha = 0.9$ and $\beta = 0.1$	86.27	86.27	75.49	89.22	86.27
$\alpha = 0.8$ and $\beta = 0.2$	89.22	83.33	89.22	83.33	80.39
$\alpha = 0.7$ and $\beta = 0.3$	89.22	83.33	86.27	80.39	83.33
$\alpha = 0.6$ and $\beta = 0.4$	86.27	80.39	86.27	89.22	89.22
$\alpha = 0.5$ and $\beta = 0.5$	94.11	83.33	92.15	89.22	89.22
$\alpha = 0.4$ and $\beta = 0.6$	80.39	78.57	86.27	92.15	88.22
$\alpha = 0.3$ and $\beta = 0.7$	80.39	86.27	89.22	86.27	83.33
$\alpha = 0.2$ and $\beta = 0.8$	89.22	83.33	86.27	80.39	78.57
$\alpha = 0.1$ and $\beta = 0.9$	89.22	78.57	80.39	89.22	83.33
$\alpha = 0.0$ and $\beta = 1.0$	80.39	80.39	83.33	75.49	83.33

matrices for each face image. The values of α and β are chosen according to

$$\begin{aligned} \beta &= 0.1 \times i, \\ \alpha &= 1.0 - \beta, \end{aligned} \quad 0 \leq i \leq 10. \qquad (9)$$

After computing the confidence matrices of all the decomposed face images, PCA is performed on these confidence matrices for further dimensionality reduction. ANN classifier (with 0.02 acceleration and 0.9 momentum) is then used to classify the face images on the basis of the extracted features. The recognition performances of the classifier on our own Database and Terravic Facial IR database are shown in Tables 3 and 4, respectively. The results are also shown graphically in Figures 11 and 12, respectively.

3.2. Harr Wavelet + PCA + Minimum Distance Classifier. In the second set of experiments, the feature set was kept the same as those in the first set of experiments, but the classifier is chosen as minimum distance classifier. The recognition performance obtained on both the thermal face databases considered here is detailed in Table 5 and also graphically compared in Figure 13.

TABLE 5: Recognition performance (on own database and benchmark database) with minimum distance classifier and the value of α and β.

Value of α and β	Recognition rate (%)	
	Own database	Terravic Facial Thermal Database
$\alpha = 1.0$ and $\beta = 0.0$	89.22	94.11
$\alpha = 0.9$ and $\beta = 0.1$	86.27	86.27
$\alpha = 0.8$ and $\beta = 0.2$	86.27	83.33
$\alpha = 0.7$ and $\beta = 0.3$	83.33	86.27
$\alpha = 0.6$ and $\beta = 0.4$	86.27	83.33
$\alpha = 0.5$ and $\beta = 0.5$	80.39	80.39
$\alpha = 0.4$ and $\beta = 0.6$	80.39	78.57
$\alpha = 0.3$ and $\beta = 0.7$	80.39	86.27
$\alpha = 0.2$ and $\beta = 0.8$	83.33	86.27
$\alpha = 0.1$ and $\beta = 0.9$	83.33	86.27
$\alpha = 0.0$ and $\beta = 1.0$	86.27	80.39

3.3. Local Binary Pattern + (PCA + ANN/Minimum Distance Classifier). In the third set of experiments, cropped face images are divided in to 161 subimages each of size 8×8

TABLE 6: Recognition performance (own database and benchmark database) with varying numbers of eigenvectors, ANN, and minimum distance classifier.

	Recognition rate (%)					
	ANN classifier					Minimum distance classifier
	10 eigenvectors	20 eigenvectors	30 eigenvectors	40 eigenvectors	50 eigenvectors	
Own database	86.27	83.33	86.27	86.27	83.33	89.22
Terravic Facial Thermal Database	86.27	89.22	83.33	92.15	89.22	94.11

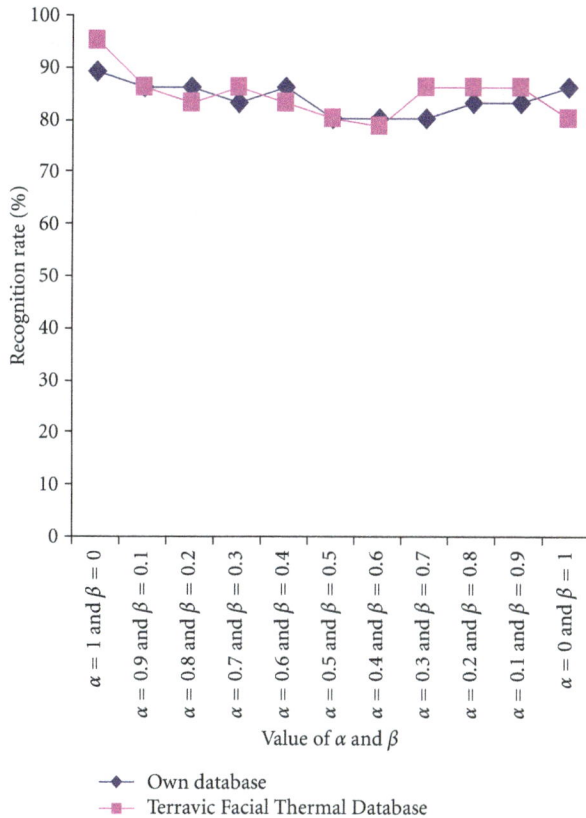

FIGURE 13: Comparative study of Recognition performance (own database and benchmark database) with minimum distance classifier and the values of α and β.

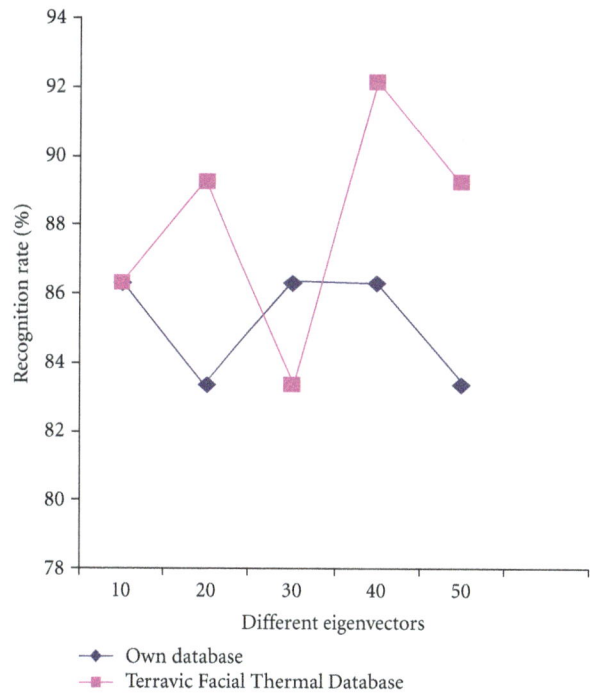

FIGURE 14: Recognition performance (own database and benchmark database) with varying numbers of eigenvectors, ANN and minimum distance classifier.

pixels. Then local binary pattern is used to extract features from each of the subimages which are concatenated in row wise manner. After performing PCA on the LBP features for dimensionality reduction, ANN and minimum distance classifier are used separately for recognition of the face images on the basis of the extracted features. The obtained recognition results are shown in Table 6. The results are also shown graphically in Figure 14.

4. Conclusions

In this paper, a comparative study of thermal face recognition methods is discussed and implemented. In this study two local-matching techniques, one based on Haar wavelet and the other based on Local Binary Pattern, are analyzed. Firstly, human thermal face images are preprocessed and cropped the face region only, from the entire face images. Then above-mentioned two feature extraction methods are used to extract features from the cropped images. Then, PCA is performed on the individual feature set for dimensionality reducation. Finally, two different classifiers are used to classify face images. One such classifier is multilayer feed forward neural network, and another is minimum distance classifier. The experiments have been performed on the database created at our own laboratory and Terravic Facial IR Database. The proposed system gave higher recognition performance in the experiments, and the recognition rate was 95.09% for $\alpha = 0.8$, $\beta = 0.2$, and number of eigenvectors is 40. This experiment was performed on our own database, which is shown in Table 3. Furthermore, no knowledge of geometry or specific feature of the face is required. However, this system is applicable to front views and constant background only. It may fail in unconstraint environments like natural scenes.

Acknowledgments

The authors are thankful to a major project entitled "Design and Development of Facial Thermogram Technology for Biometric Security System," funded by University Grants Commission (UGC), India, and "DST-PURSE Programme" at Department of Computer Science and Engineering, Jadavpur University, India, for providing necessary infrastructure to conduct experiments relating to this work.

References

[1] W. Zhao, R. Chellappa, P. J. Phillips, and A. Rosenfeld, "Face recognition: a literature survey," *ACM Computing Surveys*, vol. 35, no. 4, pp. 399–458, 2003.

[2] P. J. Phillips, A. Martin, C. L. Wilson, and M. Przybocki, "Introduction to evaluating biometric systems," *Computer*, vol. 33, no. 2, pp. 56–63, 2000.

[3] A. Samal and P. A. Iyengar, "Automatic recognition and analysis of human faces and facial expressions: a survey," *Pattern Recognition*, vol. 25, no. 1, pp. 65–77, 1992.

[4] A. Gupta and S. K. Majumdar, "Machine Recognition of Human Face," http://citeseerx.ist.psu.edu/viewdoc/download?doi=10.1.1.127.4323&rep=rep1&type=pdf.

[5] D. A. Socolinsky and A. Selinger, "A comparative analysis of face recognition performance with visible and thermal infrared imagery," in *Proceedings of the International Conference on Pattern Recognition*, vol. 4, pp. 217–222, Quebec, Canada, 2002.

[6] D. A. Socolinsky, A. Selinger, and J. D. Neuheisel, "Face recognition with visible and thermal infrared imagery," *Computer Vision and Image Understanding*, vol. 91, no. 1-2, pp. 72–114, 2003.

[7] S. Wu, Z.-J. Fang, Z.-H. Xie, and W. Liang, "Blood Perfusion Models for Infrared Face Recognition," School of information technology, Jiangxi University of Finance and Economics, China.

[8] S. G. Kong, J. Heo, B. R. Abidi, J. Paik, and M. A. Abidi, "Recent advances in visual and infrared face recognition—a review," *Computer Vision and Image Understanding*, vol. 97, no. 1, pp. 103–135, 2005.

[9] P. Buddharaju, I. T. Pavlidis, and P. Tsiamyrtzis, "Pose-invariant physiological face recognition in the thermal infrared spectrum," in *Proceedings of the Conference on Computer Vision and Pattern Recognition Workshops*, p. 133, Washington, DC, USA, June 2006.

[10] C. Manohar, *Extraction of superficial vasculature in thermal imaging [M.S. thesis]*, Department of Electrical Engineering, University of Houston, Houston, Tex, USA, 2004.

[11] A. C. Guyton and J. E. Hall, *Textbook of Medical Physiology*, W.B. Saunders Company, Philadelphia, Pa, USA, 9th edition, 1996.

[12] F. Prokoski, "History, current status, and future of infrared identification," in *Proceedings of the IEEE Workshop Computer Vision Beyond Visible Spectrum: Methods and Applications*, pp. 5–14, 2000.

[13] S. Bryan Morse, *Lecture 2: Image Processing Review, Neighbors, Connected Components, and Distance*, 1998–2004.

[14] R. C. Gonzalez and R. E. Woods, *Digital Image Processing*, Prentice Hall, 3rd edition, 2002.

[15] S. Venkatesan and S. S. R. Madane, "Face recognition system with genetic algorithm and ANT colony optimization," *International Journal of Innovation, Management and Technology*, vol. 1, no. 5, 2010.

[16] D. Hearn and M. P. Baker, Computer graphics C version.

[17] G. Beylkin, R. Coifman, and V. Rokhlin, "Fast wavelet transforms and numerical algorithms I," *Communications on Pure and Applied Mathematics*, vol. 44, no. 2, pp. 141–183, 1991.

[18] E. J. Stollnitz, T. D. DeRose, and D. H. Salestin, "Wavelets for computer graphics: a primer, part 1," *IEEE Computer Graphics and Applications*, vol. 15, no. 3, pp. 76–84, 1995.

[19] M. K. Bhowmik, D. Bhattacharjee, D. K. Basu, and M. Nasipuri, "A comparative study on fusion of visual and thermal face images at different pixel level," *International Journal of Information Assurance and Security*, no. 1, pp. 80–86, 2011.

[20] J. Heo, "Fusion of Visual and Thermal Face Recognition Techniques: A Comparative Study," The University of Tennessee, Knoxville, Tenn, USA, October 2003, http://imaging.utk.edu/publications/papers/dissertation.

[21] Z. Yin and A. A. Malcolm, "Thermal and Visual Image Processing and Fusion," SIMTech Technical Report, 2000.

[22] M. Turk and A. Pentland, "Eigenfaces for recognition," *Journal of Cognitive Neuroscience*, vol. 3, no. 1, pp. 71–86, 1991.

[23] M. A. Turk and A. P. Pentland, "Face recognition using eigenfaces," in *Proceedings of the IEEE Computer Society Conference on Computer Vision and Pattern Recognition*, pp. 586–591, June 1991.

[24] T. Ojala, M. Pietikäinen, and D. Harwood, "Performance evaluation of texture measures with classification based on Kullback discrimination of distributions," in *Proceedings of the 12th International Conference on Pattern Recognition (ICPR '94)*, vol. 1, pp. 582–585, 1994.

[25] T. Ojala, M. Pietikäinen, and D. Harwood, "A comparative study of texture measures with classification based on feature distributions," *Pattern Recognition*, vol. 29, no. 1, pp. 51–59, 1996.

[26] Lin and Lee, *Neural Fuzzy Systems*, Prentice Hall International, 1996.

[27] J. Haddadnia, K. Faez, and M. Ahmadi, "An efficient human face recognition system using Pseudo Zernike Moment Invariant and radial basis function neural network," *International Journal of Pattern Recognition and Artificial Intelligence*, vol. 17, no. 1, pp. 41–62, 2003.

12

Noise-Assisted Instantaneous Coherence Analysis of Brain Connectivity

Meng Hu and Hualou Liang

School of Biomedical Engineering, Science & Health Systems, Drexel University, 3141 Chestnut Street, Philadelphia, PA 19104, USA

Correspondence should be addressed to Hualou Liang, hualou.liang@drexel.edu

Academic Editor: Marc Van Hulle

Characterizing brain connectivity between neural signals is key to understanding brain function. Current measures such as coherence heavily rely on Fourier or wavelet transform, which inevitably assume the signal stationarity and place severe limits on its time-frequency resolution. Here we addressed these issues by introducing a noise-assisted instantaneous coherence (NAIC) measure based on multivariate mode empirical decomposition (MEMD) coupled with Hilbert transform to achieve high-resolution time frequency representation of neural coherence. In our method, fully data-driven MEMD, together with Hilbert transform, is first employed to provide time-frequency power spectra for neural data. Such power spectra are typically sparse and of high resolution, that is, there usually exist many zero values, which result in numerical problems for directly computing coherence. Hence, we propose to add random noise onto the spectra, making coherence calculation feasible. Furthermore, a statistical randomization procedure is designed to cancel out the effect of the added noise. Computer simulations are first performed to verify the effectiveness of NAIC. Local field potentials collected from visual cortex of macaque monkey while performing a generalized flash suppression task are then used to demonstrate the usefulness of our NAIC method to provide highresolution time-frequency coherence measure for connectivity analysis of neural data.

1. Introduction

To understand how brain networks process information, it is crucial to accurately quantify their connectivity patterns. For analysis of brain connectivity between two signals, current measures such as coherence [1–3] rely upon spectral estimate of each signal, which is routinely computed based on Fourier or wavelet transform. Thus, the underlying nonstationary nature of neural data presents a significant challenge for the applications of current measures. Though short-time sliding window approaches, for example, short-time Fourier transform, have been used to alleviate this problem, this issue is not completely resolved for a number of reasons. First, the stationarity of neural data within each short-time window cannot be guaranteed. Second, even though the data are stationary within each time window, the resolution of time-frequency representation is limited by Heisenberg uncertainty principle [4]. Wavelet transform [4], albeit improved, is still subject to time-frequency resolution tradeoff, that is,

frequency resolution is low at high frequencies and high at low frequencies. Moreover, wavelet analysis depends on the choice of mother wavelet, which is arbitrary and may not be optimal for time series under scrutiny.

In contrast to the aforementioned spectral estimation methods, empirical mode decomposition (EMD) method [5] adaptively decomposes nonstationary time series into a finite set of amplitude-frequency modulated components, namely, intrinsic mode functions (IMFs), without assuming any basis functions. These IMF components allow the calculation of a meaningful instantaneous frequency by virtue of Hilbert transform. As a result, a high-resolution time-frequency spectral estimation, namely, Hilbert spectrum, can be obtained, even with nonstationary time series. The last decade has witnessed the remarkable success of EMD in a large variety of applications; it is, however, limited to univariate (single-channel) data analysis. The availability of simultaneous multichannel data presents important analysis challenges and calls for multivariate extension of EMD. So far, EMD has been extended to complex EMD [6], rotation-invariant EMD [7], bivariate EMD [8], trivariate EMD [9], multidimensional ensemble EMD [10], and multivariate EMD (MEMD) [11] and its noise-assisted MEMD [12]. Of particular note is the MEMD, which is a rather generic multivariate extension and has been shown very promising in multichannel neural data analysis [13, 14].

Hence, it is natural and straightforward in this study to think of using MEMD together with Hilbert transform to perform spectral estimate of nonstationary multichannel neural data. In practice, the estimated spectra are readily computable, yet its use for subsequent coherence estimate is problematic because MEMD coupled with Hilbert transform provides high-resolution time-frequency spectra typically with many zero values, which therefore cause computational problem for estimating coherence at those zero-value positions.

In this paper, we propose a noise-assisted instantaneous coherence (NAIC) measure based on the MEMD together with the Hilbert transform to circumvent the aforementioned problems in providing high-resolution time-frequency coherence measure. First, the MEMD, together with Hilbert transform, is applied to estimate the spectra of signals. Second, we add a noise into the estimated spectra to alleviate the zero-value problem before coherence is derived. Third, we design a statistical randomization procedure to cancel out the effect of the added noise on the coherence of mixed data. We note that our procedure is not just restricted to the coherence measure demonstrated in this paper, but it can also be applied to other forms of coherence such as partial and multiple coherence as well as Granger causality [15, 16].

The paper is organized as follows. In Section 2, we briefly review the recently developed MEMD method and Hilbert transform, followed by our proposed NAIC method. In Section 3, we first conduct computer simulations to validate our NAIC method and contrast it with both Fourier-based and wavelet-based methods. Then, we apply the method to real cortical filed potential data collected from a macaque monkey while performing a generalized flash suppression task [17]. Section 4 concludes with discussions.

2. Method

2.1. Background

2.1.1. Multivariate Empirical Mode Decomposition. MEMD is a multivariate extension of EMD. The EMD [5] is a fully adaptive data-driven method which decomposes a time series into a finite set of amplitude-frequency-modulated IMFs, which represent its inherent oscillatory modes. Specifically, for a time series $x(t)$, all the local extrema are first identified, and then two envelopes $e_{min}(t)$ and $e_{max}(t)$ are obtained by interpolating between local maxima (resp., minima), and subsequently the local mean $m(t) = (e_{min}(t) + e_{max}(t))/2$ is computed. The detail $c(t) = x(t) - m(t)$ is finally iterated until it becomes an IMF, which is defined as having the symmetric envelopes and the same numbers of zero-crossing

and local extrema, differing at most by one. The residue by removing IMFs from raw signal is subject to the above procedure for the next IMF until the monotonic residue is left. Hence, a time series $x(t)$ can be expressed as: $x(t) = \sum_{j=1}^{N} c_j(t) + r(t)$, where $c_j(t)$, $j = 1, \ldots, N$ are the IMFs, and $r(t)$ is the residue.

Although the EMD has become an established tool for analysis of single-time series, mode misalignment and mode mixing are two serious problems that limit its further application for multivariate time series. The mode misalignment corresponds to a problem where the same-index IMFs across multivariate data contain different frequency modes so that the IMFs are not matched either in the scale or in the number. The mode mixing occurs when a single IMF contains multiple oscillatory modes and/or a single mode resides in multiple IMFs, which in many cases may obscure the physical meaning of IMFs.

Recently, MEMD has been proposed to alleviate the limitations of EMD and to extend the application of EMD to multivariate time series [11]. An important step in MEMD method is that the calculation of local mean as the concept of local extrema is not well defined for multivariate signals. To deal with this problem, MEMD projects the multivariate signal along different directions to generate multiple multidimensional envelopes; these envelopes are then averaged to obtain the local mean. For an n-variable signal, the MEMD algorithm is briefly summarized as follows.

(i) Construct suitable point set (e.g., the Hammersley sequence) for sampling on an $(n-1)$-sphere.

(ii) Compute a projection $\{p^{\theta_k}(t)\}_{t=1}^{T}$ of multivariate input data $\{v(t)\}_{t=1}^{T}$ along a direction vector x^{θ_k} for all k giving $\{p^{\theta_k}(t)\}_{k=1}^{K}$.

(iii) Locate the time points $t_i^{\theta_k}$ according to maxima of the set of projected signal $\{p^{\theta_k}(t)\}_{k=1}^{K}$.

(iv) Interpolate $[t_i^{\theta_k}, v(t_i^{\theta_k})]$ to acquire multivariate envelope curves $\{e^{\theta_k}(t)\}_{k=1}^{K}$.

(v) Calculate the mean $m(t)$ of the envelope curves for a set of K direction vectors, $m(t) = (1/K) \sum_{k=1}^{K} e^{\theta_k}(t)$.

(vi) Iterate on the detail $c(t) = x(t) - m(t)$ until it becomes an IMF. The above procedure is applied to the residue $r(t) = x(t) - c(t)$.

The stoppage criterion for multivariate IMF is similar to that for univariate IMFs except that the equality constraint for number of extrema and zero crossings is not imposed, as the extrema cannot be properly defined for multivariate signal.

2.1.2. Hilbert Transform. Hilbert transform [18] has been widely used to obtain analytic (complex) signal associated with a real signal $x(t)$ and consequently, instantaneous envelope, phase functions and instantaneous frequencies. Given an arbitrary time series $x(t)$, the corresponding analytic signal is defined as: $z(t) = x(t) + iH[x(t)] = a(t) \exp[i\theta(t)]$, where $a(t)$ and $\theta(t)$ are instantaneous amplitude and phase

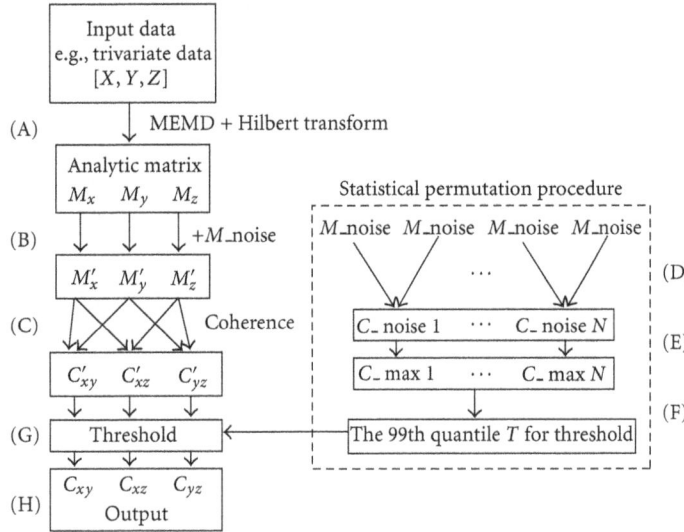

FIGURE 1: Schematic representation of the proposed noise-assisted instantaneous coherence (NAIC). A trivariate data [X Y Z] is used as an example. The first step (A) consists of transforming each time series to the corresponding analytic matrix by virtue of the MEMD and Hilbert transform. A random noise complex matrix is then added to the analytic matrix of data (B) to facilitate the calculation of coherence (C). Two random noise complex matrices are independently generated to compute their coherence. The process is repeated for N (e.g., 1000) times (D) to obtain a null distribution of the maximum coherence (E). Here, we set the P value as 0.01, thus the threshold "T" corresponds to the 10th value from the maximum of the null distribution (F). Finally, we use the "T" to threshold the coherence from (C) to be considered as statistically significant from noise (G). The output of NAIC (H) provides high-resolution time-frequency coherence spectrum.

of the analytic signal $z(t)$, and the imaginary part $H[x(t)]$ is Hilbert transform of $x(t)$: $H[x(t)] = (1/\pi)P[\int_{-\infty}^{\infty} x(u)/(t - u)du]$, where the notation P indicates the Cauchy principal value of the integral. The instantaneous frequency can then be obtained from instantaneous phase as: $f(t) = d\theta(t)/dt$.

Direct application of Hilbert transform to an arbitrary wide-band time series is of little practical value because it could produce negative frequencies, which bear no relationship to real oscillations in a time series [5, 19]. To obtain meaningful and well-behaved instantaneous frequencies, time series to be analyzed must have no riding waves and must be locally symmetrical about its mean as defined by the envelopes of local extrema. According to the definition of IMF, the IMF is an ideal candidate to take full advantage of Hilbert transform. Specifically, given an IMF $c_j(t)$, we first compute its Hilbert transform $H[c_j(t)]$ and then find its phase through the combination of $c_j(t)$ and $H[c_j(t)]$. The instantaneous frequency of IMF is finally obtained as the derivative of the instantaneous phase with respect to time. As such, we can apply Hilbert transform to the decomposed IMFs from a time series and construct a time-frequency analytic (complex) matrix, whose absolute value is the well-known Hilbert spectrum [5]. The resulting time-frequency analytic matrix makes it possible to calculate cross-spectrum between signals and autospectra of individual signals, which form the basis for coherence estimation.

2.2. Noise-Assisted Instantaneous Coherence.

Conventional coherence methods based on Fourier transform or autoregressive model assume that input signals are stationary, and their time-frequency representations suffer from the fundamental uncertainty principle. In this study, we propose a noise-assisted instantaneous coherence, which is suited to the analysis of nonstationary neural signals and offers high-resolution time-frequency coherence estimate. A schematic representation of the processing steps is shown in Figure 1.

In this method, we first employ the MEMD to adaptively decompose raw neural data into IMFs. Before applying MEMD, it should be noted that (1) neural data are often collected over certain time period from multiple channels across many trials, which can be represented as a three-dimensional matrix, that is, TimePoints × Channels × Trials, on which the MEMD cannot be directly applied, and (2) neural recordings are usually of high degree of variability, typically collected over many trials spanning from days to months, or even years, which has significant detrimental impact upon the final decomposition of MEMD when projecting the data in multidimensional space. Therefore, two important preprocessing steps [14] should be taken before applying the MEMD to neural data. First, high-dimensional neural data (e.g., TimePoints × Channels × Trials) is reshaped into such a two-dimensional time series as TimePoints × [Channels × Trials] before submitted for the MEMD analysis. It is an important step to make sure that all the IMFs be aligned not only across channels but also across trials. Second, in order to reduce the variability among neural recordings, individual time series is normalized against its temporal standard deviation before the MEMD is applied and subsequently restores the standard deviations to the corresponding IMFs after the MEMD.

Once Hilbert transform is applied to the obtained IMFs, each signal of interest yields a time-frequency analytic (complex) matrix. As described above, the analytic matrix

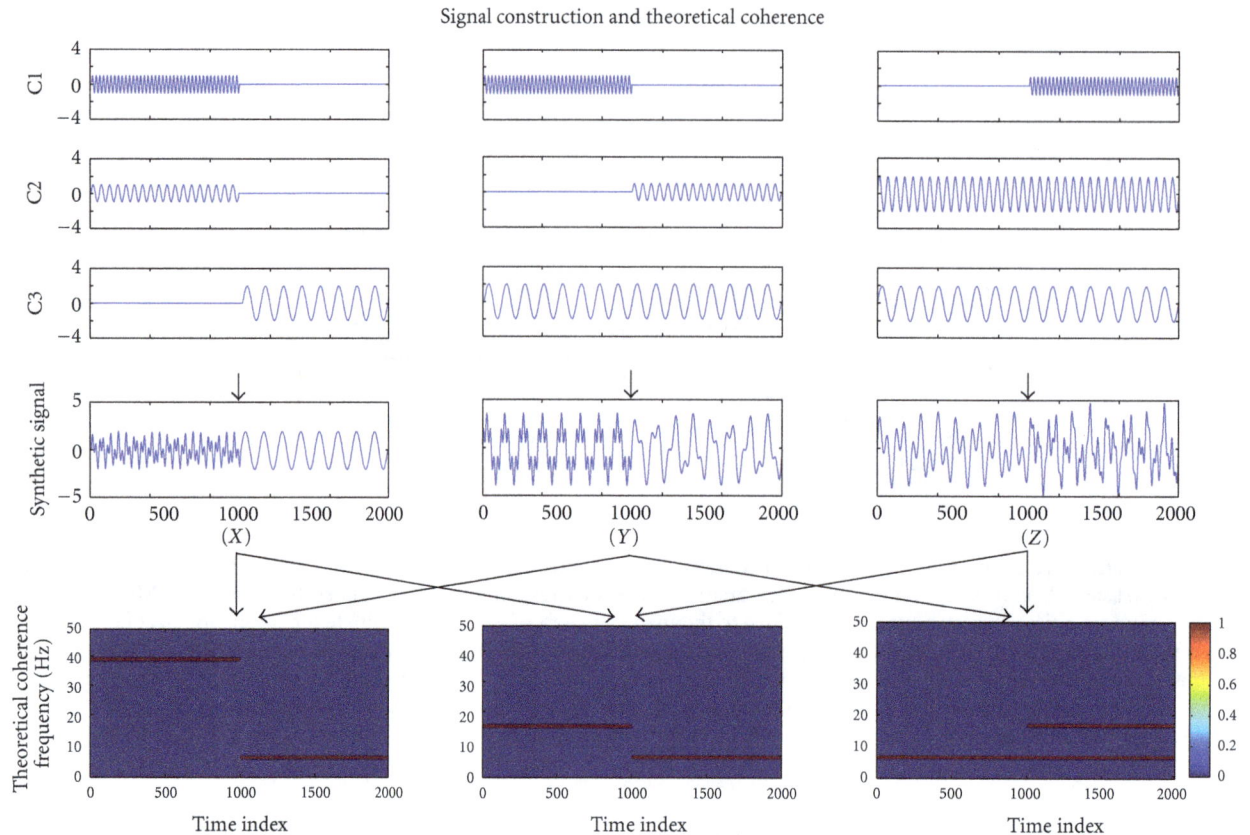

FIGURE 2: Construction of the synthetic trivariate nonstationary signal [X Y Z] and their theoretic coherence between channels. In this figure, the first three rows (C1, C2, and C3) show the components used to generate the synthetic data (the fourth row). The last row shows theoretical coherence between different channels of the synthetic data.

typically exhibits many zero values due to its high resolution, which thus cause the computational issue. As such, before computing coherence, we add a random noise complex matrix to the analytic matrix of raw data to eliminate the zero values. The real and imaginary parts of the added noise complex matrix are set to normally distributed random noise. Coherence is then estimated based on the mixed time-frequency analytic matrix. The coherence between signals i and j is defined as [20]: $C_{ij}(f) = |S_{ij}(f)|/\sqrt{S_{ii}(f)S_{jj}(f)}$, where $S_{ii}(f)$ and $S_{jj}(f)$ are the autospectra of signals, and $S_{ij}(f)$ is the cross-spectrum between signals. In this noise-assisted procedure, the added noise should be of small magnitude so as to minimize the interaction between the added noise and the original clean signal. To account for the effect of the added noise on the estimated coherence, a statistical randomization procedure is designed as follows.

(1) Time-frequency coherence of two random noise complex matrices is estimated.

(2) The maximum value of coherence across all the time and frequency in (1) is collected.

(3) Repeat Steps (1)-(2) many times, for example, 1000 times, to obtain a null distribution of the maximum coherence.

(4) From the experimental value of the estimated coherence and the aforementioned null distribution, calculate the proportion of maximum coherence in the null distribution that is larger than the experimentally observed. This proportion is called the P value.

(5) The experimental value of the estimated coherence is considered to be significant if the P value is smaller than the critical alpha level, for example, 0.05.

This procedure is a nonparametric randomization test [21], which does not need to perform statistical test at each time-frequency location, thus bypasses multiple comparison problems. By having followed all these steps, the NAIC can provide a high-resolution time-frequency coherence representation. Note that our proposed NAIC method can be readily extended to other forms of coherence such as partial and multiple coherence [20, 22] as well as Granger causality measure [16].

3. Results

3.1. Simulations. In this simulation, we generated a nonstationary three-channel signal [X, Y, Z] by concatenating and superposing three sinusoid waves, each with different frequency. Figure 2 showed how nonstationary three-channel signal was constructed and its theoretical coherence

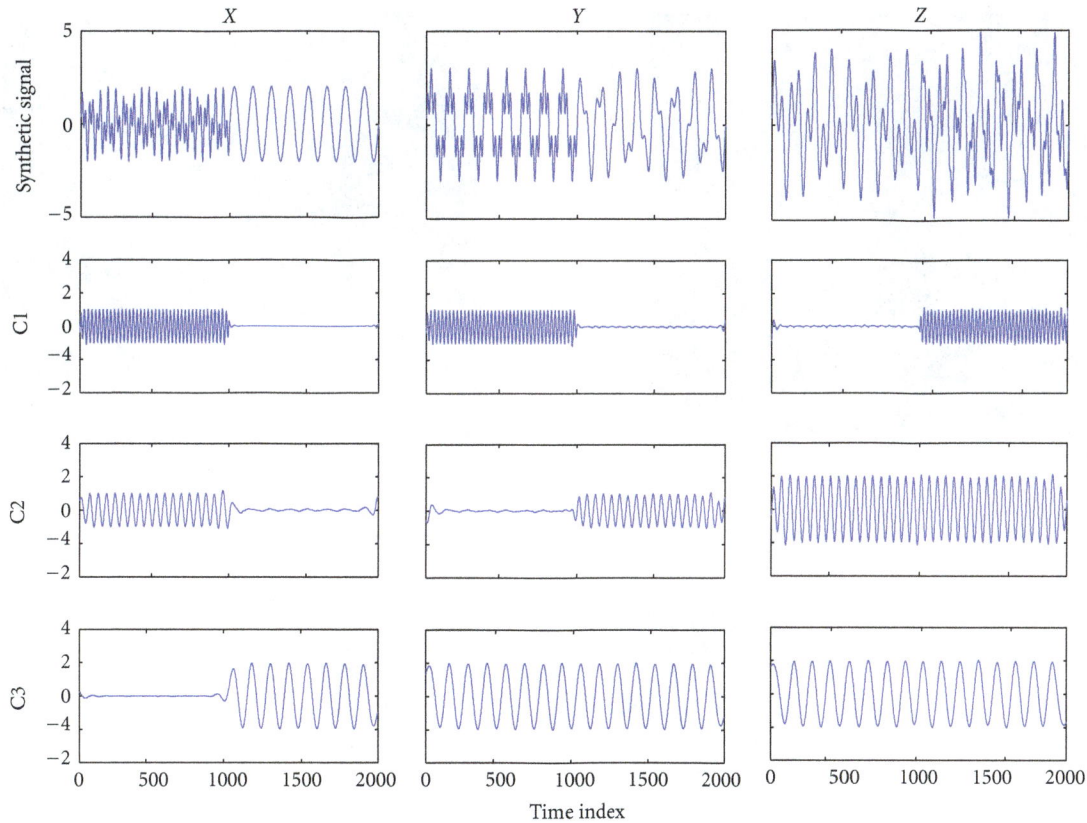

FIGURE 3: Decomposition of a synthetic trivariate nonstationary signal [X Y Z] via MEMD. The decomposed three IMFs C1–C3 correctly recover the designed components in the data (Figure 2).

between channels. We used this synthetic signal to verify the effectiveness of our NAIC in offering a high-resolution time-frequency coherence spectrum of nonstationary time series.

The MEMD was first performed to decompose the synthetic nonstationary data. Figure 3 showed that the raw data were decomposed into three IMFs, which correctly recovered the designed components in the data (Figure 2), and the common modes within the data were aligned in the IMFs with the same index. By virtue of Hilbert transform, each time series was represented by a time-frequency analytic matrix. A random noise complex matrix with noise variance of 10^{-4} was then superposed to the analytic matrix of clean data so as to facilitate the calculation of coherence between channels. Figure 4 showed time-frequency coherence based on the mixed data. From this figure, we can see that the obtained time-frequency coherence spectra reflect the designed coupling between channels. We notice, however, that the added noise induces some artifacts, shown as the bright spots scattered in the spectra. We subsequently performed the statistical randomization procedure in which the noise variance is set to 10^{-4} to identify statistically significant coherence. In Figure 5, we showed that significant coherence ($P < 0.01$) in the simulation (Figure 2) was well captured by our NAIC method. As comparisons, Fourier- and wavelet-based coherence methods were, respectively, performed to analyze the same synthetic data. For wavelet-based coherence, we used the "Morlet" as the mother wavelet

(other wavelets yield very similar results). As an example, time-frequency coherence spectra based on the Fourier and wavelet transform between channel X and Z were shown in Figure 6, in which we can see that both coherence spectra exhibit poor time-frequency resolution relative to the proposed NAIC.

In our NAIC approach, an important question is how much noise is acceptable. To examine the effect of noise on coherence estimation, we systematically varied the noise by changing its variance relative to the signal and estimated the coherence between channel X and Z in the above simulation. We measured the root mean square error (RMSE) [23] between the estimated coherence and its theoretical value as a function of noise variance. We repeated the same analysis procedure for 50 times to obtain error bars at each noise level. The result is shown in Figure 7. We can see from the figure that the RMSE declined as the noise variance decreased, and stayed constant when the noise variance approaches 10^{-4}. While the amount of noise derived from this particular simulation is empirical, it indicates that the amount of noise should be four orders of magnitude less than the signal. As a rule of thumb, we suggest that the added noise should be of infinitesimal magnitude so as to minimize the interaction between the added noise and original clean signal.

3.2. Noise-Assisted Instantaneous Coherence Analysis of Cortical Field Potential Data. In this section, we used local field

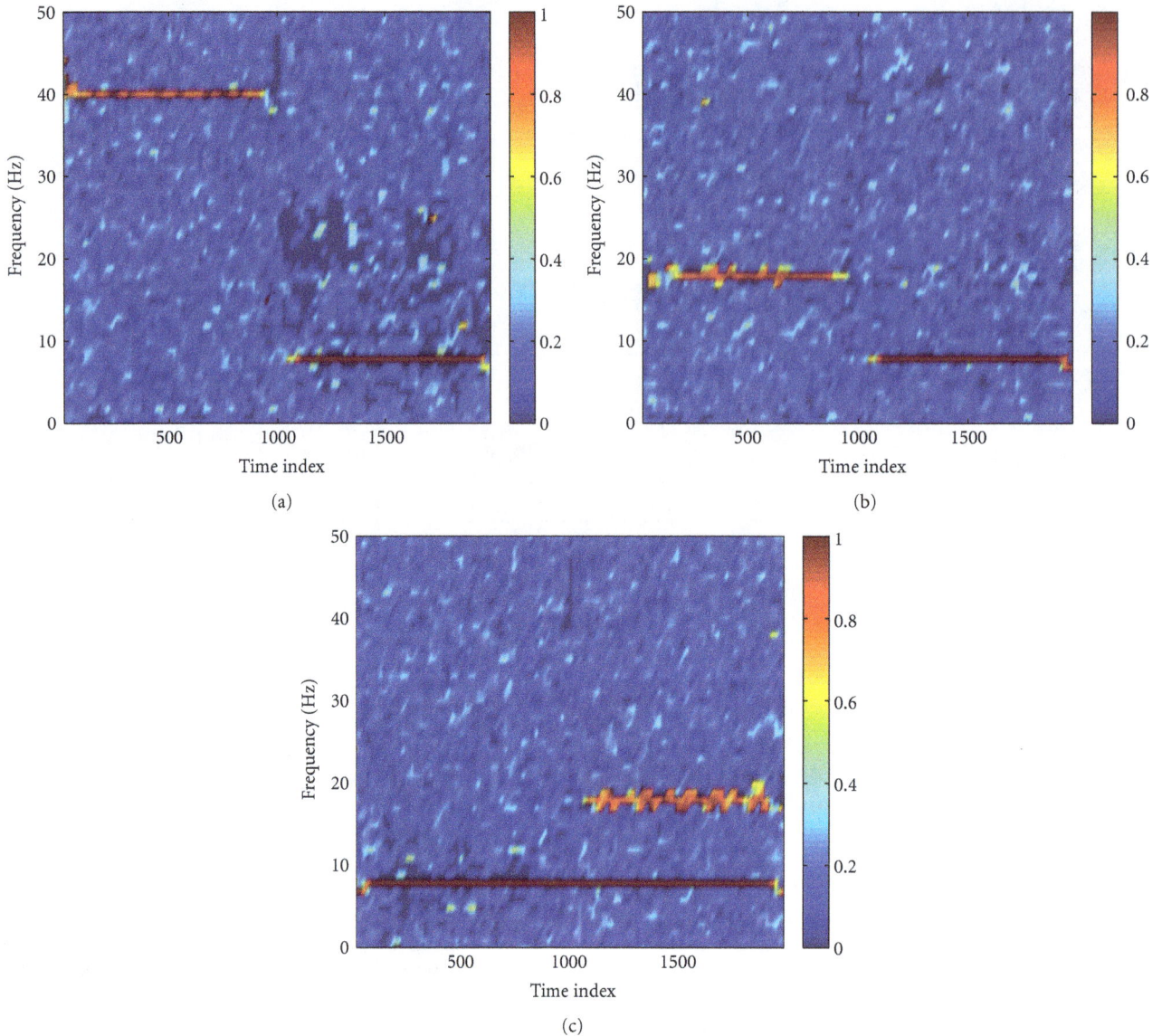

FIGURE 4: Time-frequency coherence based on the mixed analytic matrix for X-Y (a), X-Z (b), and Y-Z (c). A random noise complex matrix is superposed to the analytic matrix of data for facilitating the calculation of coherence. From these plots, we can see that the obtained coherence reflects the designed coupling patterns in the synthetic signals but contains a lot of artifacts scattered in the spectra.

potentials (LFPs) collected from visual cortex of macaque monkey while performing a visual illusion task as an example to demonstrate the usefulness of our proposed NAIC approach in providing high-resolution time-frequency coherence spectrum for nonstationary neural data.

The visual illusion task used here is called generalized flash suppression (GFS), in which a salient visual stimulus could be rendered invisible despite continuous retinal input. It provides a rare opportunity to study neural mechanisms directly related to perception [17]. In the GFS task, after the monkey gained fixation, target stimulus was presented for 1400 msec and immediately followed by the surroundings stimuli. With the presence of the surroundings, the target could be rendered subjective invisible. The monkey was trained to respond to the visibility conditions such that the

trial was classified as either "Visible" or "Invisible". Note that the stimuli in these two conditions were physically identical. Multielectrode LFP recordings were simultaneously collected from multiple cortical areas V1, V2, and V4 while monkeys performed the GFS task. The data were obtained by band pass filtering the full bandwidth signal between 1 and 500 Hz and then resampled at 1 KHz [24]. In this study, two-channel LFP from area V1 of one-second long after surrounding onset over 65 trials was used for demonstration.

As described in Method part, the MEMD was first performed on multichannel multitrial LFP data to produce the IMFs, followed by Hilbert transform to obtain the analytic matrix of data. A random noise matrix with noise variance of 10^{-4} was then added to the analytic matrix of data to facilitate the calculation of coherence. The high-resolution

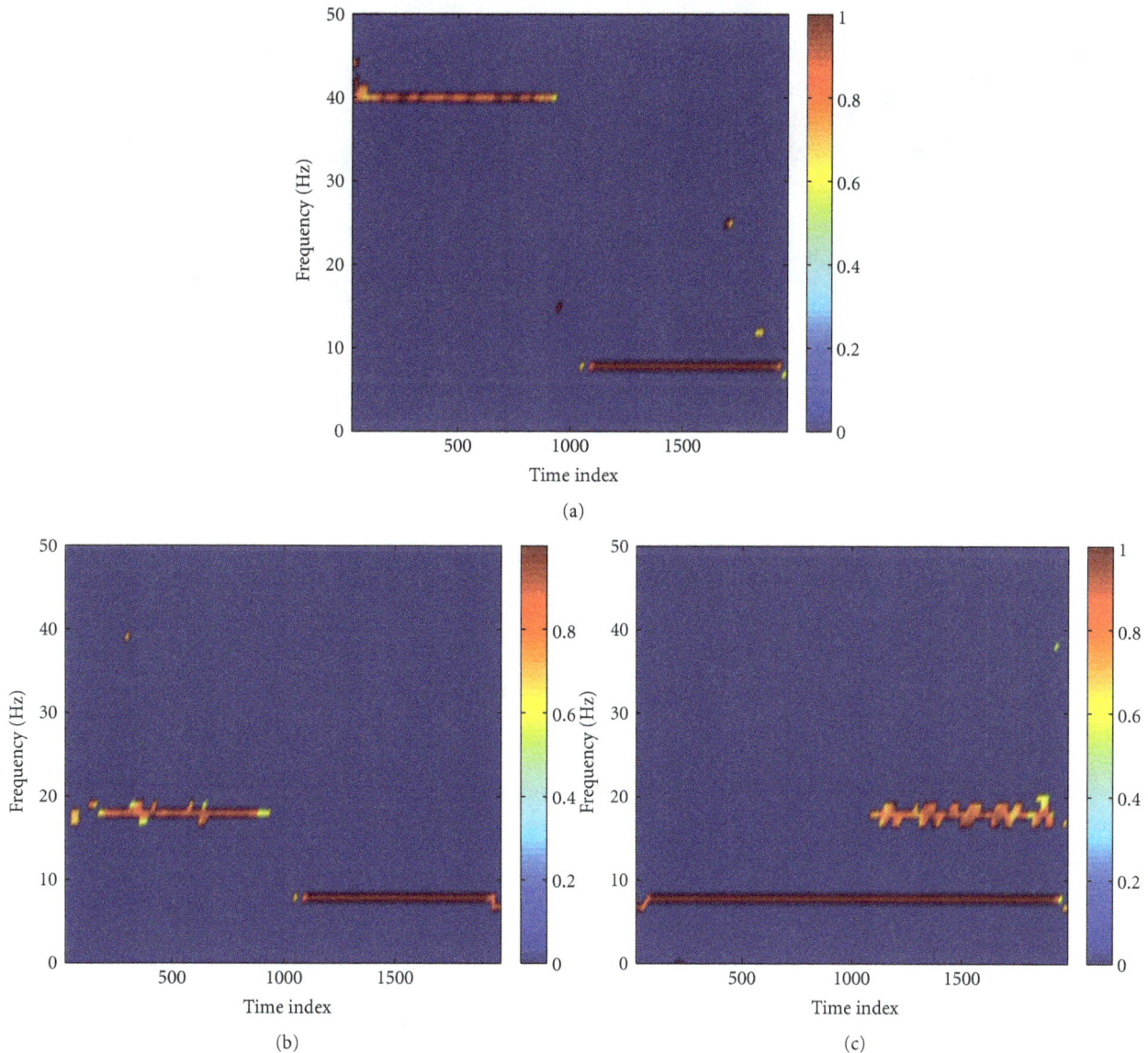

Figure 5: The noise-assisted instantaneous coherence (NAIC) for X-Y (a), X-Z (b), and Y-Z (c). From these plots, we can see that the designed coupling patterns in the simulation (Figure 2) are well captured by our NAIC method.

time-frequency coherence spectrum was finally obtained by applying the proposed statistical randomization procedure in which the noise variance was set to 10^{-4}. Figures 8(a) and 8(b) showed the grand average of the NAIC spectra in the Visible and Invisible conditions, respectively. From this figure, we can see clearly that the 10 Hz coherence initially appeared in both conditions for about 200 msec after the surrounding onset. We then observed a slightly shift of oscillatory frequency to 10–20 Hz with reduced coherence, yet the Visible condition exhibited greater coherence than the Invisible condition. As comparisons, we applied Fourier- and wavelet-based coherence methods to the same neural data, with results shown in Figures 9 and 10, respectively. Based on these figures, we can see that Fourier- and wavelet-based methods exhibited similar coherence patterns but with poor time-frequency resolution. Furthermore, we

compared the NAIC spectra between Invisible and Visible conditions to reveal how neural connectivity reflected perceptual suppression. We initially performed point-wise significance test by applying t-test to every time-frequency index between two conditions. As shown in Figure 11(a), significant perceptual suppression effect was evident in about 400 msec after the surrounding onset between 10 and 20 Hz, in which Visible condition showed significantly larger coherence than Invisible condition ($P < 0.05$, uncorrected). To deal with multiple-comparison problem, for which several methods have been proposed [21, 25, 26], we adopted a clustered-based nonparametric method [21] and found that the significant difference observed between two conditions still survived ($P < 0.05$). For comparison, we repeated the same statistical procedure to the wavelet-based coherence between two conditions. The resulting significant

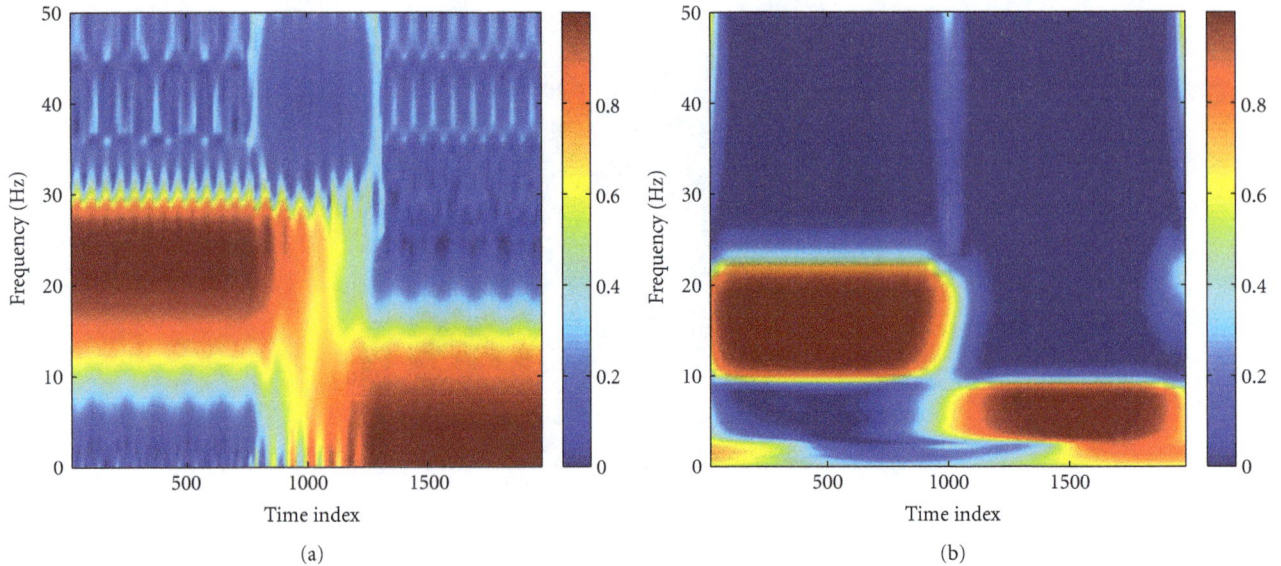

FIGURE 6: Fourier- (a) and wavelet- (b) based coherence for X-Z in the simulation. The coherence between X and Z is used as an example to demonstrate that our NAIC method (Figure 5) can provide better time-frequency resolution than Fourier- and wavelet-based coherence estimations.

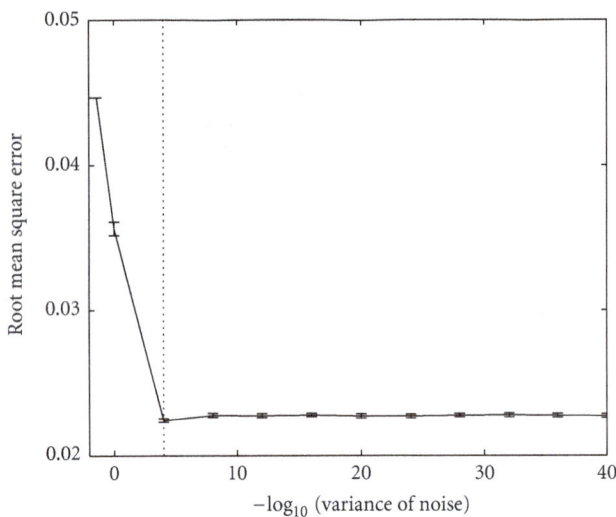

FIGURE 7: Root mean square error (RMSE) between the estimated coherence and its theoretical value as a function of the noise variance. Error bars denote standard deviations over 50 repetitions. Note that the RMSE declined as the noise variance decreased and stayed constant once the noise variance approaches 10^{-4}.

difference at both $P < 0.05$ and $P < 0.01$ is shown in Figure 11(b). Both NAIC and wavelet methods show general agreement about the concentration of significant difference in frequency. However, the NAIC is more sensitive in revealing significant difference of perceptual suppression that occurred as early as 400 msec after surrounding onset. These results together suggest that neural coherence reflects perceptual suppression, and significantly reduced coherence in Invisible condition may be associated with the reduction of brain connectivity.

4. Discussion

In this paper, we introduced a noise-assisted instantaneous coherence (NAIC) to achieve high-resolution time-frequency coherence measure. In our method, the fully data-driven MEMD, together with Hilbert transform, was first employed to provide high-resolution time-frequency spectral representation for nonstationary neural data. We then added random noise onto the spectra, which makes the calculation of coherence measure feasible. Finally, a statistical randomization procedure was designed to identify the statistically significant coherence. Computer simulations confirm that our NAIC is effective for coherence analysis of nonstationary signal. Cortical LFP data further demonstrates that our NAIC method indeed is able to provide a high-resolution time-frequency coherence representation for connectivity analysis of neural data.

The use of noise in data analysis has long been known. There are only a few that are relevant to EMD analysis. Broadly, there are two ways to utilize noise for EMD analysis. One is to assign statistical significance of information content for IMF components from any noisy data by exploiting numerical observations that (1) EMD of white noise acts essentially as a dyadic filter [27], and (2) all the IMFs of white noise follow a normal distribution [28]. Another way is to improve the EMD method by adding noise to the data. Early attempt has been made to add noise of infinitesimal amplitude to the data short of extrema in order to make the EMD operable [29]. Wu and Huang [30] explored the benefit of dyadic filter bank structure of EMD for white noise and proposed ensemble EMD (EEMD) in which multiple realizations of white noise are added to the data before applying EMD. The effect of the added white noise is to provide a uniformly distributed reference scale, which enables EMD to preserve the dyadic property and hence

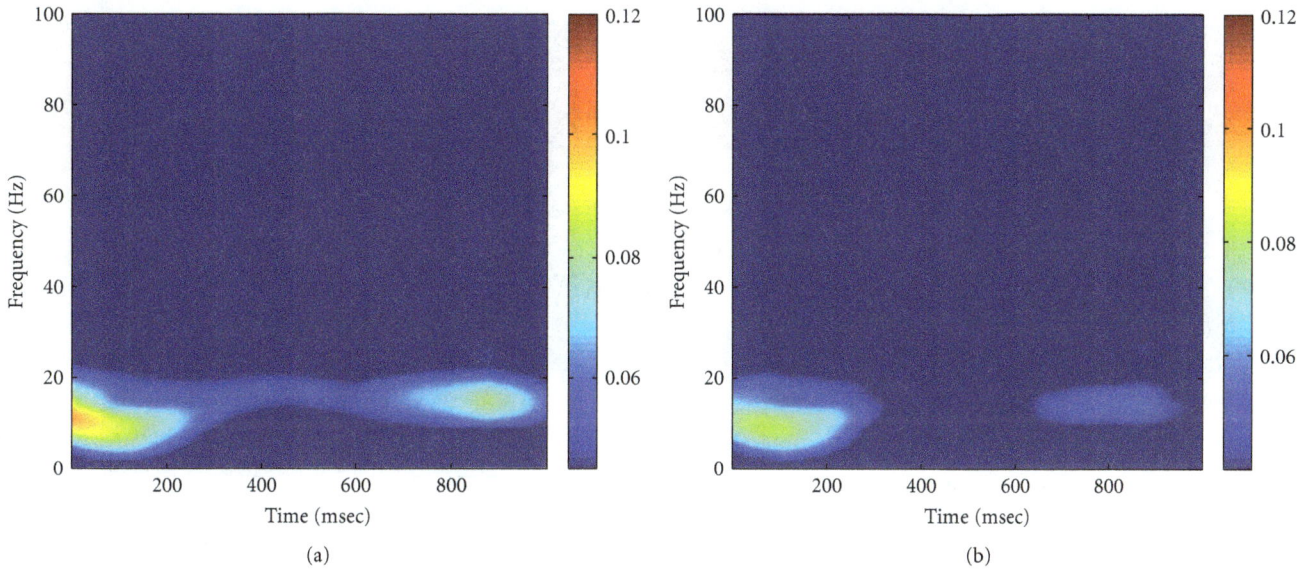

FIGURE 8: Grand average of the smoothed NAIC spectra in Visible (a) and Invisible (b) conditions. In comparison of (a) with (b), we can see that Visible condition exhibits larger coherence than Invisible condition at round 400 msec after surrounding onset.

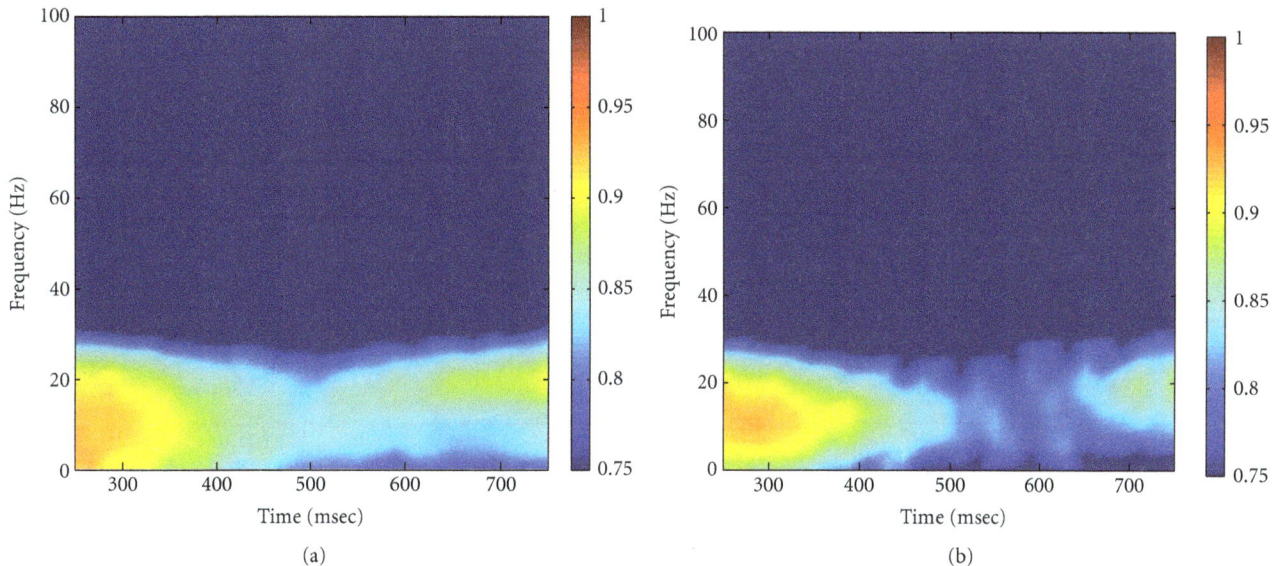

FIGURE 9: Grand average of Fourier-transform-based coherence in Visible (a) and Invisible (b) conditions. Note that the time-evolving coherence was obtained using a moving window approach, in which the window size used was 500 msec long, with a step size of 10 msec. As a result, coherence was only displayed between 250–750 msec.

reduce the chance of mode mixing. Given the random effect of noise in multiple realizations, added noise is eventually canceled out in the ensemble mean. Recently proposed noise-assisted MEMD (NA-MEMD) [12], similar to EEMD, also makes use of the dyadic property to reduce the mode-mixing problem; however, unlike EEMD, it adds white noise as separate channels and thus only a single sweep of MEMD is applied.

Our NAIC method is radically different from the above methods in that the noise is introduced *after* MEMD data decomposition. In the procedure, the noise is added to the Hilbert spectrum of data derived from MEMD to eliminate the zero values in the spectral representation and thus make the coherence estimation operable. The effect of the added noise is eliminated via a statistical randomization procedure. How much noise should be added is a crucial question in any noise-assisted methods. In our method, as a rule of thumb, we suggest that the added noise should be of infinitesimal magnitude so as to minimize the interaction between the added noise and original clean signal.

We note that our procedure is not just limited to the coherence measure demonstrated in this paper, but it can

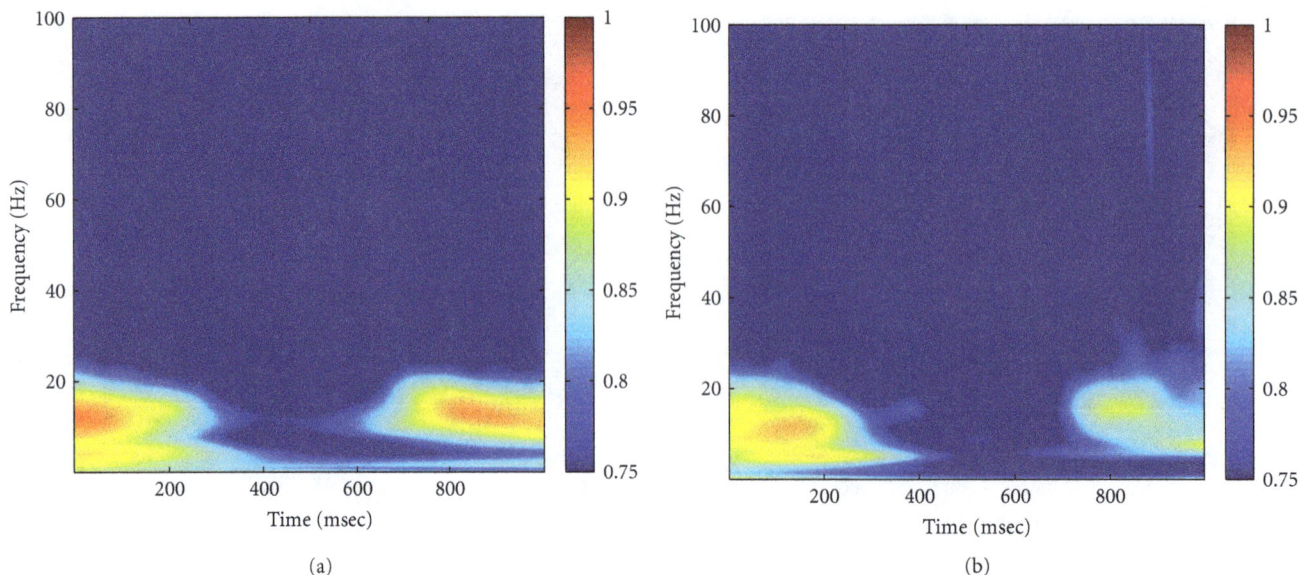

FIGURE 10: Grand average of wavelet-transform-based coherence in Visible (a) and Invisible (b) conditions. Note that we used the "Morlet" as the mother wavelet.

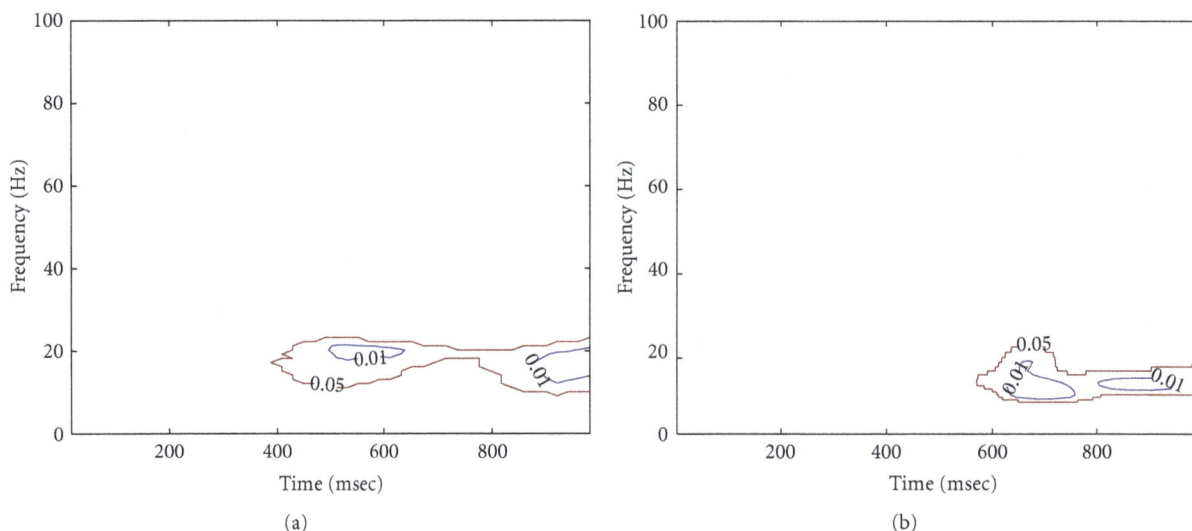

FIGURE 11: Significance test of difference between two perceptual conditions revealed by the NAIC (a) and the wavelet-based method (b). General agreement of two methods is evident, yet the NAIC is able to detect statistically significant difference of perceptual suppression occurring as early as 400 msec after surrounding onset. Level lines are depicted at $P < 0.05$ (red) and $P < 0.01$ (blue), respectively.

also be used for other forms of coherence estimation such as partial and multiple coherence as well as Granger causality [15, 16]. In addition, it is conceivable that our procedure could be applied to phase-based measures including phase synchrony [31] and even phase-based causality measure [32, 33]. Phase synchrony based on EMD [31] has clear advantage in adaptively extracting the narrow-band components (IMF) of signal and thus avoiding arbitrary preselection of frequency ranges. Importantly, phase synchrony could be used to reveal potential nonlinear coupling between IMFs of different scales, which makes the approach very attractive. We should note that Hilbert transform is usually used to

estimate instantaneous phase, which can also be calculated by alternative methods [34].

A significant strength of our coherence estimation is that the data stationarity is not required. Despite its promise, coherence is essentially a linear measure, which may fail to capture underlying nonlinear relations. As such, phase synchrony may offer a viable solution to circumvent this issue. A detailed comparison of our method with phase synchrony against well-characterized neural data would serve to identify their relative strengths and weaknesses. In addition, when signal-to-noise ratio is low, special care should be taken to interpret the estimation of coherence which could become unreliable. Furthermore, while analyzing large amount of neural data, there is a particular concern about how to reduce the computational load. Nonetheless, we have presented in the paper a noise-assisted data analysis method to achieve high-resolution coherence estimation. The analysis is supported by simulations on both synthetic and real neural data.

Acknowledgments

This work is partially supported by NIH. We thank Dr. Melanie Wilke for providing the data, which were collected at the laboratory of Dr Nikos Logothetis at Max Planck Institute for Biological Cybernetics in Germany.

References

[1] G. G. Gregoriou, S. J. Gotts, H. Zhou, and R. Desimone, "High-Frequency, long-range coupling between prefrontal and visual cortex during attention," *Science*, vol. 324, no. 5931, pp. 1207–1210, 2009.

[2] H. Liang, S. L. Bressler, M. Ding, R. Desimone, and P. Fries, "Temporal dynamics of attention-modulated neuronal synchronization in macaque V4," *Neurocomputing*, vol. 52–54, pp. 481–487, 2003.

[3] A. Brovelli, M. Ding, A. Ledberg, Y. Chen, R. Nakamura, and S. L. Bressler, "Beta oscillations in a large-scale sensorimotor cortical network: directional influences revealed by Granger causality," *Proceedings of the National Academy of Sciences of the United States of America*, vol. 101, no. 26, pp. 9849–9854, 2003.

[4] S. Mallat, *A Wavelet Tour of Signal Processing*, Academic Press, New York, NY, USA, 1998.

[5] N. E. Huang, Z. Shen, S. R. Long et al., "The empirical mode decomposition and the Hubert spectrum for nonlinear and non-stationary time series analysis," *Proceedings of the Royal Society A*, vol. 454, no. 1971, pp. 903–993, 1998.

[6] T. Tanaka and D. P. Mandic, "Complex empirical mode decomposition," *IEEE Signal Processing Letters*, vol. 14, no. 2, pp. 101–104, 2007.

[7] M. Altaf, T. Gautama, T. Tanaka, and D. P. Mandic, "Rotation invariant complex empirical mode decomposition," in *Proceedings of the IEEE International Conference on Acoustics, Speech and Signal Processing (ICASSP '07)*, pp. 1009–1012, Honolulu, Hawaii, USA, April 2007.

[8] G. Rilling, P. Flandrin, P. Goncalves, and J. M. Lilly, "Bivariate empirical mode decomposition," *IEEE Signal Processing Letters*, vol. 14, no. 12, pp. 936–939, 2007.

[9] N. Rehman and D. P. Mandic, "Empirical mode decomposition for trivariate signals," *IEEE Transaction on Signal Processing*, vol. 58, pp. 1059–1068, 2010.

[10] Z. Wu, N. E. Huang, and X. Chen, "The multi-dimensional ensemble empirical mode decomposition method," *Advances in Adaptive Data Analysis*, vol. 1, pp. 339–372, 2009.

[11] N. Rehman and D. P. Mandic, "Multivariate empirical mode decomposition," *Proceedings of the Royal Society A*, vol. 466, no. 2117, pp. 1291–1302, 2010.

[12] N. Rehman and D. P. Mandic, "Filter bank property of multivariate empirical mode decomposition," *IEEE Transactions on Signal Processing*, vol. 59, no. 5, pp. 2421–2426, 2011.

[13] M. Hu and H. Liang, "Adaptive multiscale entropy analysis of multivariate neural data," *IEEE Transactions on Biomedical Engineering*, vol. 59, no. 1, pp. 12–15, 2011.

[14] M. Hu and H. Liang, "Intrinsic mode entropy based on multivariate empirical mode decomposition and its application to neural data analysis," *Cognitive Neurodynamics*, vol. 5, no. 3, pp. 277–284, 2011.

[15] C. W. J. Granger, "Investigating causal relations by econometric models and crossspectral methods," *Econometrica*, vol. 37, no. 3, pp. 424–438, 1969.

[16] M. Dhamala, G. Rangarajan, and M. Ding, "Analyzing information flow in brain networks with nonparametric Granger causality," *NeuroImage*, vol. 41, no. 2, pp. 354–362, 2008.

[17] M. Wilke, N. K. Logothetis, and D. A. Leopold, "Generalized flash suppression of salient visual targets," *Neuron*, vol. 39, no. 6, pp. 1043–1052, 2003.

[18] A. V. Oppenheim and R. W. Schafer, *Digital Signal Processing*, Prentice Hall, Englewood Cliffs, NJ, USA, 1989.

[19] W. J. Freeman, "Origin, structure, and role of background EEG activity—part 1. Analytic amplitude," *Clinical Neurophysiology*, vol. 115, no. 9, pp. 2077–2088, 2004.

[20] M. Kaminski and H. Liang, "Causal influence: advances in neurosignal analysis," *Critical Reviews in Biomedical Engineering*, vol. 33, no. 4, pp. 347–430, 2005.

[21] E. Maris and R. Oostenveld, "Nonparametric statistical testing of EEG- and MEG-data," *Journal of Neuroscience Methods*, vol. 164, no. 1, pp. 177–190, 2007.

[22] G. M. Jenkins and D. G. Watts, *Spectral Analysis and Its Applications*, Holden-day, San Francisco, Calif, USA, 1968.

[23] Z. Wang, A. Maier, D. A. Leopold, N. K. Logothetis, and H. Liang, "Single-trial evoked potential estimation using wavelets," *Computers in Biology and Medicine*, vol. 37, no. 4, pp. 463–473, 2007.

[24] M. Wilke, N. K. Logothetis, and D. A. Leopold, "Local field potential reflects perceptual suppression in monkey visual cortex," *Proceedings of the National Academy of Sciences of the United States of America*, vol. 103, no. 46, pp. 17507–17512, 2006.

[25] D. Maraun, J. Kurths, and M. Holschneider, "Nonstationary Gaussian processes in wavelet domain: synthesis, estimation, and significance testing," *Physical Review E*, vol. 75, no. 1, Article ID 016707, 2007.

[26] D. Maraun and J. Kurths, "Cross wavelet analysis: significance testing and pitfalls," *Nonlinear Processes in Geophysics*, vol. 11, no. 4, pp. 505–514, 2004.

[27] P. Flandrin, G. Rilling, and P. Goncalves, "Empirical mode decomposition as a filter bank," *IEEE Signal Processing Letters*, vol. 11, no. 2, pp. 112–114, 2004.

[28] Z. Wu and N. E. Huang, "A study of the characteristics of white noise using the empirical mode decomposition method," *Proceedings of the Royal Society A*, vol. 460, no. 2046, pp. 1597–1611, 2004.

[29] P. Flandrin, P. Goncalves, and G. Rilling, "EMD equivalent filter banks, from interpretation to applications," in *Hilbert-Huang Transform : Introduction and Applications*, World Scientific, Singapore, 2005.

[30] Z. Wu and N. E. Huang, "Ensemble empirical mode decomposition: a noise-assisted data analysis method," *Advances in Adaptive Data Analysis*, vol. 1, pp. 1–41, 2009.

[31] C. M. Sweeney-Reed and S. J. Nasuto, "A novel approach to the detection of synchronisation in EEG based on empirical mode decomposition," *Journal of Computational Neuroscience*, vol. 23, no. 1, pp. 79–111, 2007.

[32] M. Palus and A. Stefanovska, "Direction of coupling from phases of interacting oscillators: an information-theoretic

approach," *Physical Review E*, vol. 67, no. 5, Article ID 055201, 2003.

[33] M. G. Rosenblum and A. S. Pikovsky, "Detecting direction of coupling in interacting oscillators," *Physical Review E*, vol. 64, no. 4, Article ID 045202R, 2001.

[34] M. Palus, D. Novotna, and P. Tichavsky, "Shifts of seasons at the European mid-latitudes: natural fluctuations correlated with the North Atlantic Oscillation," *Geophysical Research Letters*, vol. 32, Article ID L12805, 2005.

13

Hippocampal Anatomy Supports the Use of Context in Object Recognition: A Computational Model

Patrick Greene,[1] Mike Howard,[2] Rajan Bhattacharyya,[2] and Jean-Marc Fellous[1,3]

[1] *Graduate Program in Applied Mathematics, University of Arizona, Tucson, AZ 8572, USA*
[2] *HRL Laboratories, LLC, Malibu, CA 90265, USA*
[3] *Department of Psychology, University of Arizona, Tucson, AZ 8572, USA*

Correspondence should be addressed to Mike Howard; mdhoward@hrl.com

Academic Editor: Giorgio Ascoli

The human hippocampus receives distinct signals via the lateral entorhinal cortex, typically associated with object features, and the medial entorhinal cortex, associated with spatial or contextual information. The existence of these distinct types of information calls for some means by which they can be managed in an appropriate way, by integrating them or keeping them separate as required to improve recognition. We hypothesize that several anatomical features of the hippocampus, including differentiation in connectivity between the superior/inferior blades of DG and the distal/proximal regions of CA3 and CA1, work together to play this information managing role. We construct a set of neural network models with these features and compare their recognition performance when given noisy or partial versions of contexts and their associated objects. We found that the anterior and posterior regions of the hippocampus naturally require different ratios of object and context input for optimal performance, due to the greater number of objects versus contexts. Additionally, we found that having separate processing regions in DG significantly aided recognition in situations where object inputs were degraded. However, split processing in both DG and CA3 resulted in performance tradeoffs, though the actual hippocampus may have ways of mitigating such losses.

1. Introduction

We make sense of the world by comparing our immediate sensations with memories of similar situations. A very basic type of situation is an encounter with objects in a context. For example, objects such as a salt shaker, a glass, and a sink are expected in a kitchen. Even if these objects are encountered in an office, they suggest a kitchen-like function to the area (e.g., it is a kitchenette—not a work cubicle). In other words, the objects evoke the context in which they have been experienced in the past, and the context evokes objects that have been experienced there. The hippocampus, which is essential for the storage and retrieval of memories, is likely to play a central role in this associational process.

In rats, the hippocampus is oriented along a dorsal-ventral axis, while in primates this axis becomes an anterior-posterior axis. In both species, signals reach the hippocampus via the entorhinal cortex (EC layers II and III), which can be divided into lateral and medial portions (denoted LEC and MEC, resp.). Both the LEC and MEC can be further subdivided into caudolateral and rostromedial bands, with the caudolateral bands projecting mainly to the posterior half of the hippocampus and the rostromedial bands projecting mainly to the anterior half [1]. Within the hippocampus, these entorhinal projections reach the dentate gyrus (DG) and CA3 via the perforant path, as well as CA1. Because of the low probability of activation of its neurons, DG is thought to be responsible for producing a sparse representation of a given input which has minimal overlap with other input patterns, thereby reducing interference [2]; however the role of DG in memory is still in question [3–5]. DG projects to CA3 via the mossy fibers, a set of very strong but sparse connections. In addition to receiving inputs from DG and EC, CA3 also has many recurrent connections which are believed to serve a pattern completion purpose, allowing details lost in the sparse DG representation to be recovered in CA3 via recurrent activity and the help of EC perforant path inputs [6, 7]. The proximal region of CA3 (relative to DG) then projects to the distal portion of CA1, while the distal region of CA3 projects to the proximal portion of CA1 [8]. These connections occur in both the anterior and posterior sections of the hippocampus, with each having its own relatively independent (except in the intermediate area between anterior and posterior) DG, CA3, and CA1 subareas.

CA1 receives input from EC, with the distal portion of CA1 receiving input from LEC and proximal CA1 receiving MEC input. CA1 is essential for proper hippocampus function, since CA1 lesions result in anterograde amnesia [9]. The function of CA1 is not fully known however, although several ideas have been suggested based on theoretical [6, 7] or experimental considerations [10, 11]. We propose below a novel role for the distal and proximal areas of CA1. Each of these CA1 regions then sends output to other parts of the brain via two main pathways. The first is via the subiculum (where CA1 proximal connects to the distal part of subiculum and vice versa for CA1 distal) and to EC layers V and VI. The second pathway is via the fornix, which projects to the mammillary bodies and the thalamus.

LEC receives input mainly from perirhinal cortex and MEC receives most of its inputs from parahippocampal cortex (or postrhinal cortex in rats) which receives highly processed sensory information [12]. In this paper, we will refer to information about both the surrounding environment and spatial position within this environment, carried by the MEC, as the "context," and the information carried by LEC as the "object," which may include relational and configural information about objects [13]. It has been shown that in rats, MEC neurons display highly specific spatial grid fields, whereas LEC neurons have only weak spatial specificity [14]. This supports the notion that spatial environmental information arrives at the hippocampus primarily through MEC, whereas nonspatial information (what we call object information) is conveyed through LEC [10, 14]. Note that although our definition of context is based on the physical environment, other equally valid definitions are possible. For example, in a word list memorization task, context can refer either to the list in which a word appears (if there are multiple lists) or to a "processing context" that describes the actions done during the processing of the word, such as counting the number of vowels. It can also refer to a "temporal context" that describes, for example, whether a word was learned later or earlier during a session [15]. In the temporal context model (TCM) [12] and context maintenance and retrieval (CMR) framework [13], context is defined as an internally maintained pattern of activity different from the one corresponding to perception of the item itself. This context, consisting of background information about the object, changes over time and becomes associated with other coactive patterns.

The most obvious use of this incoming object and context information would be to associate and store object and context memories in hippocampus. However, while the necessity of hippocampus for spatial context recognition and navigation is well documented in rats [16, 17], various studies on the role of the rat hippocampus in object recognition have returned surprisingly mixed results. Several studies have found that novel object recognition in rats is impaired following hippocampal damage [18], temporary inactivation

of the dorsal region [19], or attenuation of LEC inputs to the dorsal region [10]. These experimental results suggest that detailed information about the world may indeed be represented within the dorsal hippocampus and may be dissociable from contexts, while other studies have concluded that only contextual information is stored in hippocampus [20, 21], or that the hippocampus is not required for intact spontaneous object recognition memory [22]. Analysis of neural spike data during an object recognition memory task in rats showed that hippocampal pyramidal cells primarily encode information about object location but also encode object identity as a secondary dimension [23]. Manns suggested that objects were represented mainly as points of interest on the hippocampal cognitive map, and that this map might aid the rat in recognizing encounters with particular objects [23].

In humans, the question of where memory for objects is stored is still debated, although patients such as H.M. and K.C. who have had bilateral hippocampus removals demonstrate that the hippocampus is required for the formation of new object memories and recall of most short- and medium-term memories (those formed within the last several years) [24, 25]. It is known that the human hippocampus is active during object-type recall [26]. Specifically, during successful memorization of word lists, there is significantly more activation of the posterior hippocampus than the anterior hippocampus [27]. A greater degree of posterior activation is also seen during the encoding of novel pictures [28]. However, the posterior region often responds to spatial tasks as well, particularly those concerning local spatial detail (see [29] for a review of differences in spatial and other types of processing between the anterior and posterior regions). In this study we assume that both specific object and context representations exist and are stored as memories within the hippocampus. While both regions seem to process spatial contextual information, only the posterior region has been strongly implicated in object memory as well. We therefore hypothesize that the anterior region of the primate hippocampus is primarily processing contextual information, while the posterior region is relatively more object oriented. The models that we develop in this study have explicit object recognition as a main feature and should therefore mainly be considered models of the primate hippocampus because of the evidence for explicit object representations in this case. We will discuss how our models can be related to the rat hippocampus in Section 4.

In summary, we assume that object and context memory are mainly stored in the posterior and anterior regions of hippocampus, respectively. Recall, however, that the posterior region also receives input from the caudolateral band of the MEC (which carries contextual information), and the anterior region receives input from the rostromedial band of the LEC (which carries object information). These connections raise the question of the purpose of having both object and context information reach the posterior and anterior subdivisions of the hippocampus. Recent reconsolidation experiments have shown that spatial contextual information plays a significant role in object retrieval and encoding [30, 31]. We propose that the MEC connections to the posterior stream mentioned above are vital for this. The experiments we describe next explain why context plays such a pivotal role in memory. We provide evidence that elements of hippocampal anatomy such as differentiation between the blades of DG and functional separation of the distal and proximal regions of CA1 may work together to improve the selective use of context information in object recognition, and that this can in turn improve memory performance in certain situations.

Overall, we attempt to formulate a coherent explanation for the role of several distinct anatomical features of the hippocampus and how they work together. This explanation centers on the idea that some of these anatomical differences may have evolved in order to deal with the two intrinsically different types of information that enter the hippocampus through LEC and MEC. These two types of information are "object" information (specific items within an environment, e.g., a spoon) and contextual information (the environment itself—generally less numerous than objects and related to general classes of objects, e.g., the kitchen).

Our hypothesis is that the anatomical features of the hippocampus can help manage the flow of these two types of information better than an undifferentiated hippocampus could—that they allow these two types of information to come together only in areas where it is beneficial and keep them apart otherwise. The question we are addressing in this paper is the following: can these anatomical features actually improve performance by playing the information managing role that we have proposed? We determine this by testing on a number of basic memorization tasks and find that the models with these features do indeed perform better than the baseline model on some of the tasks.

Why would we want to examine this question? There has been a large amount of work done on the theoretical aspects of how the hippocampus stores generic inputs and what role each of the main subregions (DG, CA3, and CA1) may play. In recent years, however, anatomical studies have demonstrated that there is a high degree of differentiation in terms of connectivity along multiple axes of the hippocampus (posterior-anterior and distal-proximal) and within each of the subregions. At the same time, experimental studies have shown that this differentiation has actual consequences for the memorization ability of different regions, and the studies above have shown that context plays an important role in object memorization. Thus, it is important to consider how these new findings fit into the theoretical picture of how the hippocampus works. We can no longer just consider the hippocampus or its subregions as single blocks (CA1, CA3, ...) nor consider all inputs as homogeneous if we are to have any hope of explaining existing behavioral data at the neural network level. We come at the question of how the anatomical data can explain the new experimental data with two important ideas that we believe have not been adequately expressed up to now: (1) that the anatomical features mentioned above play an information managing role whose existence only becomes necessary once we start to consider at least two different types of information converging in the hippocampus and (2) that the roles of these individual features only make sense when looking at their interaction with everything else; for example, differentiation within DG

on its own would be less useful for managing information if the rest of the upstream regions like CA1 did not also have features (like the proximal-distal distinction in our model) that make use of how DG partitions this information.

2. Methods

2.1. Model Structure and Connectivity. We use an expanded version of a model of the hippocampus developed by O'Reilly et al. [32]. The original model is a basic hippocampus consisting of a single input (EC layers II and III), a DG, CA3, and CA1 layer and a single output (EC layers V and VI). This model includes recursive connections within CA3 and DG to CA3 connections that are 10 times stronger than the EC to CA3 connections to mimic the sparse but powerful mossy fiber synapses. The smallest computational element is a "unit," which simulates a small population of neurons in a rate-coded fashion [33]. We will use the term neuron synonymously with unit in the rest of the paper. The network is trained using the Leabra algorithm, which is based on the generalized recirculation algorithm. Unlike the original model, we do not pretrain the EC $\rightarrow$ CA1 $\rightarrow$ output connection. In addition, we did not model an explicit EC output layer; we simply have an output layer. Further details of the original model can be found elsewhere [6, 34].

Our model explicitly separates the posterior and anterior halves of the hippocampus, so that the network has two CA3 regions, two DG regions, and two CA1 regions, each in the posterior and anterior poles. EC is split into lateral and medial regions (LEC and MEC, resp.), with LEC connected to all three layers on both the posterior and anterior sides to simulate the outputs of the caudolateral and rostromedial bands, respectively, and similarly for MEC. As supported by the neuroanatomy, CA3 proximal (in relation to DG) connects to CA1 distal and CA3 distal connects to CA1 proximal [8]. In order to model this distal/proximal connectivity distinction, we split each of the two CA1 regions into half again, to give four separate CA1 regions (two on the posterior side and two on the anterior side). Each CA1 receives input from the ipsilateral CA3 along with either LEC input (if it is distal) or MEC input (if it is proximal). This network will be referred to as the "Baseline" network (Figure 1).

We model inhibition in each layer as a competitive k-winner-take-all process, where only the top k most active neurons send their outputs to the next layer. Thus we can set the activity level in each region to approximately that seen in experimental results, where the activity level refers to the percentage of active neurons at any given time. EC, DG, CA3, and CA1 have experimental activity levels of 7%, 1%, 2.5%, and 2.5%, respectively [34]. In our model, these levels are set to 25%, 1.5%, 2.3%, and 2.5%, respectively. The discrepancy in EC (both LEC and MEC) is because it is serving as our input layer and does no computation; EC is just large enough to hold training patterns with 25% of the units active. The LEC and MEC layers each consist of 64 neurons. The DG, CA3, and CA1 layers on the posterior side consist of 800, 256, and 800 neurons, respectively (the distal and proximal regions of CA1 have 400 neurons each). The same numbers apply on the anterior side.

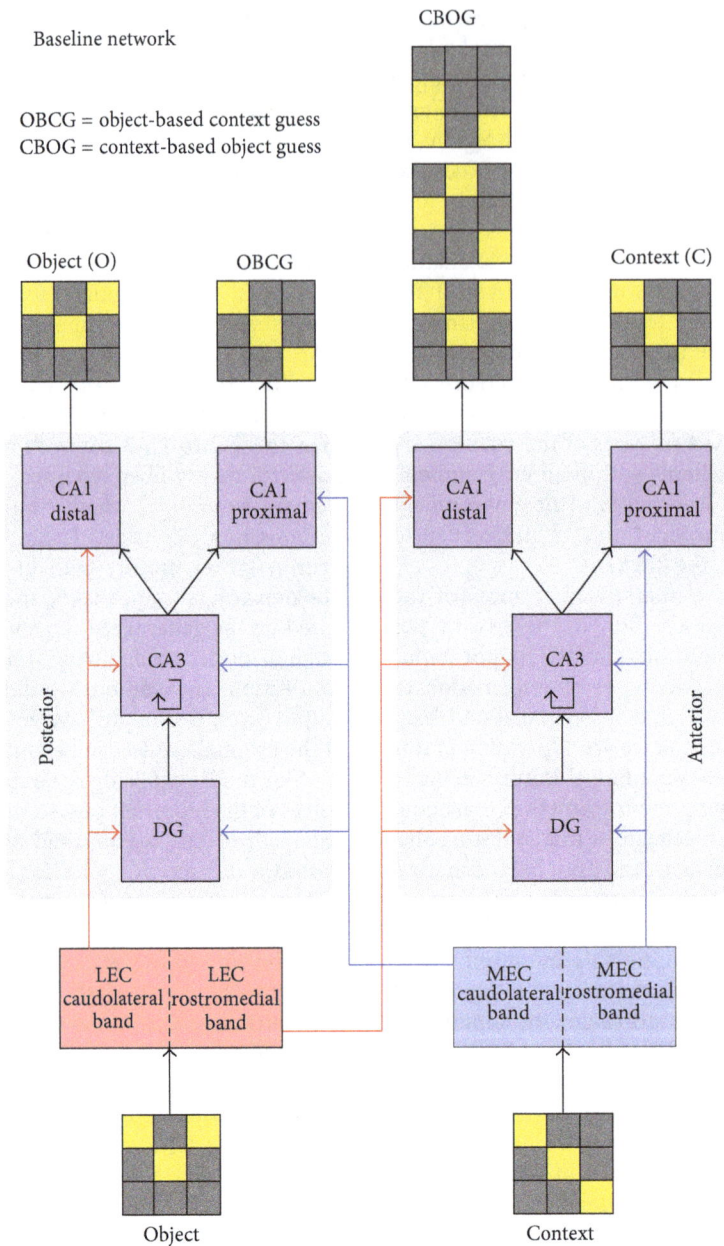

FIGURE 1: Layer and connectivity diagram of the Baseline network. Matrices representing an object and a context are the inputs to the network. The outputs are an object (O), an object-based context guess (OBCG), a context-based object guess (CBOG), and a context (C). The OBCG output is the context that the input object is associated with during training, and the CBOG output is the set of objects that were associated with the input context during training.

As discussed above, the LEC primarily carries object information while the MEC carries spatial contextual information. Hence in our model we conceptualize the LEC inputs as "objects" and MEC inputs as "context." In assigning roles to the output layers corresponding to the distal and proximal CA1 regions, we first note that these two regions lie on largely separate output pathways: CA3 proximal connects mainly to CA1 distal and CA1 distal connects mainly to the proximal part of the subiculum, which in turn projects back to the LEC [8, 35]. On the other hand, CA3 distal connects mainly to CA1 proximal and CA1 proximal connects to the distal part

of the subiculum, which in turn projects back to the MEC [8, 35]. If these pathways were both carrying the same type of information, there would be no need for such a wiring scheme to keep them separate. Since our model only contains two types of information, object and context, we assume that one of these pathways is carrying object information and the other is carrying context.

On the posterior side of hippocampus we are mainly focused on its object processing capabilities; hence we assume that the relevant outputs must be largely dependent on using object-type information from LEC. We hypothesize that these two outputs are an object guess and an object-based context guess. The object guess pathway does standard object recognition by taking the input object, matching it to the closest object in memory, and giving the best match as its output. The object-based context guess pathway uses the object input to generate the context that the object is associated with: if one gives it the object "swing set," it returns "playground," if one gives it "refrigerator," it returns "kitchen," and so forth. We emphasize that not every neuron in the given regions is doing these operations or using only one type of information to do them. But, to the extent that we have neurons that are encoding nonspatial information in these regions, we predict that there will be more of them (or alternatively, that the degree to which they are sensitive to spatial information will be lower) in the distal region of CA1 compared to the proximal region. Experimental results by Henriksen et al. provide support for this, showing that the strongest spatial modulation occurs in the proximal part of CA1, and that distal CA1 cells are less spatially tuned [36].

On the anterior side of the hippocampus, since we focus on its contextual processing capabilities, we require that its outputs be largely dependent on using context-type information from MEC. We hypothesize that these two outputs are a context guess and a context-based object guess. The context guess pathway matches the input context to the closest context in memory, and the context-based object guess uses the input context to generate a list of the set of objects associated with the given context. For example, given the context input "playground," it would output the object list "swings, sand-box, slide."

The final question is which of the distal or proximal CA1 regions is playing each of these roles. It is known that MEC projects preferentially to the proximal region of CA1, while LEC projects preferentially to the distal region [37]. Assuming that the purpose of the two CA1 streams is to keep object and context-type information largely separate, it seems unlikely that object information from LEC would then be projected to the context stream at CA1, and similarly for MEC inputs and the object stream. Thus, on the posterior side, we conclude that the object guess is output by distal CA1 and the object-based context guess is output by proximal CA1. Similarly, on the anterior side, we conclude that the context-based object guess is output by distal CA1, and the context guess is output by proximal CA1.

2.2. Model Variants.

Variants of the Baseline network were designed to investigate the effect of two additional anatomical details. The first is the differentiation between the inferior and superior blades of DG. As shown in Figure 2, the DG may be functionally separated into two parts because of the different strengths of LEC and MEC connections onto the superior and inferior blades and a postulated dendritic gating mechanism [38, 39]. Both blades receive proximal dendritic MEC input via the medial perforant path (MPP) and distal dendritic LEC input via the lateral perforant path (LPP). However, the superior blade receives stronger LPP input whereas the inferior blade receives stronger MPP input. We further hypothesize that the effect of this connectivity is

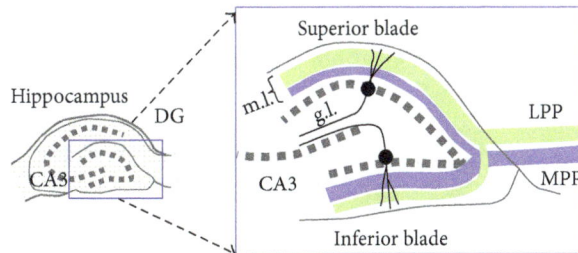

FIGURE 2: Connectivity of lateral perforant path (LPP) and medial perforant path (MPP) inputs to superior and inferior blade of DG. The LPP and MPP fiber lamina are thicker on the superior blade and inferior blades, respectively, resulting in higher effective synaptic weights (adapted from [38]).

different depending on whether the given DG region lies in the posterior or anterior hippocampus.

In the posterior hippocampus, the object information contained in the LPP input is more relevant to its task than the context information coming from the MPP input. Thus we would expect that the DG neurons in posterior hippocampus would be biased toward (or learn to weight more heavily) the LPP inputs over the MPP inputs. However, the fact remains that the MPP inputs are more proximal to the soma and thus cannot be completely ignored. The hypothesized result of this tug-of-war (more relevant LPP input but more proximal MPP input) is that, in the superior blade where the LPP object inputs are already stronger than the MPP context inputs, LPP is able to largely control the neurons' firing. In the inferior blade where LPP inputs are weaker, they are able to achieve approximate parity with the MPP input.

In anterior hippocampus the MPP contextual inputs are both more relevant and more proximal to the soma. We hypothesize that this allows the MPP inputs to control the neurons' firing, though to a greater extent in the inferior blade than the superior blade, where LPP input cannot be totally ignored.

We model the two blades of DG as separate layers in both the anterior and posterior sides of hippocampus in order to determine their effect on performance. The model with DG layers split in this way, but with all other architecture the same as in the Baseline model, will be referred to as the "SplitDG" model (Figure 3).

The second anatomical detail we consider is differentiation between the proximal and distal regions of CA3. As mentioned in the introduction, CA3 has distal and proximal regions just as in CA1 (here distal and proximal refer to distance from DG, rather than to the location on the dendrite). These regions receive different amounts of inferior and superior blade DG input and have distinct patterns of recurrent connections [8]. The amount of recurrent versus feed-forward connections is also different between the two sub-areas. Thus these two regions of CA3 may be performing functionally different roles. In order to determine the purpose of such a split and test whether it may confer some performance advantage, we construct a third network that has CA3 split into two layers on each of the posterior and anterior sides, in addition to the DG split described above.

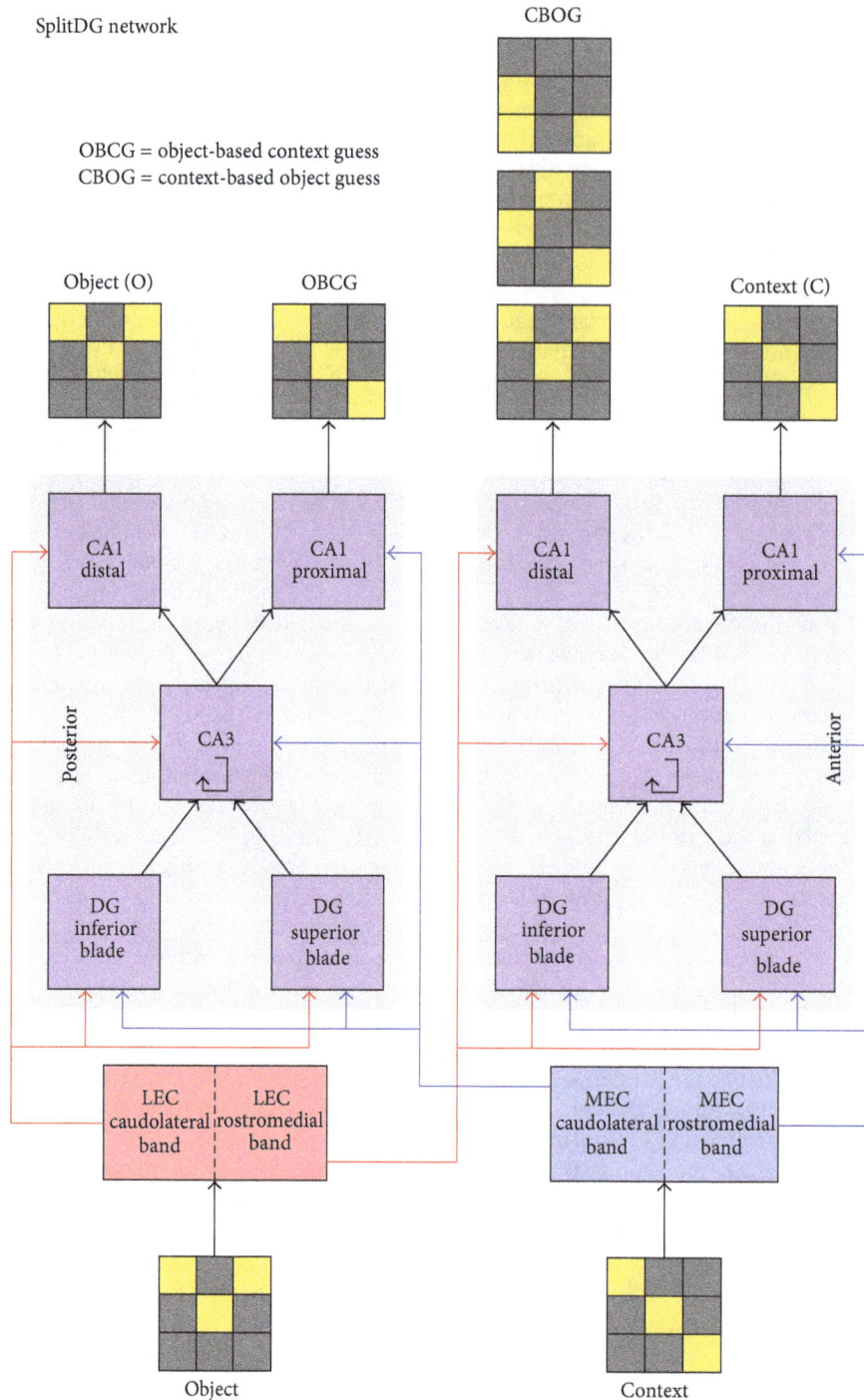

FIGURE 3: Layer and connectivity diagram of the SplitDG network.

Anatomically, the inferior blade of DG projects to proximal CA3, while the superior blade projects to both proximal and distal portions of CA3 [8]. As a modeling approximation we connect the inferior blade to proximal CA3 and the superior blade to distal CA3 only. Although our model does not capture the detailed connectivity of CA3, we believe it serves as a good starting point for understanding the purpose of having

distinct CA3 regions. We will refer to this network as the "AllSplit" network (Figure 4).

2.3. The "+" Networks. We constructed two additional networks, SplitDG+ and AllSplit+, for the purposes of comparison across networks with equal training set error. SplitDG+ is the same as SplitDG, except that each of the DG layers is

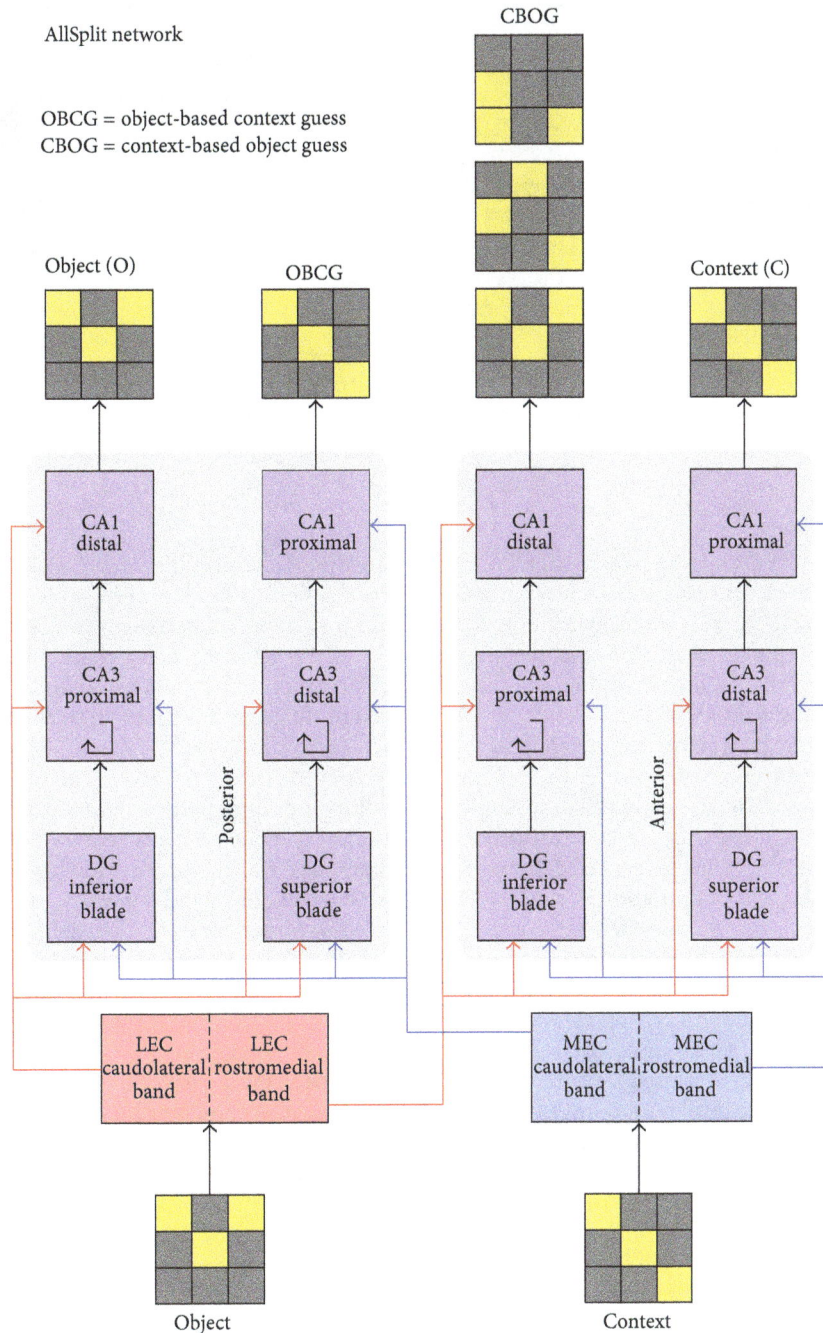

FIGURE 4: Layer and connectivity diagram of the AllSplit network.

doubled in size. Similarly, AllSplit+ is the same as AllSplit, except that both the CA3 and DG layers have been doubled in size. The relevance of these networks is addressed in more detail in the discussion.

2.4. Training and Test Sets. The training set consists of object patterns and context patterns (Figure 5). Each object is a random 8×8 matrix of zeros and ones, consisting of 16 ones (active units) and 48 zeros (inactive units). Contexts are constructed the same way. There are 120 unique objects and 40 unique contexts (3 unique objects per context).

The output layers of the network are referred to as "object" (O), "object-based context guess" (OBCG), "context-based object guess" (CBOG), and "context" (C). The correct output for the object output layer (used as a training signal and ground truth for the error metric) is the object matrix for the input object. For the OBCG layer, the correct output is the context matrix associated with the given object input. For the CBOG layer, the correct output is the three object matrices for the three objects associated with the given context. Finally, for the context output layer, the correct output is the context matrix for the input context.

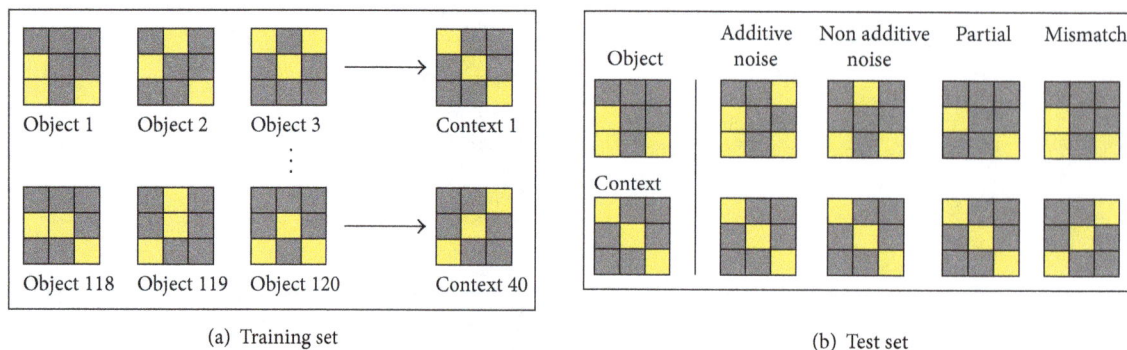

(a) Training set

(b) Test set

FIGURE 5: Training and test sets. The training set consists of 120 objects and 40 contexts, with 3 objects per context. The test sets are the same as the training set, except with either noise added (additive or nonadditive noise), part of the pattern missing (partial cue), or an object and context mismatch.

The network is trained for 20 epochs, where each epoch consists of presenting all 120 object-context pairs in a random order and applying the Leabra weight update algorithm after each presentation. Twenty epochs were chosen as the stopping point because all networks' training error had stabilized at close to their minimum value by this time.

After training, the networks' weights are frozen, and the networks' performance is measured using four test sets: additive noise, nonadditive noise, partial cue, and context mismatch (Figure 5). In additive noise tests, objects or contexts have some of the zeros in their matrix replaced by ones, simulating additional active units. In non-additive noise tests, for each zero that is replaced by a one, a one from the original pattern is replaced by a zero, so that the total number of active units remains the same. In partial cue tests, some of the ones in the original object or context pattern are replaced by zeros, resulting in a fewer number of active units overall. In the context mismatch test, an object is paired with a different context from the one it was associated with during training. The level of difficulty of each test depends on the number of units that are changed from the original pattern, which we denote by percentages in the figures.

Many experimental or real-life situations can be interpreted in terms of these simple tests or a combination of them. For example, if the object we are memorizing is a man's face, we recognize who he is even if he has grown a mustache (additive noise), is wearing a hat (non-additive noise, since it adds something but also covers his hair, which is one of his original features), or is partially turned away from us (partial cue). In addition, we recognize him even if we see the same man in a different context (mismatch), although this may be a somewhat more subtle issue than the previous ones, which we will discuss further.

3. Results

3.1. Setting the Crossconnection Weights for the Baseline Model. We will refer to the connections from LEC to the anterior side of hippocampus and from MEC to the posterior side as "crossconnections," since they bring object information into the context-dominated anterior side and context information

into the object-dominated posterior side, respectively. The first task was to determine how the relative amount of crossconnection and noncrossconnection input affects the error rate of the Baseline network and use this to maximize its performance. Since the OBCG and CBOG output layers are used in different situations from the O and C layers, we test them accordingly on a different set of tasks. The O and C layers were tested on a set with mixed additive and nonadditive noise introduced to object and context (15% noise in each layer) and a set where both object and context were incomplete (40% complete each). The OBCG layers were tested when object and context were mismatched, with noise (30%) in context only, and partial (40%) in context only. For the CBOG layer, the mismatch test was the same, but the noise and partial tests were in the object input only (30% object noise and 40% partial object) rather the context. The results can be seen in Figure 6.

To determine the optimal LEC and MEC weights for each output stream, we plot each output layer's average error over the set of relevant tests as a function of the crossconnection input it receives. This is shown in Figure 7. We use this as a guide to set the relative weights of the crossconnections for all the networks to levels which optimize their performance on the sample tests. Note that for networks such as SplitDG or AllSplit which have split layers, we optimize the crossconnection strengths for these layers independently, while for the Baseline network, we must average the optimal connection strengths over the two output types. For example, since the O output does best with a multiplier of 3 while OBCG does best with a multiplier of 0, we end up with the Baseline network having a relative weight multiplier of 1.5 for the MEC to dorsal side crossconnections. For the AllSplit network, we do not need to make this compromise and can directly use a multiplier of 3 for the MEC inputs into the DG and CA3 areas which feed into O and use a small multiplier close to 0 for the DG and CA3 areas which feed into OBCG. The SplitDG network has the same weighting for crossconnections to DG and CA1 as the AllSplit network and the same weighting to CA3 as the Baseline network, since it only has a single CA3 which the O and OBCG streams must share. These results show that there is unlikely to be a single

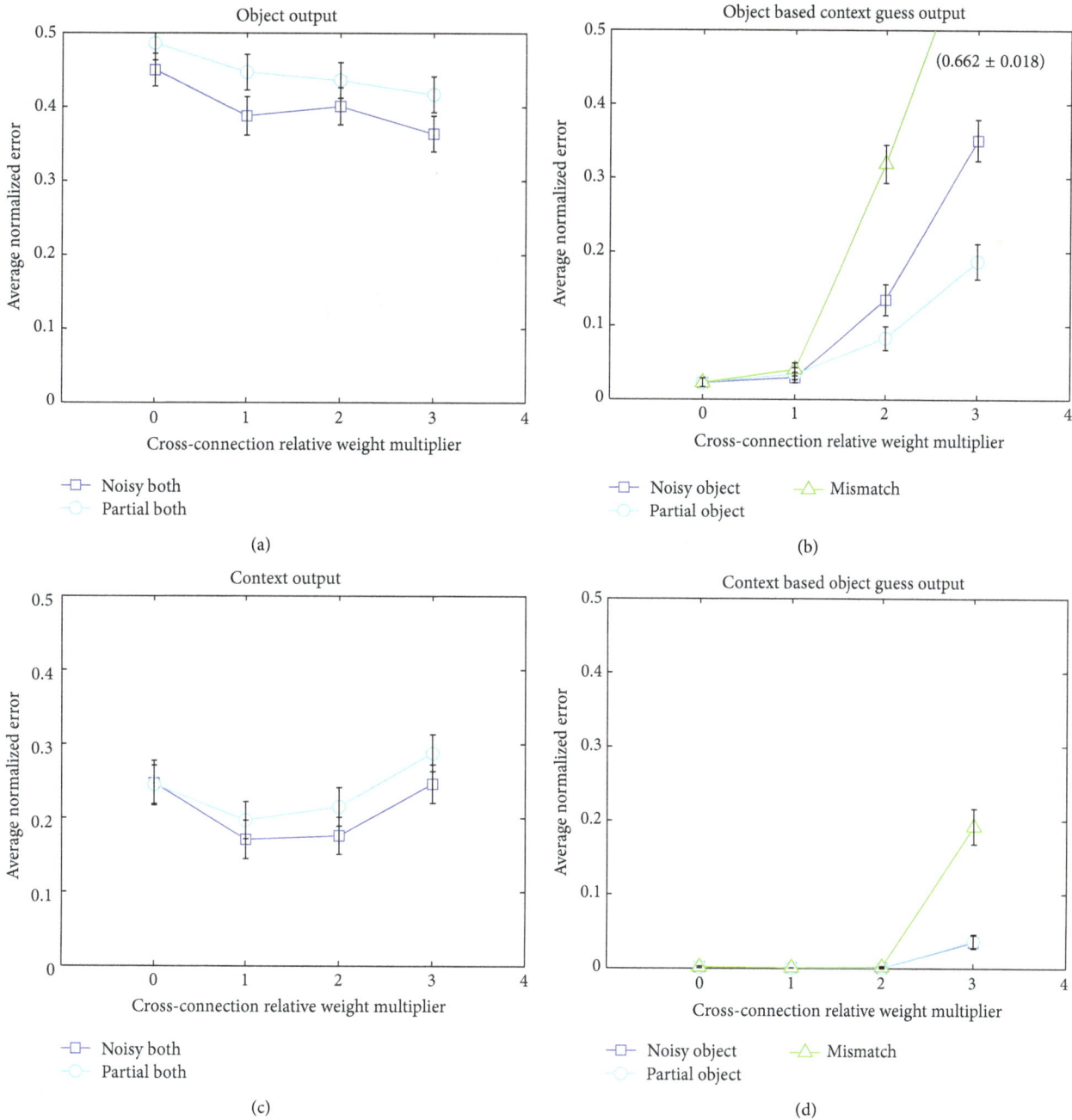

FIGURE 6: Error for each of the four output layers of the Baseline network on various sample tasks, as a function of crossconnection weighting. Crossconnection input refers to LEC input to anterior hippocampus and MEC input to posterior hippocampus. Higher relative weight multiplier values mean stronger MEC input to posterior and stronger LEC input to anterior streams. (a) Object output error on noisy and partial cue tests (where both object and context are noisy or partial, resp.) as a function of crossconnection strength. (b) OBCG output error on noisy and partial cue tests (here the noise and partial are only in the context) as a function of crossconnection strength. (c) Same as A, except the error is measured at the context output layer. (d) Same as B, except only the object is noisy or partial, and the error is measured at the CBOG output layer. Error bars are standard errors of the mean.

set of crossconnection weights that optimizes performance for the various output layers across a range of different tasks. The flexibility provided by having different DG and CA3 layers that can take different levels of crossconnection input provides an advantage and may be one of the reasons why this anatomical differentiation exists in the hippocampus.

3.2. Training Error. Having fixed the crossconnection weights in all networks to values that minimize the error over the sample test sets, we now compare the networks. First we measure the error on the training set after 20 epochs, when the error has reached its asymptotic minimum. Figure 8 shows the average error for each of the five networks,

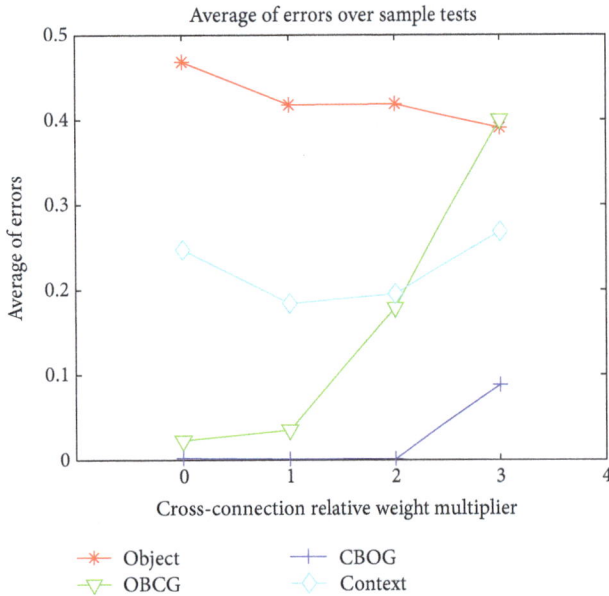

FIGURE 7: Average of errors over sample tests for each output layer, as a function of crossconnection strength.

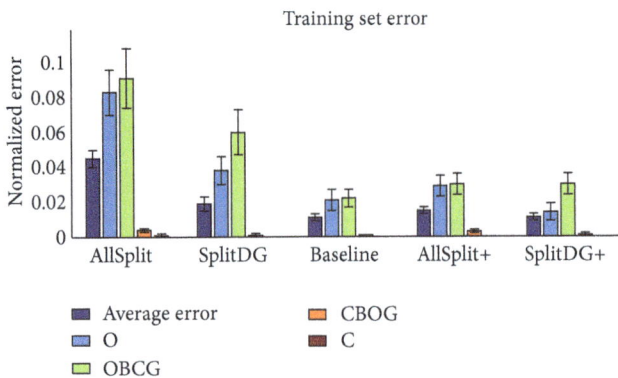

FIGURE 8: Error on the training set for each of the five networks after 20 epochs of training.

along with the error on each of the four outputs individually. The networks can be divided into two categories for further comparison: those which have the same number of neurons, consisting of AllSplit, SplitDG, and Baseline and those which have the same initial training set error, consisting of Baseline, AllSplit+, and SplitDG+. This illustrates the fact that differences in layer size may play an important role in the networks' basic memorization ability. When a layer is split, each of the halves can specialize more efficiently on the task, for example, pattern completing an object or converting an object to a context guess. On the other hand, it must hold the same number of object or context memories despite being half the size, resulting in more memorization errors. Figure 8 shows two possible outcomes of this tradeoff: for the context and CBOG streams, there is no difference in training error before and after splitting the CA3 and DG layers which lie on those streams (compare C and CBOG error between Baseline, SplitDG, and AllSplit). This is due

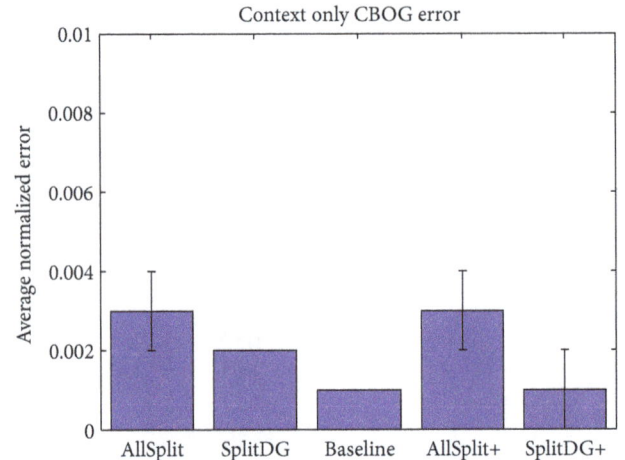

FIGURE 9: Error on the context-based object guess (CBOG) output when given only the context as input.

to the fact that these layers only need to store 40 context memories, so even when they are split in half they have no difficulty memorizing them all. However, for the object and OBCG streams, splitting their respective DG or CA3 layers results in a significant increase in training error (compare O and OBCG error between the same three networks). In this case they need to memorize 120 objects, and a CA3 or DG layer half the size is not sufficient. The results of the "+" networks show that this is no longer a problem if we simply have more neurons to start with. The question of whether it is more appropriate to compare Baseline with AllSplit+ and SplitDG+ (since they start off with the same training set error) or to compare Baseline with AllSplit and SplitDG (since they have the same number of neurons) depends on which situation is more likely to reflect biological reality and will be addressed further in the discussion. In all subsequent tests we include the results for each of the five networks.

3.3. Test Sets. We seek to determine how, and in what situations, contextual information can be used by the hippocampus to aid in object recognition and recall (and similarly how object information can aid context recognition), and what role differentiation within DG and CA3 may play in using this information. To answer these questions, we have constructed three primary networks with varying degrees of differentiation in the DG and CA3 layers and will test the ability of each of these networks to recognize objects and contexts under various conditions of degraded inputs.

A common and simple test of human memory is to have a subject memorize a list of words or set of objects, then recall them given a cue. We would like to determine if our network is capable of giving this object output even without the object input. We simulate this task in our networks by presenting a context (the cue—which would consist of the room and the experimenter) and use the CBOG output to get a list of the objects which have been memorized in the given context. Figure 9 shows that the CBOG stream performs well in this task. There is little difference between networks here since all

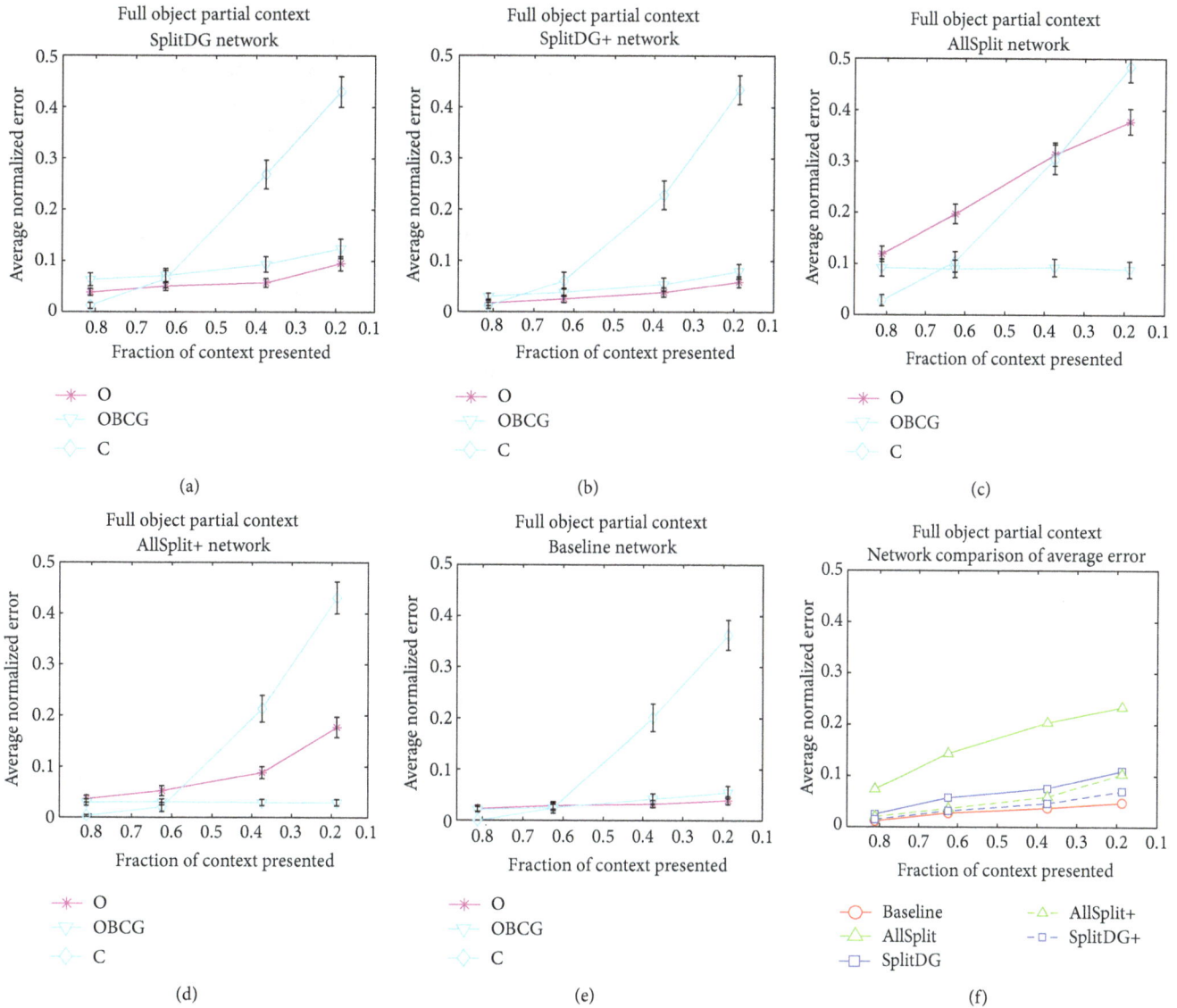

FIGURE 10: Error for each of the networks' O, C, and OBCG layers when a partial context and full object were given as input. (a) SplitDG, (b) SplitDG+, (c) AllSplit, (d) AllSplit+, (e) Baseline, and (f) average error across the object output and the lowest of the two context outputs (C or OBCG) for each network, as a function of percentage of context input presented.

use the same crossconnection strength into the anterior side, where CBOG is located.

Next we consider the case where the context, rather than the object, is missing to various degrees. This test will help us determine the degree to which relying on contextual input to recognize objects is disadvantageous when the context is degraded. Figure 10 shows the individual performance of the output layers O, OBCG, and C as a function of how much of the context is given for the various networks, illustrating the effect of having increased MEC inputs into the object stream. Because the AllSplit network's object stream uses a relatively large amount of context information, partial context input has a greater adverse effect on the AllSplit network's O output than it does on the Baseline network's O output. The same is true for SplitDG and its "+" counterpart. Thus we do not expect the AllSplit network to do well compared to the

Baseline network in this situation, and Figure 10(e), which gives the average error for each network by taking the average of the error from the O output and the best context output (either C or OBCG), confirms this. The "+" networks do relatively better since their larger CA3 sizes allow the partial context-object mix within the object stream to be pattern completed to a higher degree. This figure also shows the advantage (for all the networks) of having an OBCG output when context is difficult to discern. When the fraction of context drops below 60%, the networks can rely on OBCG for their context guess rather than the context stream output C.

The analogous situation on the object side is to present a partial object and a full context. This test helps us determine how well the various network architectures can utilize context to aid object recognition. At first glance it seems that we ought to make use of the CBOG output to generate an object guess

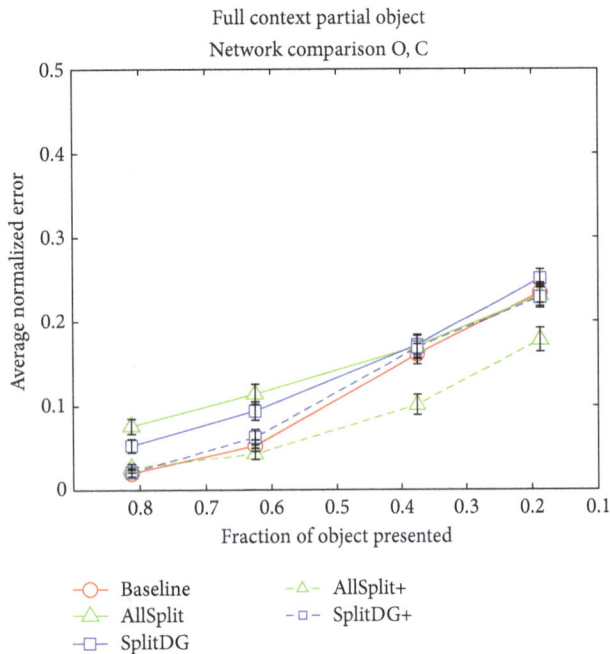

Full context partial object
Network comparison O, C

FIGURE 11: Average error on the O and C output layers when full contexts and partial objects were given as input.

using the clean context, just as we used the OBCG layer in the partial context case above. However, the problem is that the CBOG layer activates multiple possible objects rather than a single object, and thus we would need a way of picking the correct object out of this list. Cortical areas outside of hippocampus could conceivably accomplish this by picking the closest match either to the original input or the O output; however, since we restrict our model to the hippocampus proper, we have not attempted to implement such a scheme and instead use the O output as our exclusive object guess. We consider this issue further in the discussion. Figure 11 shows that, when the object is partially given, the increased amount of context information that the AllSplit network uses via the MEC to posterior crossconnections becomes an advantage rather than a liability, as it now has an error rate similar to that of the Baseline network. When the initial training set memorization disadvantage is accounted for under the AllSplit+ network, a consistent advantage for all partial conditions is seen. Surprisingly, neither SplitDG nor SplitDG+ is able to do better than the Baseline network, suggesting that some degree of heterogeneity within CA3 is necessary to take advantage of the additional context information.

Figure 12 illustrates the effect of having additive-only noise in the object or context input layers. These tests are of the same nature as the partial input tests done previously and are designed to determine if there is any difference in how the networks deal with noise, and whether this allows more or less effective use of the crossconnection inputs. As with the partial object case, the AllSplit network performs well with object noise by using the additional context information available to its object stream to help it guess the object. In this case, the SplitDG and SplitDG+ networks also do better

than the Baseline network and about the same as their AllSplit counterparts, though slightly worse in high noise situations. When the noise is in the context input, AllSplit does worse since it must deal with additional noise in its object representation. The larger DG and CA3 areas of the SplitDG and "+" networks clearly help with this task and bring performance on par with or even better than the Baseline network (in the case of SplitDG+), indicating that even if the context input is highly noisy, a large CA3 can extract enough additional context information to aid in object identification.

Figure 13 shows the results of the non-additive noise task. As in the additive-only task, the split networks perform better than the Baseline network when the object is noisy, with the AllSplit network performing better than SplitDG. When the context is noisy, the pattern is reversed, although SplitDG does just as well as the Baseline network.

4. Discussion

4.1. Anterior-Posterior Crossconnections. The results in Figure 6 suggest that a split network provides performance advantages compared to the Baseline network. Each output layer requires a different object to context input ratio in order to perform optimally on the relevant tasks. The object output layer gives the network's best guess as to what the actual object is, meaning it needs to perform well in low to medium noise and partial situations where either the object or context input (or both) is degraded. Surprisingly, additional contextual information is helpful even when that context is as noisy/incomplete as the object. This can be thought of as providing a "bigger picture" for the network to look at, and thus making it more likely that it can find some relevant clue which it can use to decipher the entire input. For example, suppose one is looking at a photograph of a person taken from a side angle so it is difficult to determine who it is (partial cue). If a wider-angle photo is now given which includes some of the person's body or clothing (partial context), this information gives a clue as to who the person is, even if the full context is unavailable. The same idea applies to noisy objects and contexts.

However, since each context contains several possible objects, the context input gives less information than the object input, and therefore its value (as far as the object output is concerned) decreases rapidly to zero with the amount of signal degradation. It is not a case when more information is beneficial regardless of how noisy it is. At some point, the error introduced by the noise outweighs the value of having additional information. If the object is presented noiselessly, then additional contextual information is not very useful, particularly if it itself contains noise. For the CA3 size used in our AllSplit network, this point of zero benefit occurs approximately when the context begins to have more noise or be more incomplete than the object. This is why, in the "partial context" and "noisy context" tests, we see the AllSplit network perform rather poorly with its relatively large amount of context input into the object stream (via the strong MEC connection). As we would expect, the more degraded the input context compared to the input object, the worse the AllSplit network performance. On the other hand,

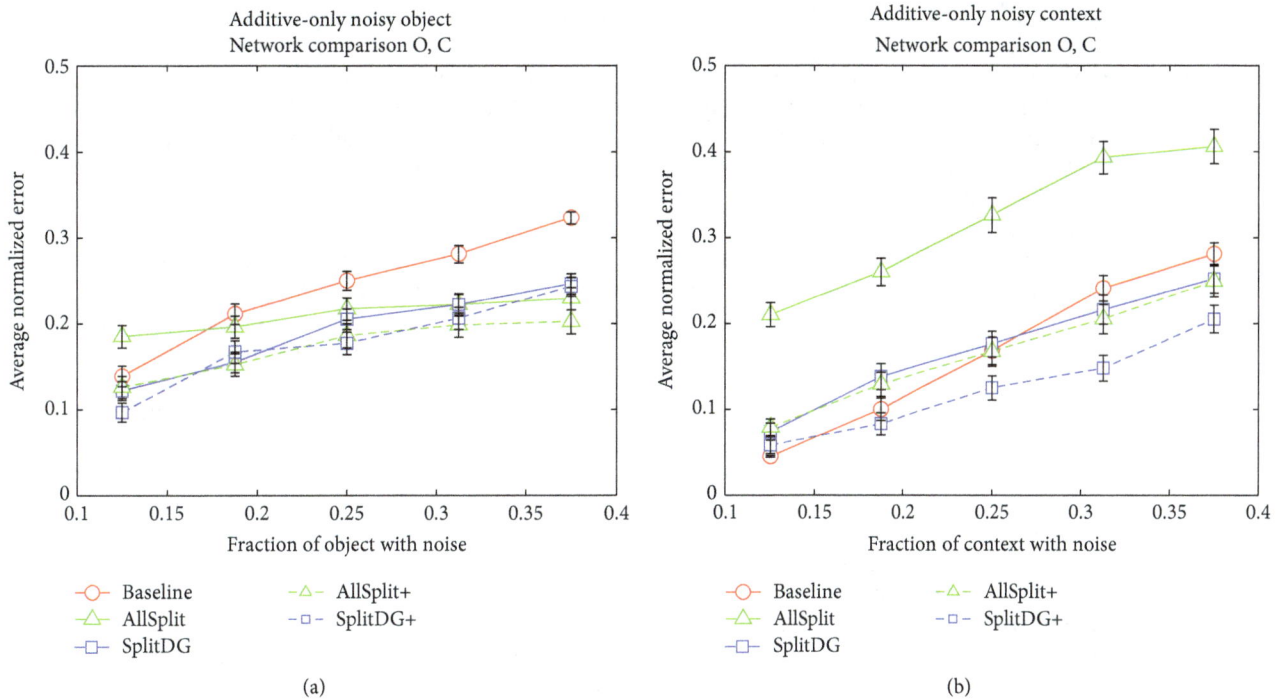

FIGURE 12: Additive-only noise tests. (a) Error across networks, averaged over the O and C output layers, when noisy objects and noiseless contexts were presented as input. (b) Error across networks, averaged over the O and C output layers, when noiseless objects and noisy contexts were presented as input.

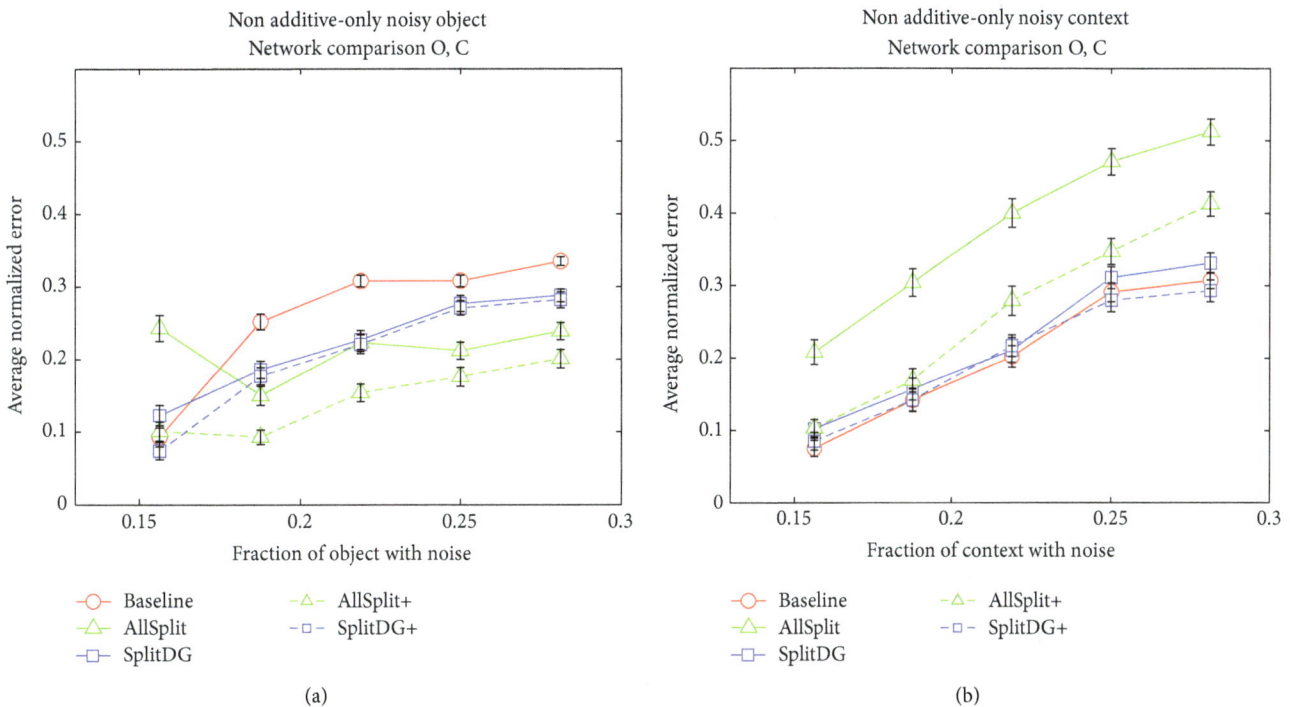

FIGURE 13: Nonadditive noise tests. (a) Error across networks, averaged over the O and C output layers, when noisy objects and noiseless contexts were presented as input. (b) Error across networks, averaged over the O and C output layers, when noiseless objects and noisy contexts were presented as input.

when the input context is less degraded than the input object, as in the "partial object" and "noisy object" tests, the AllSplit performance increases above that of the other networks. Again, because the context inputs have less absolute predictive value than the object inputs (for the object output layer) to begin with, the beneficial effect of noiseless context is less than the detrimental effect of degraded context, and as a result the noiseless context benefit does not come into play until object noise/partial levels are slightly higher. However, the beneficial effects can clearly be seen at moderate object noise levels, and for low noise levels the error is near the training threshold.

In all the networks, in the case of the context being particularly noisy/incomplete, the context output from the anterior stream may be too noisy for use. The hippocampal network would then turn to the object-based context guess output to deliver a context prediction, provided that the object input is relatively noiseless. Thus the OBCG layer needs to be effective in noisy/partial context and mismatch situations, which is what we test in Figure 6(b). In order to achieve good performance, the output must not use the MEC context input, since this layer will only be called on when the context is particularly noisy or incomplete. In addition, if the output relies too much on context, it begins to duplicate the functionality of the anterior context stream. Fortunately for the AllSplit network, this highly degraded or mismatched context situation in which C must be substituted with OBCG is also exactly the situation in which the object output fails; hence it may be able to conveniently rely on the OBCG layer's output to give it a reliable context to use. We have not implemented this backup functionality in our network.

The context output layer is similar to the object output layer in that it must be able to deal with noise in both object and context, and dealing with object noise is of higher priority (as it is with the object output) because the OBCG layer provides a backup in the case of high context noise. For the context layer, this means that it should have a small amount of object input relative to context input. Figure 6(c) shows that this naturally occurs thanks to the fact that there are much fewer contexts than objects, and thus the context stream is very effective at determining context even when they are noisy/incomplete. As a result additional object information is of little use to it, so the LEC to context stream input has less influence than the MEC to object stream input.

As with the context stream, the CBOG stream has fewer input-output associations to store; hence it relies less on the object input from LEC crossconnections. It is important that it depends mostly on context for the same reason that OBCG depends mostly on object, although the CBOG list may get called on even when the object input is usable, since it provides additional information that the object output cannot give. This layer provides a mechanism by which a list of objects can be recalled given only a single contextual cue. Networks consisting of only a single object and context output would not be able to model this task. One artificial feature of this output is that it is N times as large as the object output, where N is the number of objects per context (here 3). We are not implying that in the actual hippocampus, the region that distal CA1 on the anterior side projects to is N times as

large or N times as active as the regions all the other CA1 areas project to. In the actual hippocampus these object outputs may come out one at a time, as the network activity has a time component in spiking networks. Since our model is strictly a rate-based connectionist model, the only way we can represent this output is as a single matrix in which all objects are represented at once. The OBCG output could also be represented this way, in the case where objects are allowed to appear in more than one context.

The temporal dynamics of context-based object retrieval in free recall situations have been given a theoretical foundation in the TCM (temporal context model) and CMR (context maintenance and retrieval) frameworks [40, 41]. Our model explicitly represents the biological structures and connections that make possible the basic multiple object to context associations (referred to as source clustering) assumed by these frameworks, but we do not attempt to provide a realization of any of the temporal aspects of memory (temporal clustering) which TCM and its generalizations also deal with, such as associations between successively presented contexts and the recency effect. However, allowing objects to be associated with more than one context (as they are in the case of the temporal context), our model could conceivably provide a starting point for a biological realization of the TCM framework. The varying internal context of TCM could be produced within our model by having objects output by CBOG feed back into the OBCG stream to produce an associated set of contexts, which would then be used as inputs into the CBOG stream to produce the next object to be recalled, in a repeated cycle.

4.2. Effects of Layer Size. There are two ways to approach the interpretation of the other test results, beginning with the training set error. The first way is to ignore the size of the network and compare only those networks that have similar amounts of error on the training set. In this view, a fair comparison would be between those networks that start out with equal amounts of knowledge on the training set, regardless of how many epochs it took them to get their error to that level or how many neurons they have. Here, splitting a layer into two separate sublayers has little to no disadvantage, because each sublayer is still large enough to do its task at the same level as the full layer. This has precedent in the cognitive psychology literature, where, for example, subjects being tested on recall of a list over time or in different contexts may be allowed as many trials as they need to memorize the list in the first place, so that all participants start out with the same low training error rate. This assumes that humans have enough neurons available to memorize the training list to whatever degree of accuracy is required, given enough time. In addition, it is known that in rats, during the course of a particular spatial task, only a small fraction of the hippocampal CA1 neurons fire during the entire duration of the task. This suggests that the hippocampus has many more neurons than necessary for any given task.

Of course, neurons cannot be added to actual test subjects, but in our test networks this provides an effective way to accomplish the same goal of reducing the error on the training set, so that all networks start with the same baseline error

Object error on partial object (40%) in baseline network

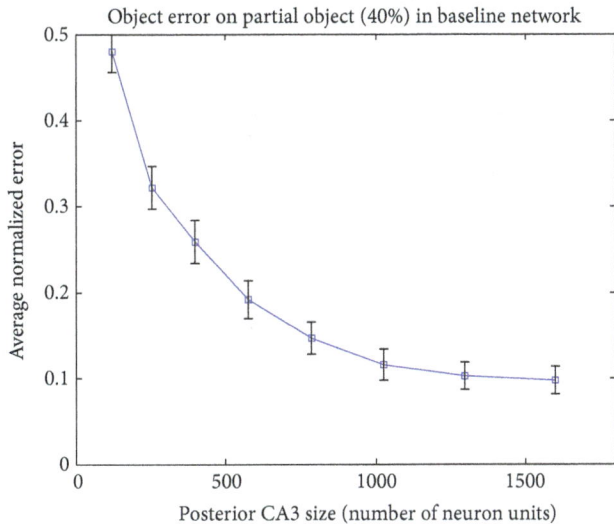

FIGURE 14: Error on object output layer in a 40% partial object task as a function of the size of posterior CA3 in the Baseline network.

rate. From the biological standpoint, this way of comparing networks essentially says that, in the actual brain, the memorization ability of the hippocampus for any particular task is not limited by the number of neurons, but rather the way in which they are connected. From this point of view, the basic AllSplit and SplitDG networks should be ignored, and the results of Baseline should only be compared against AllSplit+ and SplitDG+, since all three of these networks have the same error rate on the initial training set.

The second way to interpret the results is to take the neuron-limited view, where a fair comparison would be between networks which have the same number of neurons, regardless of how well they are able to store the initial training set. In this view, splitting a layer into two separate sublayers incurs the penalty of each sublayer now being half the size. Biologically, this means that neurons are costly in terms of energy required to build and maintain, and that the brain has as few neurons as possible while still being able to perform its required tasks. From this point of view, Baseline should be compared with AllSplit and SplitDG since they have the same number of neurons, and AllSplit+ and SplitDG+ should be ignored.

In the biological hippocampus, the answer probably lies somewhere between the two extremes. Figure 14 shows that increasing the size of CA3 in the Baseline network results in lower error rates, but that eventually the error stops decreasing with layer size. If the hippocampus is in the rightmost region of the graph, then it has enough neurons and there is little cost to splitting a layer, so it is best approximated by the "+" models. On the other hand, if it is near the leftmost region of the graph, it is severely neuron constrained, and splitting a layer results in a dramatic decrease in performance on each of the streams. In this case it would be better approximated by the normal (non-+) models.

Overall, the test results show that the AllSplit network is best for noisy or partial object situations and worst when given noisy or partial context. AllSplit+ has uniformly better

performance as expected but follows the same general pattern as AllSplit. On the other hand, the Baseline network is relatively better at noisy or partial context situations than with noisy or partial object. Rarely is it the best network at any particular task, however, with the exception of partial context. It is most similar to the SplitDG network, which is what we expected based on its architecture. The SplitDG network has good all-around performance. Compared to Baseline, it does consistently better in noisy or partial object tests, about the same in noisy context, but noticeably worse when presented with partial context. SplitDG+ is generally about the same as SplitDG on noisy or partial object tests, but its larger DG seems to aid in the incorporation of context information when it is noisy or partial. This allows it to do significantly better than SplitDG in such tasks and puts it on par or better than Baseline. Our results thus suggest that differentiation within DG provides uniformly better performance over a nondifferentiated DG if it is large enough (SplitDG+), and generally better performance with the exception of partial context tasks if DG is size constrained (SplitDG). Additional differentiation within CA3 (AllSplit and AllSplit+) may work to further increase noisy and partial object task performance, but at the cost of the corresponding degraded context task performance.

4.3. Object Noise versus Context Noise. These results raise the question of whether it is better for the object stream to be able to deal with noisy objects (AllSplit) or noisy contexts (Baseline), where we will use the term "noise" to refer to partial cues as well. We argue that there is inherently less noise in contexts than in objects; hence dealing with object noise is more important. To make things concrete, consider the case of an animal in search of food. It has to find edible plants and insects and has to memorize a large amount of object-related information. Depending on the time of year and the time of day, the types of plants or insects it can eat and their appearance change (noise). On the other hand, the season and spatial environment are contextual cues that change slowly, and there are only a relatively small number of different contexts it must identify: its dwelling, its scavenging grounds, what season it is, and so forth. In general, the much larger number of objects in existence makes it likely that interference and noise are much more likely to occur between objects than between object and contexts, which are few in number and change only slowly over time.

The second argument is that, given some recurrent support structures, noise in context is easier for the hippocampus to deal with than noise in object. The context stream deals with context noise relatively well since the contexts are few and well memorized. Thus getting a clean context to the object stream requires only taking the context stream output (C) and feeding it back into the object stream. If the context is very noisy or absent (to the point that the context stream output is no longer useful), the output of the OBCG layer can be used instead. Thus there are two independent ways for the object stream to not have to deal with context noise, each involving only a recurrent loop.

With object noise, the situation is different. The object stream is itself responsible for determining the object; thus the only place it can turn to for additional object information is the CBOG output, which uses context to make object guesses. However, since the CBOG stream uses mainly context information, the best it can do is to give a list of possible objects that are associated with that context. Choosing one object out of this list would then require a separate calculation where the input object is compared with the CBOG output list and the best match selected. This would not be an easy task when the input object is noisy, although it would be significantly easier than the object stream's original task, which is to compare the input object to a list of 120 possible objects and choose the closest match. Thus the object noise problem can certainly be overcome with the help of additional structures, but it may be more judicious to simply use context information in the object stream from the beginning, which is exactly the solution that the AllSplit and SplitDG networks use. They then trade the object noise problem for a context noise problem, but this seems to be a much easier issue to deal with.

4.4. Mismatches. Mismatches, consisting of an object appearing in a different context from that it was learned into, are by definition rare events. If they happened frequently, the object would simply be associated with the new context and it would no longer be considered a mismatch. On the posterior side, a mismatch means that the incoming context information does not match the primary object input from LEC, thus putting it in a situation similar to having a very noisy context but noiseless object. On the anterior side, where MEC context information is primary, the incoming object input introduces uncertainty, and the situation is similar to a very noisy object but noiseless context. Due to the smaller number of inputs it needs to store and the fact that LEC input is relatively weak, mismatches have little effect on the anterior stream— if we see someone from the office at the mall, we do not have any trouble recognizing our context as the mall. On the other hand, the large amount of MEC input into the posterior stream means that a mismatched context can significantly affect object recognition—it may take us several seconds to recognize a colleague if we unexpectedly encounter them at the mall, whereas the recognition is nearly instantaneous when we see them at the office.

Any encoding and retrieval scheme which uses contextual information to recognize objects, as we believe the hippocampus does, will naturally have problems in mismatch situations. However, this is only the case if we believe that a familiar object in a different context from usual ought to still be recognized as the same familiar object. In many situations it may make sense to consider object A in context A as effectively different from object A in context B [42]. The large amount of error that a mismatch produces may be beneficial for signaling that something is wrong or unexpected and deserves our attention.

4.5. Relation to Rat Hippocampus. Our model is not explicitly a place field model, and in the way we have conceptualized it and in its current form our model better reflects the primate hippocampus. However, with some minor modifications the model would be consistent with the observation of higher-resolution place fields in dorsal compared to ventral

hippocampus. We will switch to using the appropriate terminology for the rat anatomy in this discussion, so that anterior and posterior in our model are now ventral and dorsal, and the caudolateral and rostromedial bands of MEC and LEC are now dorsolateral and ventromedial, respectively.

In our model, for simplicity's sake, we make no distinction between the dorsolateral and ventromedial bands of the MEC, modeling both as carrying the same context information, albeit to different parts of hippocampus (dorsal versus ventral, resp.). However, it is known that neurons in the dorsolateral band of MEC are more spatially tuned than those in the ventromedial band [43], and thus we would expect that the dorsal hippocampus, receiving higher-resolution spatial information from the dorsolateral band, would have the tighter place fields that are seen experimentally. If we wanted to extend our model to cover this additional aspect of the anatomy, we could do this by having two different types of contextual inputs, a "local" context and a less precise "global" context which might represent the context at a larger spatial scale or contain some other nonspatial information, with the local context being carried by the dorsolateral MEC and the global context being carried by the ventromedial MEC.

Note that both the dorsal and ventral subdivisions of the hippocampus receive the nonspatial LEC inputs to some extent. However, we refer to the dorsal hippocampus as the more object-oriented layer in our model compatible with human fMRI studies and our set of sample tests (shown in Figures 6 and 7) which led us to set the relative weighting of the LEC input larger than that of the MEC input for optimal performance (and the reverse is true on the ventral hippocampus for context information). Of course, the set of "tests" that the rat hippocampus has evolved to do could be different from the basic tests that we proposed. For example, the performance on the mismatch test (where the presented object and context were not associated) was a significant factor in determining how strong the MEC to dorsal hippocampus connections should be. A strong MEC to dorsal connection results in a large amount of error on the OBCG output, and as a result those connections were kept very weak. In the rat hippocampus, however, it could be the case that it simply just does badly on mismatches because they are so rare that they do not need to be protected against with weak MEC to dorsal weighting, or it could be that in the case of mismatches, additional cortical processing is involved. In either case, the MEC to dorsal signal could well be just as strong or stronger than the LEC to dorsal signal.

In conjunction with the dorsolateral versus ventromedial band differences mentioned above, the dorsal and ventral streams of our rat-modified model would not contradict the general conception of the dorsal stream as being context oriented and more finely spatially tuned than the ventral side. In summary, the degree to which the MEC's spatial contextual information is relevant in the dorsal side of the rat hippocampus is probably much higher than that indicated in our model, where we look at objects, rather than context, as the primary information the hippocampus is storing and view context as information that can contribute to object recognition.

5. Conclusion

We constructed hippocampus models that include anatomical and functional details such as the distinction between the posterior and anterior subdivisions of the hippocampus, connections from the medial and lateral entorhinal cortex to both the posterior and anterior regions, differences between the superior and inferior blades of the dentate gyrus, and connectivity differences between distal and proximal (relative to DG) portions of CA3 and CA1. We hypothesized distinct roles for each of the CA1 areas on the proximal and distal sides and attempted to show how these anatomical details work together to increase performance on certain tasks. In particular, we showed that object and context require different treatment in terms of how much one is used to help recognize the other. This is simply due to the greater number of objects compared to the number of contexts rather than intrinsic differences in representation. In addition, we showed how the hippocampal anatomy supports the use of contextual information to help object recognition and proposed ways in which the tradeoffs inherent to this could possibly be mitigated.

Our models make several predictions that may be experimentally tested. We predict that the inferior blade of DG and proximal CA3 in the posterior region of hippocampus receives more MEC innervation, or that these neurons are more sensitive to MEC inputs, than is the case with LEC inputs into the anterior side of hippocampus. Blocking MEC input into posterior hippocampus should have a significant negative effect on object recognition when the object is noisy or only partially shown, assuming that the object was associated with a specific context, but should have only a mildly negative or even a positive effect if the context is noisy or obscured. Blocking LEC input into anterior hippocampus should have much less of an effect on context recognition in either case, assuming that there are many more objects than contexts. If the number of contexts and the number of objects are roughly equal, then we should see effects similar to those seen on the posterior stream with MEC input. Our assumptions about the two different types of information being carried along the output pathways can also be experimentally tested by comparing the information content of proximal CA1 and distal CA1 neurons. We predict that distal CA1 neurons on both the posterior and anterior sides will be more likely to carry object-type information, while proximal CA1 neurons will tend to carry primarily context-type information.

We found that the models that have only DG split (SplitDG and SplitDG+) did the best overall on our test sets, generally doing about the same as the Baseline model when the context input was degraded, and significantly better when the object input was degraded. The models with both DG and CA3 split (AllSplit and AllSplit+) did even better in noisy or incomplete object situations, but at a cost in performance on the corresponding degraded context tasks. As we mentioned in the discussion, it may be the case that degraded context situations are relatively rare compared to degraded object situations, and thus the performance tradeoff of the AllSplit networks may in fact be optimal. However, it is probably also the case that the hippocampus does not make

as severe a tradeoff as we have in our models, where CA3 is either completely unified or completely split. For instance, both regions of CA3 in the actual hippocampus receive superior blade input from DG, rather than just the distal region. In our model, the superior blade on the posterior side of hippocampus carries mainly LEC object information, so including this feature may change the ratio of object to context information within proximal CA3 in favor of object information and thereby reduce some of the deleterious effects of noisy context that we observed in the AllSplit network. The two regions of CA3 also communicate to an extent, although they have different connectivity patterns in terms of the proportion of projections they send within CA3 and onward to CA1. Exactly how these differences affect hippocampal function remains a topic for future research.

To date, much of the computational literature on the hippocampus has either focused on only object memorization or only spatial context memorization and has not attempted to identify how these different types of information may mutually support each other within the hippocampus or elucidate specific anatomical details within the hippocampus that may allow this to occur. On the other hand, experimental literature that addresses details such as the LEC and MEC cross-connections has often assigned them only the vague role of allowing a mixing or integration of object and context information. We have hypothesized specific ways that object and context information may be used in the posterior and anterior regions of the hippocampus, shown that the connectivity of hippocampus supports and enables these uses, and identified specific situations in which these object-context interactions have a beneficial or deleterious effect. Our results thus suggest new ways of thinking about the sort of computations that the hippocampus may do, and how it uses both object and context to perform them.

Acknowledgments

This paper is supported by the Intelligence Advanced Research Projects Activity (IARPA) via Department of the Interior (DOI) contract no. D10PC20021. The US Government is authorized to reproduce and distribute reprints for governmental purposes notwithstanding any copyright annotation thereon. The views and conclusions contained hereon are those of the authors and should not be interpreted as necessarily representing the official policies or endorsements, either expressed or implied, of IARPA, DOI, or the US Government. Additional support is provided by the National Institutes of Health (NIH) Grant GM084905: Computational and Mathematical Modeling of Biomedical Systems and by and NSF Robust Intelligence Grant IIS 1117388 (JMF).

References

[1] M. P. Witter, G. W. Van Hoesen, and D. G. Amaral, "Topographical organization of the entorhinal projection to the dentate gyrus of the monkey," *Journal of Neuroscience*, vol. 9, no. 1, pp. 216–228, 1989.

[2] J. K. Leutgeb, S. Leutgeb, M. B. Moser, and E. I. Moser, "Pattern separation in the dentate gyrus and CA3 of the hippocampus," *Science*, vol. 315, no. 5814, pp. 961–966, 2007.

[3] C. B. Alme, R. A. Buzzetti, D. F. Marrone et al., "Hippocampal granule cells opt for early retirement," *Hippocampus*, vol. 20, no. 10, pp. 1109–1123, 2010.

[4] W. Deng, M. Mayford, and F. H. Gage, "Selection of distinct populations of dentate granule cells in response to inputs as a mechanism for pattern separation in mice," *ELife*, vol. 2, Article ID e00312, 2013.

[5] L. M. Rangel and H. Eichenbaum, "What's new is older," *ELife*, vol. 2, Article ID e00605, 2013.

[6] R. C. O'Reilly and J. W. Rudy, "Conjunctive representations in learning and memory: principles of cortical and hippocampal function," *Psychological Review*, vol. 108, no. 2, pp. 311–345, 2001.

[7] A. Treves and E. T. Rolls, "Computational analysis of the role of the hippocampus in memory," *Hippocampus*, vol. 4, no. 3, pp. 374–391, 1994.

[8] M. P. Witter, "Intrinsic and extrinsic wiring of CA3: indications for connectional heterogeneity," *Learning & Memory*, vol. 14, no. 11, pp. 705–713, 2007.

[9] S. Zola-Morgan, L. R. Squire, and D. G. Amaral, "Human amnesia and the medial temporal region: enduring memory impairment following a bilateral lesion limited to field CA1 of the hippocampus," *Journal of Neuroscience*, vol. 6, no. 10, pp. 2950–2967, 1986.

[10] M. R. Hunsaker, G. G. Mooy, J. S. Swift, and R. P. Kesner, "Dissociations of the medial and lateral perforant path projections into dorsal DG, CA3, and CA1 for spatial and nonspatial (visual object) information processing," *Behavioral Neuroscience*, vol. 121, no. 4, pp. 742–750, 2007.

[11] M. R. Hunsaker, P. M. Fieldsted, J. S. Rosenberg, and R. P. Kesner, "Dissociating the roles of dorsal and ventral CA1 for the temporal processing of spatial locations, visual objects, and odors," *Behavioral Neuroscience*, vol. 122, no. 3, pp. 643–650, 2008.

[12] R. D. Burwell, "The parahippocampal region: corticocortical connectivity," *Annals of the New York Academy of Sciences*, vol. 911, pp. 25–42, 2000.

[13] C. B. Cave and L. R. Squire, "Equivalent impairment of spatial and nonspatial memory following damage to the human hippocampus," *Hippocampus*, vol. 1, no. 3, pp. 329–340, 1991.

[14] E. L. Hargreaves, G. Rao, I. Lee, and J. J. Knierim, "Neuroscience: major dissociation between medial and lateral entorhinal input to dorsal hippocampus," *Science*, vol. 308, no. 5729, pp. 1792–1794, 2005.

[15] S. Dennis and M. S. Humphreys, "A context noise model of episodic word recognition," *Psychological Review*, vol. 108, no. 2, pp. 452–478, 2001.

[16] S. Gaskin, A. Gamliel, M. Tardif, E. Cole, and D. G. Mumby, "Incidental (unreinforced) and reinforced spatial learning in rats with ventral and dorsal lesions of the hippocampus," *Behavioural Brain Research*, vol. 202, no. 1, pp. 64–70, 2009.

[17] M. B. Moser and E. I. Moser, "Functional differentiation in the hippocampus," *Hippocampus*, vol. 8, no. 6, pp. 608–619, 1998.

[18] R. E. Clark, S. M. Zola, and L. R. Squire, "Impaired recognition memory rats after damage to the hippocampus," *Journal of Neuroscience*, vol. 20, no. 23, pp. 8853–8860, 2000.

[19] M. N. De Lima, T. Luft, R. Roesler, and N. Schröder, "Temporary inactivation reveals an essential role of the dorsal hippocampus in consolidation of object recognition memory," *Neuroscience Letters*, vol. 405, no. 1-2, pp. 142–146, 2006.

[20] O. Hardt, P. V. Migues, M. Hastings, J. Wong, and K. Nader, "PKMζ maintains 1-day- and 6-day-old long-term object location but not object identity memory in dorsal hippocampus," *Hippocampus*, vol. 20, no. 6, pp. 691–695, 2010.

[21] D. G. Mumby, S. Gaskin, M. J. Glenn, T. E. Schramek, and H. Lehmann, "Hippocampal damage and exploratory preferences in rats: memory for objects, places, and contexts," *Learning and Memory*, vol. 9, no. 2, pp. 49–57, 2002.

[22] J. A. Ainge, C. Heron-Maxwell, P. Theofilas, P. Wright, L. De Hoz, and E. R. Wood, "The role of the hippocampus in object recognition in rats: examination of the influence of task parameters and lesion size," *Behavioural Brain Research*, vol. 167, no. 1, pp. 183–195, 2006.

[23] J. R. Manns and H. Eichenbaum, "A cognitive map for object memory in the hippocampus," *Learning and Memory*, vol. 16, no. 10, pp. 616–624, 2009.

[24] R. S. Rosenbaum, S. Köhler, D. L. Schacter et al., "The case of K.C.: contributions of a memory-impaired person to memory theory," *Neuropsychologia*, vol. 43, no. 7, pp. 989–1021, 2005.

[25] S. Corkin, "Lasting consequences of bilateral medial temporal lobectomy: clinical course and experimental findings in H.M," *Seminars in Neurology*, vol. 4, no. 2, pp. 249–259, 1984.

[26] K. A. Paller and G. McCarthy, "Field potentials in the human hippocampus during the encoding and recognition of visual stimuli," *Hippocampus*, vol. 12, no. 3, pp. 415–420, 2002.

[27] G. Fernández, H. Weyerts, M. Schrader-Bölsche et al., "Successful verbal encoding into episodic memory engages the posterior hippocampus: a parametrically analyzed functional magnetic resonance imaging study," *Journal of Neuroscience*, vol. 18, no. 5, pp. 1841–1847, 1998.

[28] C. E. Stern, S. Corkin, R. G. González et al., "The hippocampal formation participates in novel picture encoding: evidence from functional magnetic resonance imaging," *Proceedings of the National Academy of Sciences of the United States of America*, vol. 93, no. 16, pp. 8660–8665, 1996.

[29] J. Poppenk, H. Evensmoen, M. Moscovitch, and L. Nadel, "Long-axis specialization of the human hippocampus," *Trends in Cognitive Sciences*, vol. 17, no. 5, pp. 230–240, 2013.

[30] A. Hupbach, O. Hardt, R. Gomez, and L. Nadel, "The dynamics of memory: context-dependent updating," *Learning and Memory*, vol. 15, no. 8, pp. 574–579, 2008.

[31] B. Jones, E. Bukoski, L. Nadel, and J. M. Fellous, "Remaking memories: reconsolidation updates positively motivated spatial memory in rats," *Learning & Memory*, vol. 19, no. 3, pp. 91–98, 2012.

[32] R. C. O'Reilly, R. Bhattacharyya, M. D. Howard, and N. Ketz, "Complementary learning systems," *Cognitive Science*, pp. 1–20, 2011.

[33] B. Aisa, B. Mingus, and R. O'Reilly, "The emergent neural modeling system," *Neural Networks*, vol. 21, no. 8, pp. 1146–1152, 2008.

[34] R. C. O'Reilly and Y. Munakata, *Computational Explorations in Cognitive Neuroscience: Understanding the Mind By Simulating the Brain*, The MIT Press, Cambridge, Mass, USA, 2000.

[35] V. Cutsuridis, B. Graham, S. R. Cobb, and I. Vida, *Hippocampal Microcircuits: A Computational Modelers' Resource Book*, Springer, 2010.

[36] E. J. Henriksen, L. L. Colgin, C. A. Barnes, M. P. Witter, M. B. Moser, and E. I. Moser, "Spatial representation along the proximodistal axis of CA1," *Neuron*, vol. 68, no. 1, pp. 127–137, 2010.

[37] G. Shepherd and S. Grillner, *Handbook of Brain Microcircuits*, Oxford Univ Press, Oxford, UK, 2010.

[38] H. Hayashi and Y. Nonaka, "Cooperation and competition between lateral and medial perforant path synapses in the dentate gyrus," *Neural Networks*, vol. 24, no. 3, pp. 233–246, 2011.

[39] P. Poirazi, T. Brannon, and B. W. Mel, "Pyramidal neuron as two-layer neural network," *Neuron*, vol. 37, no. 6, pp. 989–999, 2003.

[40] P. B. Sederberg, M. W. Howard, and M. J. Kahana, "A context-based theory of recency and contiguity in free recall," *Psychological Review*, vol. 115, no. 4, pp. 893–912, 2008.

[41] S. M. Polyn, K. A. Norman, and M. J. Kahana, "A context maintenance and retrieval model of organizational processes in free recall," *Psychological Review*, vol. 116, no. 1, pp. 129–156, 2009.

[42] L. Nadel, *The Hippocampus and Context Revisited*, Hippocampal Place Fields. Oxford Scholarship Online Monographs, 2008.

[43] T. Hafting, M. Fyhn, S. Molden, M. B. Moser, and E. I. Moser, "Microstructure of a spatial map in the entorhinal cortex," *Nature*, vol. 436, no. 7052, pp. 801–806, 2005.

14

Augmenting Weak Semantic Cognitive Maps with an "Abstractness" Dimension

Alexei V. Samsonovich and Giorgio A. Ascoli

Krasnow Institute for Advanced Study, George Mason University, 4400 University Drive MS 2A1, Fairfax, VA 22030-4444, USA

Correspondence should be addressed to Alexei V. Samsonovich; asamsono@gmu.edu

Academic Editor: Rajan Bhattacharyya

The emergent consensus on dimensional models of sentiment, appraisal, emotions, and values is on the semantics of the principal dimensions, typically interpreted as valence, arousal, and dominance. The notion of weak semantic maps was introduced recently as distribution of representations in abstract spaces that are not derived from human judgments, psychometrics, or any other a priori information about their semantics. Instead, they are defined entirely by binary semantic relations among representations, such as synonymy and antonymy. An interesting question concerns the ability of the antonymy-based semantic maps to capture all "universal" semantic dimensions. The present work shows that those narrow weak semantic maps are not complete in this sense and can be augmented with other semantic relations. Specifically, including hyponym-hypernym relations yields a new semantic dimension of the map labeled here "abstractness" (or ontological generality) that is not reducible to any dimensions represented by antonym pairs or to traditional affective space dimensions. It is expected that including other semantic relations (e.g., meronymy/holonymy) will also result in the addition of new semantic dimensions to the map. These findings have broad implications for automated quantitative evaluation of the meaning of text and may shed light on the nature of human subjective experience.

1. Introduction

The idea of representing semantics geometrically is increasingly popular. Many mainstream approaches use vector space models, in which concepts, words, documents, and so forth are associated with vectors in an abstract multidimensional vector space. Other approaches use manifolds of more complex topology and geometry. In either case, the resultant space or manifold together with its allocated representations is called a *semantic space* or a *semantic (cognitive) map*. Examples include spaces constructed with Latent Semantic Analysis (LSA) [1] and Latent Dirichlet Allocation (LDA) [2], as well as many related techniques, for example, ConceptNet [3, 4]. Other examples of techniques include Multi-Dimensional Scaling (MDS) [5], including Isomap [6], and related manifold-learning techniques [7], Gardenfors' conceptual spaces [8], very popular in the past models of self-organizing feature maps, and more.

The majority of these approaches are based on the idea of a dissimilarity metrics, which is to capture semantic dissimilarity between representations (words, documents, concepts, etc.) with a geometrical distance between associated space elements (points or vectors). In other words, the metrics that determines the allocation of representations in space is a function of their semantic dissimilarity. In this case, two representations allocated at close points in space must have similar semantics and vice versa: two representations with similar semantics must be close to each other in space. Conversely, representations unrelated to each other must be separated by significant distance.

We introduced the term "weak semantic cognitive mapping" to denote an alternative approach, exploited here, which is not based on dissimilarity [9–11]. The idea is not to separate all different meanings from each other (like in MDS), nor to allocate them based on their individual semantic characteristics given a priori (as in LSA), but rather to arrange them in space based on their mutual semantic relations. The notion of weak semantic cognitive maps was originally introduced in a narrow sense, where these relations were limited to synonymy and antonymy only [9–11]. In a more general sense, as discussed below, weak semantic cognitive maps may capture other binary semantic relations as well, including hypernymy-hyponymy, holonymy-meronymy, troponymy, causality, and dependence.

While the understanding of dissimilarity as the basis of antonymy is widespread, many examples of the dictionary antonym pairs used in our analysis suggest that dissimilarity and antonymy are distinct notions. Most unrelated words may be considered dissimilar (e.g., "apple" and "inequality"), yet do not constitute antonym pairs. In contrast, antonym pairs include words that are related to each other and in a certain sense are similar to each other in their meaning and usage, for example, king and queen, major and minor, and ascent and descent. It appears that most antonym pairs (at least in the dictionaries that we used) are consistent with the notion of "opposite" rather than "dissimilar."

More generally, the method of weak semantic mapping is essentially different from most vector-space-based approaches including LSA, LDA, MDS, and ConceptNet [1–4], primarily because there is no a priori attribution of semantic features to representations in the constructive definition of the map. Only relations, but not semantic features, are given as input. As a result, semantic dimensions of the map that are not predefined to emerge naturally, starting from a randomly generated initial distribution of words in an abstract space with no a priori given semantics and following the strategy to pull synonyms together and antonyms apart [10, 11] (see Section 2: Methods). In contrast to LSA, principal component analysis is used here to reveal the main emergent semantic dimensions at the final stage only. The advantage of the antonymy-based weak semantic cognitive map compared to "strong" maps based on dissimilarity metrics is that its dimensions have clearly identifiable semantics (naturally given by the corresponding pairs of antonyms) that are domain-independent. For example, the notion of "good versus bad" that corresponds to the first principal component applies to all domains of human knowledge.

Interestingly, semantics of the emergent dimensions of antonym-based weak semantic cognitive maps are closely related to those of another broad category of "dimensional models" of affects [12] that attempt to capture human emotions, feelings, affects, appraisals, sentiments, and attitudes. Examples range from original classical models such as Osgood's semantic differential [13], Russell's circumplex [14], and Plutchik's wheel [15] to many more recent derivative integrated frameworks, like PAD (pleasure, arousal, and dominance) [16], ANEW (Affective Norms for English Words) [17], EPA (evaluation, potency, and arousal) [18], and a recent 3D model linking emotions to main neurotransmitters [19]. These dimensional models are usually derived from human experimental studies involving psychometrics or introspective judgment evaluated on the Likert scale [20]. While these models provide the most common bases for opinion mining or sentiment analysis [21], the weak semantic map is more complete in the sense that (i) it assigns values to all words, not only to emotionally meaningful words, (ii) it measures semantics associated with all antonym pairs, not only emotionally meaningful antonym pairs, and therefore is applicable to all domains of knowledge, and (iii) its

dimensions are orthogonal and independent of each other. The combination of these features makes weak semantic maps extremely valuable for numerous applications.

It is surprising that the well-known dimensions of the semantic differential, PAD, EPA, and related models can be recognized in the main principal components (PC) of the above cited weak semantic map, where PC1 is related to valence, PC2 to arousal, and PC3 to dominance [11]. (This correspondence is approximate, because the principal components have zero correlations with each other, while the variables of, e.g., ANEW are strongly correlated.) For example, "love" and "joy" have top values of valence in the affective database ANEW and also top values of PC1 of weak semantic cognitive map. Words like "anger" and "excitement" have top values of arousal in the affective database ANEW and also top values of PC2 in weak semantic cognitive map. This correspondence is consistent in weak semantic maps constructed based on different corpora in several major languages [11]. The observation is unexpected, because the weak semantic map is not derived from any semantic features of words given a priori, and is not explicitly related to emotions and feelings by its construction. In fact, any pair of antonyms defines a map dimension, including antonym pairs that are not associated with affects, for example, "abstract-specific." It is also surprising that the weak semantic map is low-dimensional: the number of PCs that account for 95% of the variance of the multidimensional distribution typically varies from 4 to 6, depending on the corpus [11].

How complete is the weak semantic map narrowly defined only by antonym pairs? Certainly at least some semantic differences cannot be captured by antonymy relations, because not all concepts have antonyms (e.g., the number 921714083). Here we address a different question: whether all universal semantic dimensions can be captured by antonymy relations. For example, it may seem obvious that causality cannot be captured by antonymy. However, the issue is nontrivial, as there are many examples of causally related antonyms (e.g., attack-defend, begin-end, send-receive, and even cause-effect). Thus, two logical possibilities stand.

(1) Antonym-based semantic maps separate representations along all semantic dimensions that make sense for all domains of knowledge. Thus, if there is a semantic characteristic X that makes sense for all domains of knowledge such that some concepts can be characterized as having more X than others, then there is a direction on the narrow weak semantic map along which those concepts are separated based on their value of X.

(2) The alternative: there is at least one general semantic characteristic X defined for all domains that is ignored by the antonym-based weak semantic map. In other words, the variance in X measured across all concepts is not accounted by the map coordinates of concepts, and vice versa, no significant part of the variance of the map can be accounted by X.

Here we argue for (2), quantifying the notion of "abstractness" (or ontological generality) as an example of X. Our

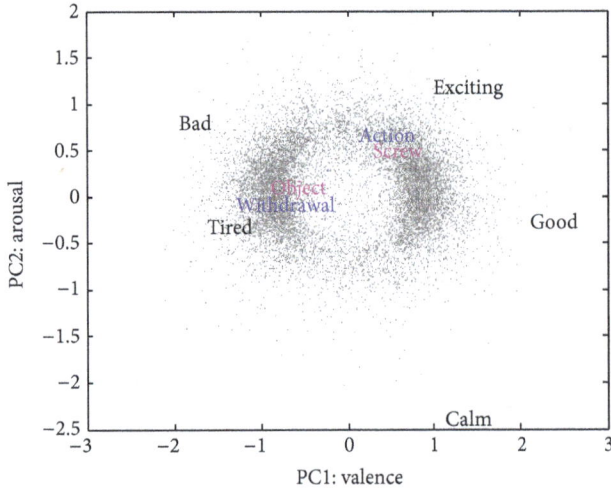

FIGURE 1: A sample from the antonymy-based weak semantic cognitive map constructed by Samsonovich and Ascoli [11]. Grey dots show all 15,783 words from the MS Word English dictionary. Similar results were obtained by WordNet. Words shown in color are examples of hypernym-hyponym pairs: "action-withdrawal" and "object-screw." Selected examples illustrate that there is no clear separation of hypernyms and hyponyms on the map.

technical definition of "abstractness" is based on hyponym-hypernym relations among words.

Before presenting results of the computational study, we briefly discuss the hypothesis at an intuitive level. While "abstract-specific" is a pair of antonyms, which corresponds to a direction on the narrow weak semantic map, the two antonyms "abstract" and "specific" themselves have approximately the same measure of "abstractness" (the X value) associated with them. Intuitively, this observation must hold for most antonym pairs, because antonyms pairs do not typically constitute a hypernym-hyponym couple. Therefore, it is unlikely that there is a hyperplane on the map that separates more abstract from more specific words. Therefore, we do not expect to find a dimension of the map based on synonyms and antonyms that could separate words by "abstractness" (see Figure 1). In contrast, there is a hyperplane (PC1 = 0) that separates "good" and "bad" words and a hyperplane (PC2 = 0) that separates "calming" and "exciting" words. That is to say, "good words" tend to be synonyms of the word "good," but "abstract words" are not synonyms of the word "abstract" or of each other.

2. Methods

2.1. Weak Semantic Cognitive Mapping. The general idea of semantic cognitive mapping is to allocate representations (e.g., words) in an abstract space based on their semantics. This paradigm is common for a large number of techniques overviewed in Introduction. While most studies in semantic cognitive mapping are based on the notion of a dissimilarity metrics and/or on a set of semantic features given a priori, weak semantic mapping ignores dissimilarity as well as any individually predefined semantics.

The algorithm for antonymy-based weak semantic mapping is described in our previous work [11]. The semantic space is created by minimization of the "energy" of the entire distribution of words on the map, starting from a random distribution. Then, the emergent semantics of the map dimensions are defined by the entire distribution of representations on the map and typically are best characterized by the pairs of antonyms that separated by the greatest distance along the given dimension. The main semantic dimensions are defined by the principal components of the emergent distribution of words on the map. Semantics associated with the first three PCs can be characterized as "good" versus "bad" (PC1), "calming, easy" versus "exciting, hard" (PC2), and "free, open" versus "dominated, closed" (PC3) [11]. When limited to affects, these semantics approximately correspond to the three PAD dimensions: pleasure, arousal, and dominance.

More precisely, the narrow weak semantic cognitive map is a distribution of words in an abstract vector space (with no semantics preassociated with its elements or dimensions) that minimizes the following energy function [11]:

$$H(\mathbf{x}) = -\frac{1}{2}\sum_{i,j=1}^{N} W_{ij}\mathbf{x}_i \cdot \mathbf{x}_j + \frac{1}{4}\sum_{i=1}^{N} |\mathbf{x}_i|^4, \quad \mathbf{x} \in \mathfrak{R}^N \otimes \mathfrak{R}^D. \quad (1)$$

Here $\mathbf{x}_i$ is a D-vector representing the ith word (out of N). The W_{ij} entries of the symmetric relation matrix equal $+1$ for pairs of synonyms, -1 for pairs of antonyms, and zero otherwise. D is set to any integer (e.g., 100) that is substantially greater than the number of resulting significant principal components of the distribution, which typically ranges from 4 to 6 and determines the dimensionality of the map. In this case the choice of D does not change the outcome. The energy function (1) follows the principle of parsimony: it is the simplest analytical expression that creates balanced forces of desired signs between synonyms and antonyms, preserves symmetries of semantic relations, and increases indefinitely at the infinity, keeping the resultant distribution localized near the origin of coordinates.

The procedure is that the initial coordinates of all words are sampled by a random number generator. Then the energy (1) minimization process starts that pulls synonym vectors together and antonym vectors apart. Then principal component analysis is used to reveal the main emergent semantic dimensions of the optimized map [10, 11]. Thus, the initial space coordinates are not associated with any semantics a priori: instead, words are allocated randomly in an abstract multidimensional space. In contrast, the starting point of traditional techniques based on LSA [1, 22] is a feature space, where dimensions have definite semantics a priori.

The representative weak semantic map shown in Figure 1 includes $N = 15,783$ words and was constructed based on the dictionary of English synonyms and antonyms available as part of Microsoft Word (MS Word) [11]. A similar map was also constructed using WordNet in the same work [11] and is also used in this study, together with maps constructed in [11] for other languages. Figure 1 represents the first two PCs of the distribution of words on the map constructed using the English MS Word thesaurus. The axes of the map are defined by the PCs. Selected words shown on the map in black at

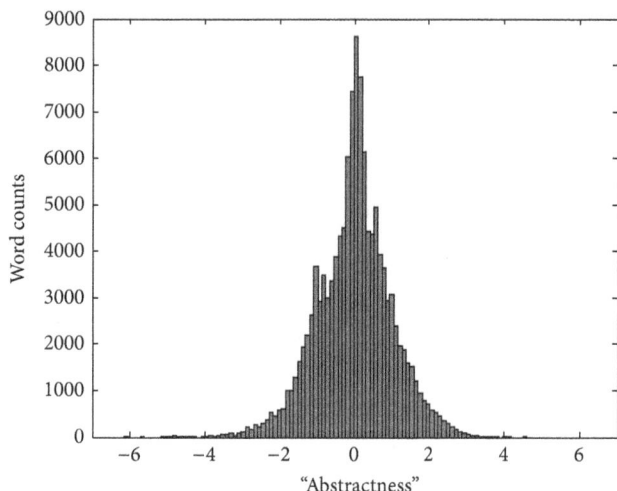

FIGURE 2: Distribution histogram of the 124,408 WordNet 3.0 words along the "abstractness" dimension.

TABLE 1: The tails of the list of 124,408 words sorted by "abstractness" in descending order.

The beginning of the list	The end of the list
Entity	Chain wrench
Physical entity	Francis turbine
Psychological feature	Tricolor television tube
Auditory communication	Tricolor tube
Unmake	Tricolour television tube
Cognition	Tricolour tube
Knowledge	Edmontonia
Noesis	*Coelophysis*
Natural phenomenon	*Deinocheirus*
Ability	*Struthiomimus*
Social event	*Deinonychus*
Craniate	*Dromaeosaur*
Vertebrate	*Mononychus olecranus*
Higher cognitive process	*Oviraptorid*
Physiological property	*Superslasher*
Mammal	*Utahraptor*
Mammalian	*Velociraptor*

their map locations characterize the semantics of the map. The two hypernym-hyponym pairs, "object-screw" (shown in pink) and "action-withdrawal" (in blue), illustrate the map inability to capture the "abstractness" dimension, confirmed quantitatively by correlation analysis in the next section. It should be pointed out here that the negative valence of "object" can be attributed to the meaning of the verb "object" that is merged with the noun "object" on this string-based semantic map.

2.2. Measuring the "Abstractness" of Words. Here we refer to the "abstractness" of a concept as its ontological generality. The WordNet database contains information that allows us to arrange English words on a line according to their

"abstractness" (or ontological generality). This information is contained in the hyponym-hypernym relations among words. The goal is to separate hypernym-hyponym pairs in one dimension tentatively labeled "abstractness," so that each hyponym has a lower "abstractness" value compared to its hypernyms. Given a consistent hierarchy, a solution would be, for example, to interpret the order of a word in the hierarchy as a measure of its "abstractness." Unfortunately, the system of hyponym-hypernym relations among words available in WordNet is internally inconsistent: it has numerous loops and conflicting links. Therefore, we use an optimization approach analogous to the antonymy-based weak semantic mapping based on (1). The underlying idea is to give each word i its "abstractness" coordinate x_i in such a way that the overall correlation between the difference in word "abstractness" coordinates x and the reciprocal hypernym-hyponym relations of the two words is maximized. Unfortunately, an energy function similar to H (1) cannot be used here, because the symmetry of hypernym-hyponym relations is different from the symmetry of antonym and synonym relations. Nevertheless, we showed in previous work [23] that the goal can be achieved by using the following definition of word "abstractness" values $\{x\}$:

$$\vec{x} = \operatorname*{argmin}_{\mathbb{R}^n} \left[\sum_{i,j=1}^{n} W_{ij} \left(x_i - x_j - 1 \right)^2 + \mu \sum_{i=1}^{n} x_i^2 \right], \quad (2)$$

where n is the number of words, μ is a regularization parameter, and $W_{ij} = 1$ if the word i is a hypernym of the word j and zero otherwise. Here the first sum is taken over all ordered hyponym-hypernym pairs.

The publicly available WordNet 3.0 database (http://wordnet.princeton.edu/) was used in this study. The hypernym-hyponym relations among $n = 124,408$ English words were extracted from the database as a connected graph defining the matrix W, which was used to compute the energy function (2). Optimization was carried out with standard MATLAB functions, as described in [23].

3. Results

3.1. Measuring Correlations of Augmented Map Dimensions. The one-dimensional semantic map of "abstractness" was computed as described in Section 2. The resultant distribution of 124,408 WordNet words in one dimension is shown in Figure 2. The two ends of the sorted list of words along their "abstractness" are given in Table 1.

This map was then combined with several antonymy-based weak semantic maps that are previously constructed [11]. The "abstractness" map was merged with any given narrow weak semantic map as the following. First, the set of words was limited to those that are common for both maps. Then, the augmented map was defined as a direct sum of the two vector spaces; that is, the "abstractness" dimension was added as a new word coordinate.

The resultant augmented maps were used to compute the correlation between "abstractness" and other map

(a)

(b)

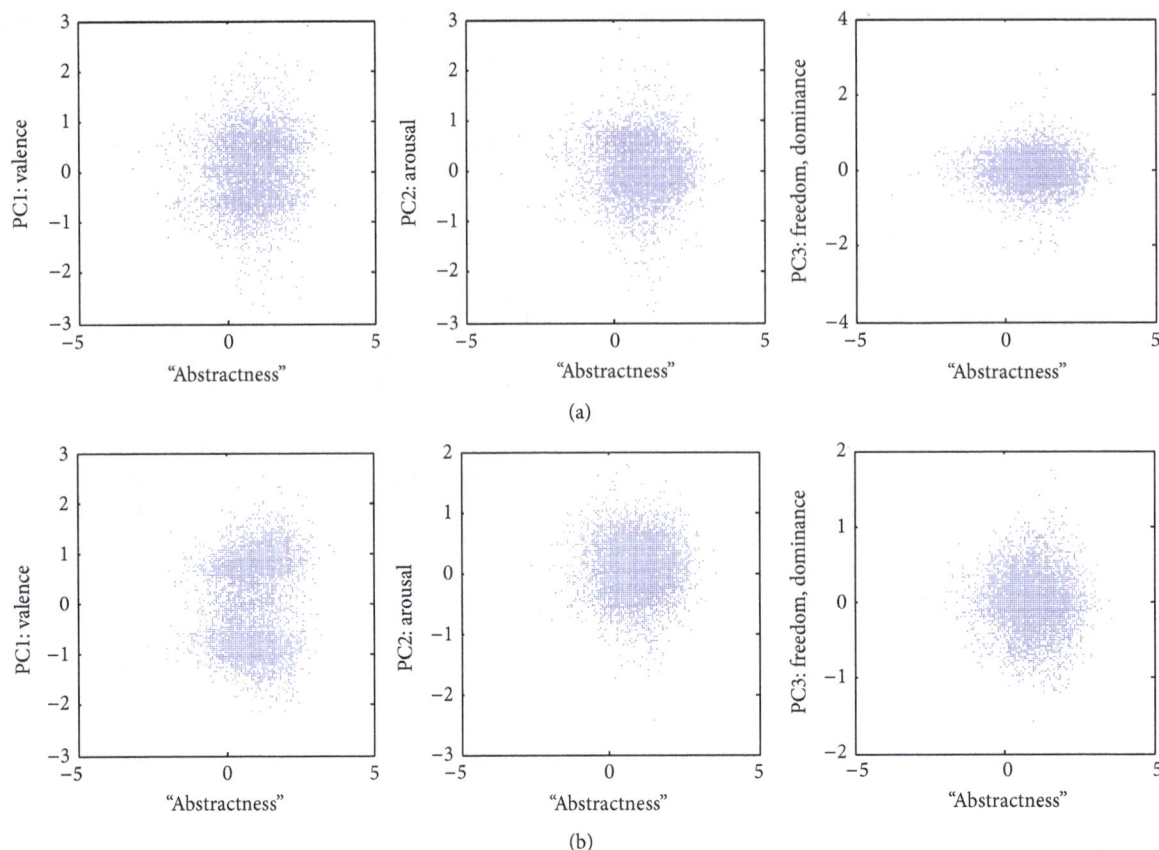

FIGURE 3: Correlations of "abstractness" with principal components of the antonymy-based weak semantic cognitive maps. (a) The map constructed using WordNet 3.0; (b) the map constructed using the Microsoft Word thesaurus.

TABLE 2: Pearson correlation coefficient R and the corresponding accounted variance (R^2) of "abstractness" with PC1: valence, PC2: arousal, and PC3: freedom/dominance, measured in four augmented maps constructed based on WordNet 3.0 and the MS Word English, French, and German thesauri.

	PC1: valence		PC2: arousal		PC3: freedom, dominance	
	R	R^2	R	R^2	R	R^2
WordNet	0.09	0.8%	−0.07	0.5%	−0.01	0% (NS)
MS Word English	0.12	1.4%	0.01	0% (NS)	−0.03	0.1% (NS)
MS Word French	0.11	1.2%	0.02	0% (NS)	0.01	0% (NS)
MS Word German	0.14	2.0%	−0.02	0% (NS)	0	0% (NS)

dimensions. The main question was how, if at all is the new "abstractness" dimension related to the principal components of the antonymy-based weak semantic map? Figure 3 illustrates the scatterplots of word "abstractness" values derived from WordNet with the dimensions of narrow weak semantic maps derived from WordNet data (Figure 3(a)) and from MS Word (Figure 3(b)). The Pearson correlation coefficient R and the corresponding accounted variance R^2 are given in Table 2 for each PC.

Similar results were obtained for augmented weak semantic maps in other languages (constructed based on the MS Word thesaurus as described in [11]): French (Figure 4(a))

and German (Figure 4(b)). Automated Google translation was used to merge maps in different languages.

In all cases "abstractness" is only positively correlated with valence ($P < 10^{-8}$ in all corpora), while none of the correlation coefficients with the other two dimensions (arousal and freedom) are statistically significant in a consistent way across corpora. Even in the case of valence, the correlation coefficient remains small (Table 2). This finding is further addressed in Section 4.

Overall, the results (Figures 3 and 4) show that the new "abstractness" dimension is practically orthogonal to the narrowly defined weak semantic map dimensions. Indeed,

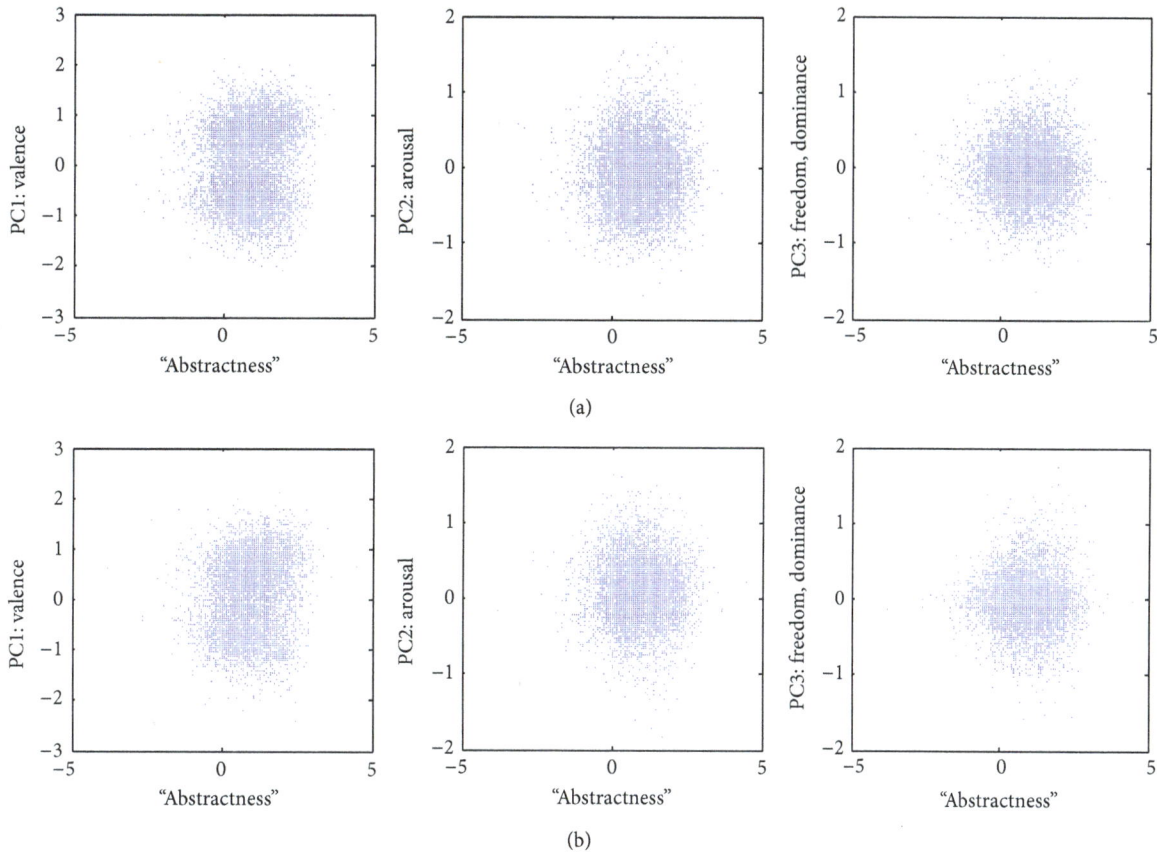

FIGURE 4: Correlations of "abstractness" with principal components of the antonymy-based weak semantic cognitive maps in other languages. (a) The map constructed using the French dictionary of MS Word. (b) The map constructed using the German dictionary of MS Word.

in most cases the correlation is not significant. In the minority of the cases where the correlation is statistically significant, the correlation coefficient is sufficiently small as to become marginal. Specifically, little information is lost by disregarding the fraction of the variance of the distribution of words on the weak semantic map accounted by the word "abstractness" or, vice versa, the fraction of the variance in the word "abstractness" accounted by the weak semantic map dimensions (Table 2).

In conclusion, the previous weak semantic map dimensions do not account for a substantial fraction of variance in "abstractness," and word "abstractness" values do not account for a substantial fraction of variance in the distribution of words on antonymy-based weak semantic maps.

3.2. Examples of Document Mapping with the Augmented Semantic Map. Traditionally, only the valence dimension is used in sentiment analysis. At the same time, other dimensions including "abstractness" are frequently indicated as useful (e.g., [24]). We previously applied the weak semantic map to analysis of Medline abstracts [25]. As an extension of that study, we now applied the augmented semantic map to analyze various kinds of documents.

Using the MS Word English narrow weak semantic map merged with the WordNet-based "abstractness" map, this part of the study asked the following key research questions: how informative is the new dimension compared to familiar dimensions at the document level? Specifically, how well are different kinds of documents separated from each other on the augmented map compared to the narrow weak semantic map? How capable is the new "abstractness" dimension compared to antonymy-based dimensions in terms of document separation? Being aware of more advanced methods of sentiment analysis [21, 26], here we adopted the simplest "bag of words" method (computing the "center of mass" of words in the document, not to be confused with LSA). This parsimonious choice is justified because at this point we are interested in assessing the value of the new dimension compared to familiar dimensions of the narrow weak semantic map, rather than achieving practically significant results.

For each document, the average augmented map coordinates of all words were computed, together with the standard error in each dimension. The results are represented in Figure 5 by crossed ovals, with the center of the cross representing the average and the size of the oval representing the standard error (i.e., the standard deviation divided by the square root of the number of identified words). The large black crosses in each panel represent the average of all words in the dictionary weighted by their overall usage

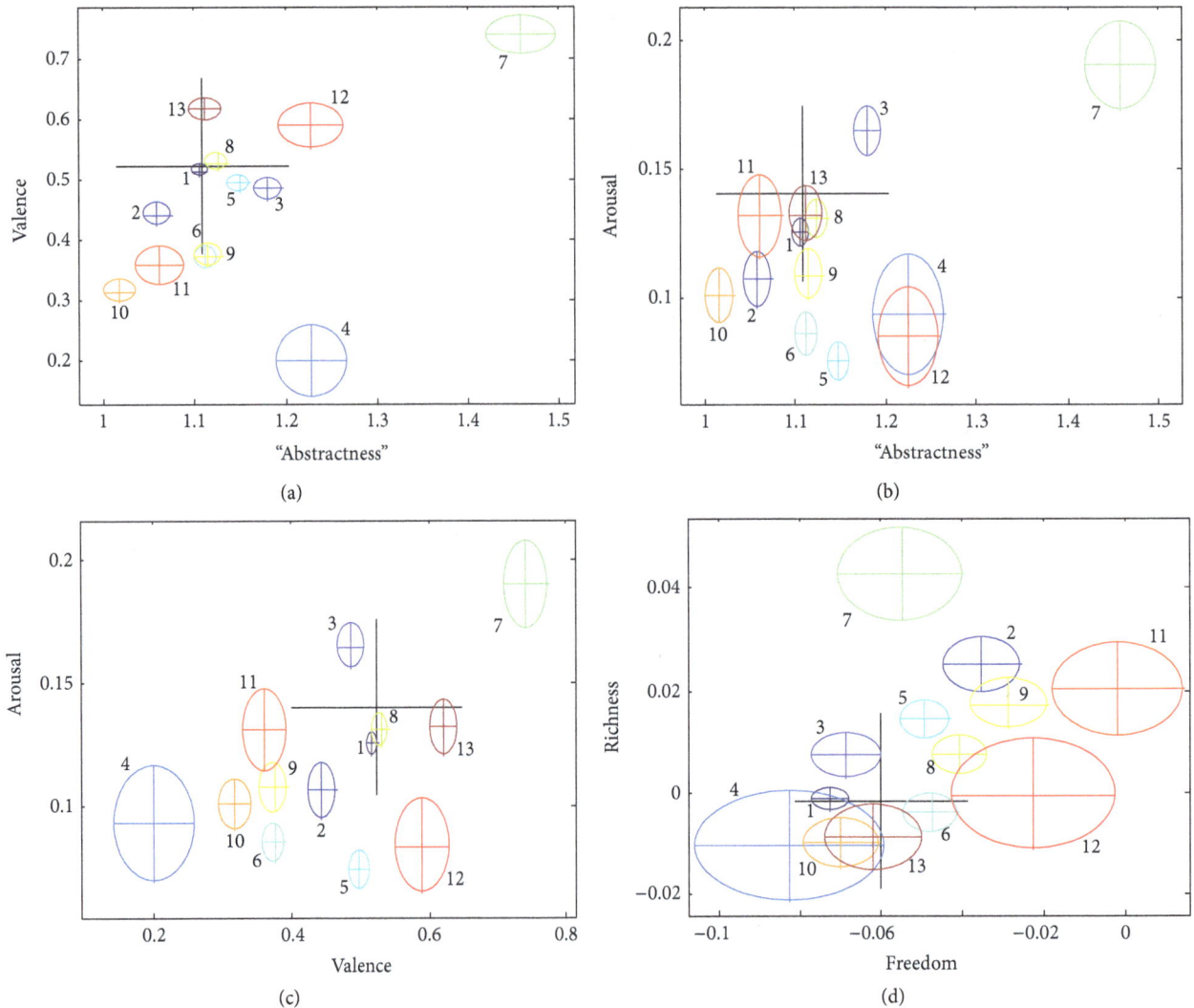

FIGURE 5: Representations of 13 documents (details in the text) on the augmented semantic map. The Pearson correlation coefficient R and the corresponding P value were computed for each panel. None of the correlations are significant. (a) Valence versus "abstractness," $R = 0.54$, $P = 0.06$. (b) Arousal versus "abstractness," $R = 0.50$, $P = 0.09$. (c) Arousal versus valence, $R = 0.54$, $P = 0.057$. (d) Richness (PC4) versus freedom (PC3), $R = 0.46$, $P = 0.12$.

frequency, not limited to materials of this study and derived as in [11]. Colors and numbers of ovals in Figure 5 correspond to RGB values and item numbers given in the following list of corpora:

(1) Project Gutenberg's A Text-Book of Astronomy, by George C. Comstock (http://www.gutenberg.org/files/34834/34834-0.txt), 9626 words, rgb = (0, 0, 6);

(2) Martha Stewart Living Radio Thanksgivings Hotline Recipes 2011 (http://www.hunt4freebies.com/free-martha-stewart-thanksgiving-recipes-ebook-download), 2091 words, rgb = (0, 0, 9);

(3) Al Qaida Inspire Magazine Issue 9 (http://www.en.wikipedia.org/wiki/Inspire_(magazine)), 2555 words, rgb = (0, 2, 10);

(4) A suicide blog (http://www.tumblr.com/tagged/suicideblog), 387 words, rgb = (0, 5, 10);

(5) 152 Shakespeare sonnets [27], 4170 words, rgb = (0, 8, 10);

(6) The Hitchhiker's Guide to the Galaxy, by Douglas Adams (http://www.paulyhart.blogspot.com/2011/10/hitchhikers-guide-to-galaxy-text_28.html), 4187 words, rgb = (1, 10, 9);

(7) 10 abstracts of award-winning NSF grant proposals (downloaded from http://www.nsf.gov/awardsearch), 585 words, rgb = (4, 10, 6);

(8) 196 reviews of the film "Iron Man", 2008 (http://www.mrqe.com/movie_reviews/iron-man-m100052975/), 3902 words, rgb = (8, 10, 2);

(9) 170 reviews of the film "Superhero Movie", 2008 (http://www.mrqe.com/movie_reviews/superhero-movie-m100071304/), 2204 words, rgb = (10, 9, 0);

(10) 160 reviews of the film "Prom Night", 2008 (http://www.mrqe.com/movie_reviews/prom-night-m100076394/), 2114 words, rgb = (10, 6, 0);

(11) 47 anecdotes of/about famous scientists (retrieved from http://jcdverha.home.xs4all.nl/scijokes/10.html), 919 words, rgb = (10, 3, 0);

(12) transcript of Obama's speech at the DNC on September 6, 2012 (http://www.foxnews.com/politics/2012/09/06/transcript-obama-speech-at-dnc), 491 words, rgb = (10, 0, 0);

(13) "Topological strings and their physical applications," by Andrew Neitzke and Cumrun Vafa (http://www.arxiv.org/abs/hep-th/0410178v2), 1909 words, rgb = (7, 0, 0).

The selected documents are mostly well separated in 3 dimensions, including valence (PC1), arousal (PC2), and "abstractness" (Figure 5). At the same time, the ovals more frequently overlap on the plane freedom-richness (PC3-PC4). Visually, "abstractness" is approximately as efficient as valence (PC1) in its ability to separate documents and appears to be more efficient than other dimensions; however, the oval separation on the valence-arousal projection (Figure 5(c)) looks slightly better than on the valence-"abstractness" projection (Figure 5(a)). This observation suggests that disregarding "abstractness" may not significantly affect the quality of results, while disregarding valence would substantially impair the quality of document separation (e.g., on the "abstractness-"arousal plane, Obama's speech overlaps substantially with the suicide blog, while valence separates the two documents significantly).

Differences between the above 13 documents along these 5 dimensions were quantified with analysis of variance. Specifically, the MANOVA P value was 0.027, suggesting that all five semantic dimensions are mutually independent in characterizing the selected 13 corpora. Moreover, in order to compare how informative different semantic dimensions are relative to each other, two sets of characteristics were computed (Table 3), namely, (i) the ANOVA P values to reject the null hypothesis that all 13 corpora have the same mean in each selected semantic dimension and (ii) the MANOVA P values to reject the null hypothesis that the means of all 13 corpora belong to a low-dimensional hyperplane within the space of all but one semantic dimensions.

These results can be interpreted as follows. The lower the P value for ANOVA is, the more informative the selected semantic dimension is. On the contrary, the lower the P value for MANOVA is, the less informative the selected semantic dimension is, because MANOVA was computed in the space of all semantic dimensions except the one selected. Therefore, results represented in Table 3 indicate that "abstractness" (dimension 0) is nearly as informative as valence (dimension 1) and could be more informative than arousal (dimension 2, based on ANOVA only), freedom (dimension 3), and richness (dimension 4). More data are needed to verify this interpretation.

4. Discussion

Statistical analysis indicates that "abstractness" is positively (if marginally) correlated with valence consistently across corpora, which is not the case with other semantic dimensions. On the one hand, the amount of variance in the distributions of words that can be attributed to interaction between valence and "abstractness" is not substantial (only 2% of variance or less); therefore, the two dimensions can be considered orthogonal for practical purposes. On the other hand, the consistent significance of this negligibly small correlation across datasets and languages indicates that there may be a universal factor responsible for it. This factor could be the usage frequency of words that affects the probability of word selection for dictionaries. Stated simply, abstract positive words and specific negative words are used more frequently than abstract negative words and specific positive words. Specifically, our previous study [11] showed that the mean valence (normalized to unitary standard deviation) of all words weighted by their usage frequency is significantly positive (0.50 using frequency data from a database of Australian newspapers and 0.59 using frequency data from the British National Corpus). Using the results in the present study, the mean normalized "abstractness" is between 0.99 (weighted by "Australian" frequency) and 1.39 (weighted by "British" frequency). An equivalent explanation is that abstract words and positive words are both used more frequently than specific words and negative words. Specifically, the correlation with frequency is small but significantly positive both for valence (0.064 Australian, 0.061 British) and for "abstractness" (0.036 Australian, 0.019 British). This interpretation is consistent with data at the level of documents (Figure 5(a)), where the correlation coefficient is even higher, yet not significant (not shown). Another potential source of correlation is the selection of words for inclusion in dictionaries. It seems, however, counterintuitive that the overall picture should be affected by marginal inclusions of rare words. Nevertheless, it would be interesting to check elsewhere how the correlation changes across sets of words found in various types of documents.

The method of weak semantic mapping is an alternative to other vector-space-based approaches including LSA, LDA, MDS, and ConceptNet [1-4], primarily because (i) no semantic features of words are given as input and (ii) the abstract space of the map has no semantics associated a priori with its dimensions. It is therefore not surprising that emergent semantic features (dimensions) in weak semantic mapping are substantially different from emergent semantic dimensions obtained by LSA and related techniques: the latter are typically domain specific and harder to interpret [22].

From another perspective, it is interesting that emergent semantic dimensions of a weak semantic map are so familiar. All generally accepted dimensional models of sentiment, appraisal, emotions and values, attitudes, feelings, and so forth converge on semantics of their principal dimensions, typically interpreted as valence, arousal, and dominance [12-14, 16-18]. Antonymy-based weak semantic mapping appears to be consistent with this emergent consensus [9-11], despite the stark difference in methodologies (human

TABLE 3: ANOVA and MANOVA P values for selected semantic dimensions characterizing the means of the 13 corpora. Dimensions are numbered as follows: 0, "abstractness", 1, PC1 (valence), 2, PC2 (arousal), 3, PC3 (freedom/dominance), and 4, PC4 (richness).

Semantic Dimension	0	1	2	3	4
One dimension, ANOVA	$1.2e-36$	$5.9e-57$	$3.1e-15$	$2.1e-7$	$6.2e-11$
All but one, MANOVA	0.018	0.040	0.041	$5.1e-7$	$1.4e-7$

judgment or psychometrics versus automated calculations based on subject-independent data). The number of semantic dimensions, or factors, used in the literature varies from 2 to 7, which roughly corresponds to the variability in the number of significant principal components of the narrow weak semantic map [11]. Why do antonyms relating to the "dimensional models" of affect, and not others, make for good PCs? This interesting question remains open and should be addressed by future studies.

The present study unambiguously demonstrates the inability of narrow weak semantic maps to capture all universal semantic dimensions. Here we presented one dimension, "abstractness," that is not captured by "antonymy-" defined weak semantic maps. This is due to the fact that, in general, hypernym-hyponym pairs are not antonym pairs and vice versa. Therefore, hypernym-hyponym relations cannot be captured with the map defined by antonym relations, and the map needs to be augmented. The example of "abstractness" that we found is probably not unique: we expect a similar outcome for the holonym-meronym relation, which will be addressed elsewhere. Our previous results indicated that antonym relations are essential for weak semantic mapping, while synonym relations are not [28].

Thus, the present work shows that narrow weak semantic maps (and related dimensional models of emotions) are not complete in this sense and need to be augmented by including other kinds of semantic relations in their definition. A question remains open as to whether any augmented semantic map may be considered complete—or there will always be new semantic dimensions that can be added to the map. We speculate that there exists a complete finite-dimensional weak semantic map. Moreover, the number of its dimensions can be relatively small. This is because the number of distinct semantic relationships in natural language is limited, as is the number of primary categories [29], or the number of primary semantic elements of metalanguage known as semantic primes [31, 32]. This notion of "completeness," however, may only be applicable to a limited scope, for example, all existing natural languages.

We found that hyponym-hypernym relations induce a new semantic dimension on the weak semantic map that is not reducible to any dimensions represented by antonym pairs or to the traditional PAD or EPA dimensions. Its tentative labeling as "abstractness" or ontological generality, however, remains speculative. In any case, it is not our ambition here to define the notion of "abstractness" or to establish a precise connection between the real notion of abstractness and our new "abstractness" dimension, a topic that should be addressed elsewhere.

Findings of this study have broad implications for automated quantitative evaluation of the meaning of text,

including semantic search, opinion mining, sentiment analysis, and mood sensing, as exemplified in Figure 5 and Table 3. While multidimensional approaches in opinion mining are nowadays popular, the problem is finding good multidimensional ranking of all words in the dictionary. Traditional bootstrapping methods (e.g., based on cooccurrence of words) to extend the ranking of positivity from a small subset of words to all words may not work, for example, for "abstractness." The approach presented here should be useful for such applications.

Finally, we speculate that this approach may shed light on the nature of human subjective experience [30] by revealing fundamental semantics of qualia as PCs of the weak semantic cognitive map. In addition, we suggest other connections of our findings, for example, to semantic primes [31, 32].

Acknowledgments

The authors are grateful to Mr. Thomas Sheehan for help with extraction of the relation data from WordNet. Part of Giorgio A. Ascoli's time was supported by the Intelligence Advanced Research Projects Activity (IARPA) via Department of the Interior (DOI) Contract no. D10PC20021. The US Government is authorized to reproduce and distribute reprints for governmental purposes notwithstanding any copyright annotation thereon. The views and conclusions contained herein are those of the authors and should not be interpreted as necessarily representing the official policies or endorsements, either expressed or implied, of IARPA, DOI, or the US Government.

References

[1] T. K. Landauer and S. T. Dumais, "A solution to Plato's problem: the Latent Semantic Analysis theory of acquisition, induction, and representation of knowledge," *Psychological Review*, vol. 104, no. 2, pp. 211–240, 1997.

[2] T. L. Griffiths and M. Steyvers, "Finding scientific topics," *Proceedings of the National Academy of Sciences of the United States of America*, vol. 101, no. 1, pp. 5228–5235, 2004.

[3] C. Havasi, R. Speer, and J. Alonso, "ConceptNet 3: a flexible, multilingual semantic network for common sense knowledge," in *Proceedings of the International Conference on Recent Advances in Natural Language Processing (RANLP '07)*, Borovets, Bulgaria, 2007.

[4] E. Cambria, T. Mazzocco, and A. Hussain, "Application of multidimensional scaling and artificial neural networks for biologically inspired opinion mining," *Biologically Inspired Cognitive Architectures*, vol. 4, pp. 41–53, 2013.

[5] R. F. Cox and M. A. Cox, *Multidimensional Scaling*, Chapman & Hall, 1994.

[6] J. B. Tenenbaum, V. de Silva, and J. C. Langford, "A global geometric framework for nonlinear dimensionality reduction," *Science*, vol. 290, no. 5500, pp. 2319–2323, 2000.

[7] L. K. Saul, K. Q. Weinberger, J. H. Ham, F. Sha, and D. D. Lee, "Spectral methods for dimensionality reduction," in *Semisupervised Learning*, O. Chapelle, B. Schoelkopf, and A. Zien, Eds., pp. 293–308, MIT Press, Cambridge, Mass, USA, 2006.

[8] P. Gärdenfors, *Conceptual Spaces: The Geometry of Thought*, MIT Press, Cambridge, Mass, USA, 2004.

[9] A. V. Samsonovich, R. F. Goldin, and G. A. Ascoli, "Toward a semantic general theory of everything," *Complexity*, vol. 15, no. 4, pp. 12–18, 2010.

[10] A. V. Samsonovich and G. A. Ascoli, "Cognitive map dimensions of the human value system extracted from natural language," in *Advances in Artificial General Intelligence: Concepts, Architectures and Algorithms. Proceedings of the AGI Workshop 2006*, B. Goertzel and P. Wang, Eds., vol. 157 of *Frontiers in Artificial Intelligence and Applications*, pp. 111–124, IOS Press, Amsterdam, The Netherlands, 2007.

[11] A. V. Samsonovich and G. A. Ascoli, "Principal semantic components of language and the measurement of meaning," *PLoS ONE*, vol. 5, no. 6, Article ID e10921, 2010.

[12] E. Hudlicka, "Guidelines for designing computational models of emotions," *International Journal of Synthetic Emotions*, no. 1, pp. 26–79, 2011.

[13] C. E. Osgood, G. Suci, and P. Tannenbaum, *The Measurement of Meaning*, University of Illinois Press, Urbana, Ill, USA, 1957.

[14] J. A. Russell, "A circumplex model of affect," *Journal of Personality and Social Psychology*, vol. 39, no. 6, pp. 1161–1178, 1980.

[15] R. Plutchik, "A psychoevolutionary theory of emotions," *Social Science Information*, vol. 21, no. 4-5, pp. 529–553, 1982.

[16] A. Mehrabian, *Nonverbal Communication*, 2007.

[17] P. J. Lang, M. M. Bradley, and B. N. Cuthbert, "Emotion and motivation: measuring affective perception," *Journal of Clinical Neurophysiology*, vol. 15, no. 5, pp. 397–408, 1998.

[18] C. E. Osgood, W. H. May, and M. S. Miron, *Cross-Cultural Universals of Affective Meaning*, University of Illinois Press, Urbana, Ill, USA, 1975.

[19] H. Lövheim, "A new three-dimensional model for emotions and monoamine neurotransmitters," *Medical Hypotheses*, vol. 78, no. 2, pp. 341–348, 2012.

[20] R. Likert, "A technique for the measurement of attitudes," *Archives of Psychology*, vol. 22, no. 140, pp. 1–55, 1932.

[21] B. Pang and L. Lee, "Opinion mining and sentiment analysis," *Foundations and Trends in Information Retrieval*, vol. 2, no. 1-2, pp. 1–135, 2008.

[22] T. K. Landauer, D. S. McNamara, S. Dennis, and W. Kintsch, Eds., *2007Handbook of Latent Semantic Analysis*, Lawrence Erlbaum Associates, Mahwah, NJ, USA.

[23] A. V. Samsonovich, "A metric scale for "Abstractness" of the word meaning," in *Intelligent Techniques for Web Personalization and Recommender Systems: AAAI Technical Report WS-12-09*, D. Jannach, S. S. Anand, B. Mobasher, and A. Kobsa, Eds., pp. 48–52, The AAAI Press, Menlo Park, Calif, USA, 2012.

[24] D. McNamara, Y. Ozuru, A. Greasser, and M. Louwerse, "Validating coh-metrix," in *Proceedings of the 28th Annual Conference of the cognitive Science Society*, R. Sun and N. Miyake, Eds., Erlbaum, Mahwah, NJ, USA, 2006.

[25] A. V. Samsonovich and G. A. Ascoli, "Computing semantics of preference with a semantic cognitive map of natural language: application to mood sensing from text," in *Multidisciplinary Workshop on Advances in Preference Handling, Papers from the 2008 AAAI Workshop, AAAI Technical Report WS-08-09*, J. Chomicki, V. Conitzer, U. Junker, and P. Perny, Eds., pp. 91–96, AAAI Press, Menlo Park, Calif, USA, July 2008.

[26] R. Moraes, J. F. Valiati, and W. P. G. Neto, "Document-level sentiment classification: an empirical comparison between SVM and ANN," *Expert Systems with Applications*, vol. 40, no. 2, pp. 621–633, 2013.

[27] W. Shakespeare, *A Lover's Complaint*, The Electronic Text Center, University of Virginia, 1609/2006.

[28] A. V. Samsonovich and C. P. Sherrill, "Comparative study of self-organizing semantic cognitive maps derived from natural language," in *Proceedings of the 29th Annual Cognitive Science Society*, D. S. McNamara and J. G. Trafton, Eds., p. 1848, Cognitive Science Society, Austin, Tex, USA, 2007.

[29] I. Kant, *Critique of Pure Reason*, vol. A 51/B 75, Norman Kemp Smith, St. Martins, NY, USA, 1781/1965.

[30] G. A. Ascoli and A. V. Samsonovich, "Science of the conscious mind," *The Biological Bulletin*, vol. 215, no. 3, pp. 204–215, 2008.

[31] A. Wierzbicka, *Semantics: Primes and Universals*, Oxford University Press, 1996.

[32] C. Goddard and A. Wierzbicka, "Semantics and cognition," *Wiley Interdisciplinary Reviews: Cognitive Science*, vol. 2, no. 2, pp. 125–135, 2011.

15

Modeling Spike-Train Processing in the Cerebellum Granular Layer and Changes in Plasticity Reveal Single Neuron Effects in Neural Ensembles

Chaitanya Medini,[1] Bipin Nair,[1] Egidio D'Angelo,[2,3] Giovanni Naldi,[4] and Shyam Diwakar[1]

[1] *Amrita School of Biotechnology, Amrita Vishwa Vidyapeetham (Amrita University), Amritapuri, Clappana, Kollam 690525, Kerala, India*
[2] *Department of Physiology, University of Pavia, Via Forlanini 7, 21000 Pavia, Italy*
[3] *Brain Connectivity Center, IRCCS C. Mondino, Via Mondino 2, 27100 Pavia, Italy*
[4] *Department of Mathematics, University of Milan, Via Saldini 50, 27100 Milan, Italy*

Correspondence should be addressed to Shyam Diwakar, shyam@amrita.edu

Academic Editor: Steven Bressler

The cerebellum input stage has been known to perform combinatorial operations on input signals. In this paper, two types of mathematical models were used to reproduce the role of feed-forward inhibition and computation in the granular layer microcircuitry to investigate spike train processing. A simple spiking model and a biophysically-detailed model of the network were used to study signal recoding in the granular layer and to test observations like center-surround organization and time-window hypothesis in addition to effects of induced plasticity. Simulations suggest that simple neuron models may be used to abstract timing phenomenon in large networks, however detailed models were needed to reconstruct population coding via evoked local field potentials (LFP) and for simulating changes in synaptic plasticity. Our results also indicated that spatio-temporal code of the granular network is mainly controlled by the feed-forward inhibition from the Golgi cell synapses. Spike amplitude and total number of spikes were modulated by LTP and LTD. Reconstructing granular layer evoked-LFP suggests that granular layer propagates the nonlinearities of individual neurons. Simulations indicate that granular layer network operates a robust population code for a wide range of intervals, controlled by the Golgi cell inhibition and is regulated by the post-synaptic excitability.

1. Introduction

Decoding neural activity is the key to understand spatiotemporal patterns that the brain receives as sensory information regarding the world. Time-scale of operation is closely correlated to the activity of the neural circuit and decoding such activity reveals principles regarding the function. One of the main circuits in the cerebellum is the large input layer circuit formed of granule and Golgi cells. Spatiotemporal information is one of the unique functional characteristics observed in the cerebellar input layer network [1, 2]. Cerebellar granular layer forms the input stage of the cerebellum in which information coming from the peripheral and central systems converge through the mossy fibers.

The granular layer has by far the smallest ($\sim 5\,\mu$m) and the most numerous neurons ($\sim 10^{11}$) in humans. Understanding how the granular layer process information appears critical to understand the cerebellar function, since signals coming into upper cortical layers are provided by the granular layer. The granule cells form the largest neuronal population in the mammalian brain and regulate information transfer along the major afferent systems to the cerebellum. The granule layer receives excitatory input primarily from mossy fibers and inhibition via synapses from interneurons like Golgi cell. Mossy fiber input excites both the granule cell and inhibitory interneurons like the Golgi cell. The granule cell is a small neuron with 3–5 dendrites. Timing in the cerebellar granular layer plays a key role via passage-of-time

Modeling Spike-Train Processing in the Cerebellum Granular Layer and Changes in Plasticity Reveal Single
Neuron Effects in Neural Ensembles

175

representation (POT), learning or adaptation to movements [3, 4], modulation of information transfer to Purkinje cells (activation of granule cell subsets with respect to time). Knockout and lesion studies revealed that disruption of the granular layer leads to abnormal functioning of the cerebellar mossy fiber-granule cell relay [5], affects the learning-dependent timing of conditioning eyelid responses [6], loss of rapid spike processing in the cerebellum (results in ataxia) [7] which in turn affects the plasticity of granule-Purkinje cell synapses. Prion protein (PrP) knockout mice showed a large proportion of granule cells (~40%) with slow non-overshooting nonrepetitive action potential, slow EPSPs, and no inward rectification [5]. Likewise, FHF1-FHF4 mutants showed impaired granule cell excitability which prevented rapid burst transmission in the cerebellum [7].

The objective of the current paper is to study how excitation operates in the granular layer network [1, 2, 8–11]. The focus is also on understanding the modulatory role of inhibition [1, 2] in the granule cells [12] and underlying ensemble activity in terms of combinatorial operations on granular layer network [8].

In order to estimate spiking behaviour and to test reliability in modeling, we tried using detailed and simple models of neurons in our network models. The use of detailed multicompartmental models was focused towards reproducing the spatiotemporal dynamics of normal cerebellar activity. The detailed models allow reconstructing information including local field potentials [13], which were not seen while using less detailed models. We also used a spiking neuron model (modified from [14]) for reconstructing network activity in order to understand the contributions of individual spikes in the cerebellar cortex. The necessity of these spiking models was to retain the biologically plausibility of Hodgkin-Huxley-type dynamics, while maintaining the low computational cost. It was also an attempt to validate whether such simple spiking models also allow us to create computationally simpler yet large-scale models of cerebellum. Using properties of the granule cell [9] and Golgi neuron [15, 16], we developed simple spiking models [17] to represent the spiking behavior in a network. Estimates of spiking and reproducibility of spiking could be very useful for computationally efficient and large network models. Spike modulation due to the effect of feed-forward inhibition has been known to play an important role in time-windows hypothesis [18]. The simulations quantify how spikes pass through the granular layer network and the role of feed-forward inhibition in the neuronal microcircuit. This gives a new paradigm on the functional relevance of patterns in the cerebellar granular layer circuitry. The main objectives of the paper were also to reconstruct the center-surround excitation patterns and observe role of combined inhibition and excitation geometry in frequency-dependent transmission of spike information.

The objective was also to understand the effect of combinatorial operations on the granular layer network [8]. Combinatorial operations included combined excitation and inhibition which forms the spatiotemporal pattern in granular layer network *in vitro* and *in vivo*. Another objective behind the simulations was to understand information flow

in granule neurons via burst-burst transmission and feed-forward inhibition [2]. Together they suggest the role of granule cells in expansion recoding and sparse activation via mossy fiber-granule cell synapses. The paper reports potential reconstruction of network activity in the form of center-surround structures [8], spike properties of underlying cells, and modulation of spikes due to changes induced by synaptic plasticity and due to inhibition. A special case of NMDA receptor blocking was simulated since GABA (γ-aminobutyric acid)-ergic inhibition is especially effective in controlling NMDA (N-methyl-D-aspartate) receptor-dependent depolarization in the granular layer [18]. Network simulations predict specific computational roles of granule cells in processing bursts and overall spike processing in the cerebellar granular layer.

Understanding population code via comparisons of spatiotemporal properties of simulated neural activity and with experimental measurements using multielectrode recordings [18] is useful to identify how information encoding happens in microcircuits. Therefore, evoked local field potential (LFP) responses from granular layer *in vitro* [19] were reconstructed computationally [20]. The main intention of local field reconstruction on the network *in vitro* was to study the role of inhibition in generating the N_{2b} wave [4, 20]. The paper also investigates the combined role of excitation-inhibition affecting the granular layer clusters.

2. Methods

The study carried out in this paper involved the use of computational models of neurons based on experimental data from p17–23 Wistar rat cerebellum [9]. Mathematical neuron models of granule cell (GrC) [9, 19] and Golgi cell model (GoC) [15, 16] were used in this network study. Modeling reliability for spiking models was based on the extensive characterization of membrane currents and the compact electrotonic structure of cerebellar granule cells [9, 19]. The models used AMPA (2-amino-3-(5-methyl-3-oxo-1,2-oxazol-4-yl) propanoic acid) and NMDA receptor components as excitatory mossy fiber (MF)-GrC synapses and GABAergic synapses for the GoC-GrC relay [1, 15, 19, 21]. On an average, each granule cell receives excitatory connections from 4-5 mossy fibers [9].

2.1. Simple Neuronal Models. The objective of using simple models in the study was to understand how spatiotemporal patterns integrated over time to produce responses that are selective to specific patterns and to reconstruct the representation of spiking behavior in networks especially to study how inhibition affects intrinsic electroresponsiveness. We also wanted to see how the high variability in MF spike trains affects firing behavior. A simple spiking model [14] was used to study the neuronal spiking activity. A good model should be feasible with Hodgkin-Huxley dynamics and be computationally efficient [14] to reproduce the firing behavior of biorealistic model. The simple spiking model [14] of neuron primarily used two equations one regulating the membrane potential (V) and the other regulating adaptation current (w)

$$C\frac{dV}{dt} = -g_L(V - E_L) + g_L\Delta T \exp\left(\frac{V - V_T}{\Delta T}\right) - w + I \quad (1)$$

$$t_w\frac{dw}{dt} = a(V - E_L) - w, \quad (2)$$

where C is the membrane capacitance, g_L is the leak conductance, V_T is the voltage threshold, ΔT is the slope factor which quantifies the sharpness of the spikes, w is an adaptation current, and I is the injected current. The membrane time constant is

$$t_m = \frac{C}{g_L}. \quad (3)$$

The reset values were activated when the membrane potential reached the desired peak voltage:

$$\text{if } (V > 30\,\text{mV}), \text{then}$$

$$V = V_r \quad (4)$$

$$w = w + b.$$

The change in parameters (as shown in (4)) allowed the simple spiking model to replicate amplitude of granule cell firing and Golgi cell firing (see Table 1). Capacitance values of Golgi and granule cells were modelled to simulate similar frequency and firing patterns matching experimentally observed data, as observed in biophysical models [9, 15].

While replicating the basic firing behavior, the Adaptive Exponential Integrate-and-fire (AdEx, see [14]) granule cell model was tested with various combinations of synaptic connections. AMPA [17, 22, 23] synaptic kinetics was used as the excitatory synaptic dynamics and GABA [17, 24, 25] as the inhibitory synaptic kinetics as observed via experiments [19]:

$$g_{\text{AMPA}} = g_{\text{AMPA,max}} * e^{-t/18} * \frac{1 - e^{-t/2.2}}{0.68} \quad (5)$$

$$I_{\text{AMPA}} = (V_m - 0.0) * g_{\text{AMPA}}.$$

Likewise, GABA synaptic kinetics was modeled using the GABA-A equation [17]:

$$g_{\text{GABA}} = g_{\text{GABA,max}} * e^{-t/25} * \frac{(1 - e^{-t/1.0})}{0.84} \quad (6)$$

$$I_{\text{GABA}} = (V_m + 75) * g_{\text{GABA}}.$$

The maximal conductance of AMPA and GABA was adjusted to suit the observed biophysical firing pattern (see Table 2). For simulating the case without inhibition, maximal conductance value of GABA synapses is set to zero (see Table 2). To model inhibition, varying values of maximal conductance for different number of inhibitory synapses were used. With these two receptor kinetics, we were able to match the number of spikes and amplitude to that of biophysical model. The models also reproduced a close correlation to the biophysical model while varying the intrinsic excitability and release probability (data not shown). By changing conductance and dynamic parameters, we simulated LTP and LTD, matching to experimental and computational values [26].

To understand network dynamics in terms of firing and temporal processing, we used a simple spiking network with 1680 AdEx granule cell models and 1 AdEx Golgi cell model. The properties of the network model were matched (see Tables 1 and 2 and [9, 15, 16, 27, 28]) to experimental data.

2.2. Multicompartmental Models. A detailed multicompartmental GrC model [19] was used, and simulations were performed by varying the excitatory (E) and inhibitory (I) synaptic inputs. The model of the granule cell was based on multicompartmental cable theory and included soma, axon, hillock and dendritic compartments. The model consisted of 52 active compartments connected to each other via the 3/2-power law [29]. For each of the compartments, membrane voltage V_m had to be estimated separately:

$$\frac{dV_m}{dt} = \frac{1}{\tau_m}\left(V - \frac{\sum_i g_i(V - V_i) + \sum_{\text{syn}} g_{\text{syn}}(V - V_{\text{syn}}) + \sum_{\text{br}} g_{\text{br}}(V - V_{\text{br}})}{g_{\text{tot}}}\right), \quad (7)$$

where g is the conductance corresponding to i (ion channel), syn (synaptic dynamics), br (neighboring attached branch), and tot (total). Here, $\tau_m = R_m C_m$ which is the time constant of oscillation of the membrane based on its membrane resistance, R_m and membrane capacitance, C_m. The calcium current in the model was included as

$$\frac{d[\text{Ca}]}{dt} = -\frac{I_{\text{Ca}}}{(2F \cdot A \cdot d)} - (\beta_{\text{Ca}}([\text{Ca}] - [\text{Ca}]_O)), \quad (8)$$

where d is the depth of a shell adjacent to the cell surface of area A, β_{Ca} determines the loss of calcium ions from the shell approximating the effect of fluxes, ionic pumps, diffusion, and buffers, $[\text{Ca}]_O$ is resting [Ca], and F is the Faraday's constant. [Ca] is the calcium channel dynamics as reported in [19].

The model GrC has 1–4 excitatory (one for each dendrite) and 0 (no inhibition)–4 inhibitory connections (one for each dendrite) [19]. The detailed explanations of

Modeling Spike-Train Processing in the Cerebellum Granular Layer and Changes in Plasticity Reveal Single Neuron Effects in Neural Ensembles

177

TABLE 1: Parameter values used for AdEx spiking models.

Model	C (pF)	E_L (mV)	V_T (mV)	V_r (mV)	b (pA)	g_L (nS)	ΔT (mV)	a (nS)	t_w (ms)
Granule spiking model	150	−70	−50	−64	250	10	4	9	13
Golgi spiking model	350	−58	−60	−50	1460	12	7	12	7

TABLE 2: Maximal synaptic conductance values used for AMPA and GABA receptor kinetics in the simple spiking network.

Number of excitatory connections	Number of inhibitory connections	Without inhibition		With inhibition	
		Excitatory maximal conductance (nS)	Inhibitory maximal conductance (nS)	Excitatory maximal conductance (nS)	Inhibitory maximal conductance (nS)
1	4	0.14	0	0.1	0.05
2	3	0.25	0	0.11	0.1
3	2	0.256	0	0.256	0.11
4	1	0.27	0	0.3	0.18

ionic channel dynamics, compartmental localization of ion channels and electronic structure of this granule neuron model are described elsewhere [9, 19, 21]. Since granule cell is one of the rarest neurons where the ionic channel densities can be accurately determined using whole-cell patch clamp, the ion channel dynamics that was modeled previously [7, 9, 21] is not repeated here. Also, excitatory and inhibitory synaptic inputs to the dendrites were located in dendritic tips although in neighboring dendritic compartments. Presynaptic dynamics for the MF-GrC was modeled separately as in [19, 21] due to components such as facilitation and depression. Excitatory postsynaptic mechanisms were shown as AMPA and NMDA postsynaptic receptor components as seen in granule neurons. AMPA receptor dynamics was modeled using a three-state scheme and a 2D diffusion model, whereas the NMDA receptors used Boltzmann equation as seen in [30]. Both the excitatory presynaptic and excitatory postsynaptic mechanisms are described in detail elsewhere [21]. The GoC-GrC inhibitory synapse model was based on the following presynaptic dynamics: release probability = 0.35, τ_{REC} = 36 ms, τ_{facil} = 58.5 ms, and τ_I = 0.1 ms, respectively and as described in [31]. Effects of blocking inhibition by adding gabazine were also simulated by setting GABAergic conductance in inhibitory fibers to zero.

The Golgi neuron was adapted from [15, 16]. All simulations were performed with NEURON environment [32] running on HP Blade C3000 node. Timing and initial time-window modulations are mainly affected by the role of feed-forward inhibition as it happens with only a slight delay from the mossy-fiber input and hence the role of feed-back inhibition was not simulated.

2.3. Granular Layer Network.

Granular layer spiking network model consisted of 140 homogenous mossy fibers (MF) rosettes, 1680 granule cells (GrC), and 1 Golgi cell (GoC). In this network, about 48 GrC receive 1 excitatory input from the same mossy fiber, and each granule cell receives four excitatory connections from four different mossy fibers. Along with these excitatory inputs given to GrC, mossy fibers also provide excitatory input to GoC whose ratio was set

in this model to about 78 : 1 (see [15, 16], each dendrite had 26 synapses in the GoC model and assuming a total of 3 dendrites, we approximated to 78 synapses) providing an overall glomeruli connectivity pattern [1]. The network topology is illustrated in Figure 1(a).

Modeling responses in brain slices *in vitro* were simulated by giving single spike as input via mossy fiber (MF) terminals. Anaesthetized rat brain recordings *in vivo* showed bursts as inputs through mossy fibers [33]. Therefore, *in vivo* inputs to GrC were simulated as bursts of (5 or 9) spikes via the MF input.

2.4. Center-Surround "Spot" Pattern.

Stimulating mossy fibers with an electrode at a particular point activates granule cells in the network in a center-surround activation pattern [8]. Within a "spot," cells which are in close proximity to the electrode will receive high excitation and the periphery layer cells receive less excitation. In the network model, we simulated the center-surround pattern (see Figure 4A in [8]) defined as a "spot," showing decreasing strengths of excitation spreading from the center to the periphery. In each spot, 48 cells in the center received 4 mossy fiber (MF) inputs, 144 cells received 3 MF inputs, 48 cells had 2 MF inputs, and 144 cells received 1 MF input.

2.5. Simulating LTP/LTD.

The granule cell model was modeled based on data from Wistar rats [9, 19]. By modifying intrinsic excitability and release probability [34, 35], we simulated plasticity in the GrC. Intrinsic excitability was modified by changing ionic current channel gating dynamics. On-off gating characteristics of sodium channel were altered to modify sodium activation and inactivation parameters [13, 26] for higher and lower intrinsic excitability. Three cases were studied where the intrinsic excitability of the GrC is low (low intrinsic excitability), normal (control), and high (high intrinsic excitability). The release probability (U) of MFs varied from 0.1 to 0.4 for cells with low intrinsic excitability, and from 0.5 to 0.8 for cells with high intrinsic excitability while the control value remained as 0.416 in simulations for normal cells [36].

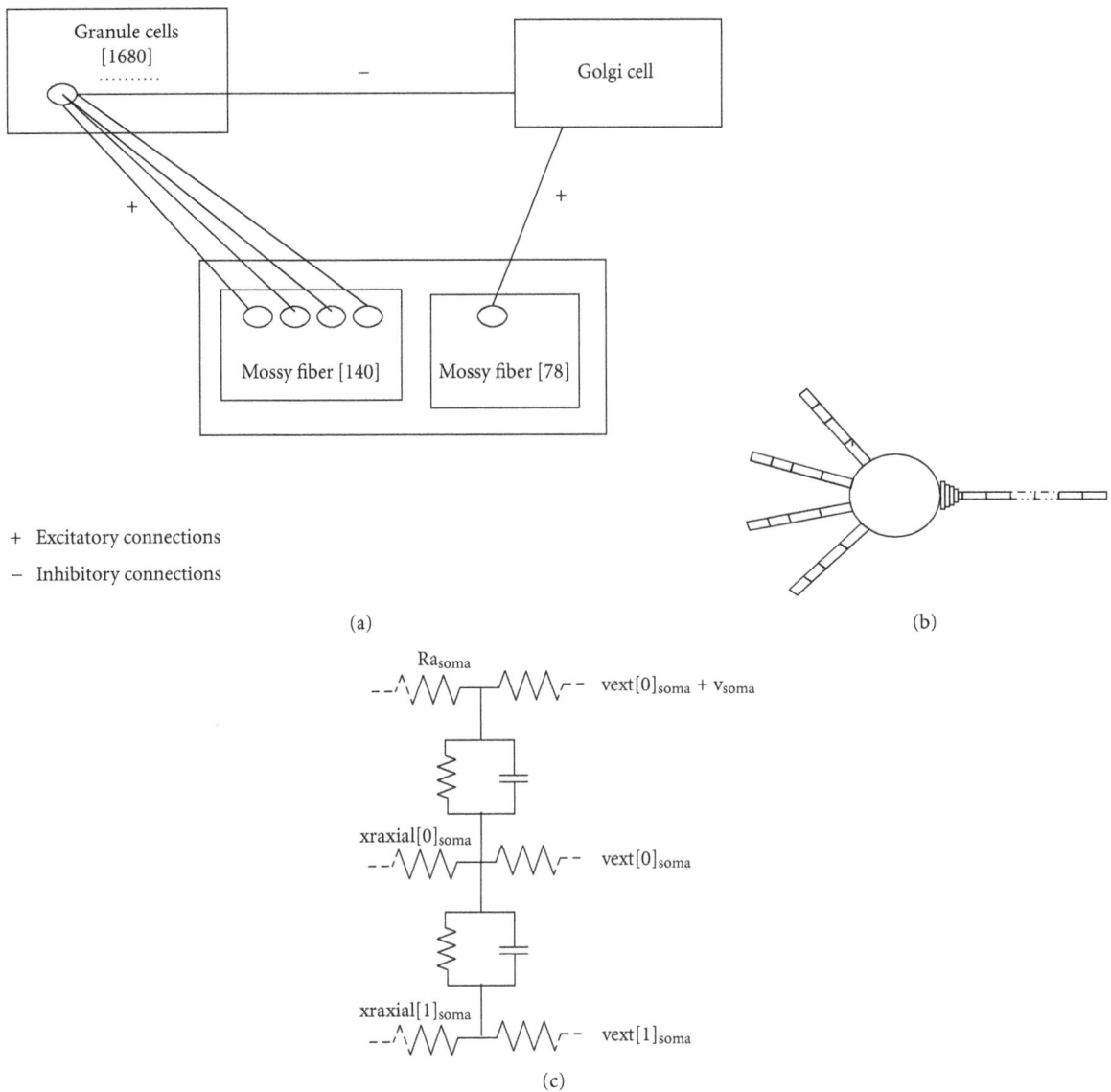

FIGURE 1: Granular layer network topology. (a) Network topology map. Granule cells (GrCs) receive 1–4 excitatory inputs from mossy fibers. GrC receive 0 (no inhibition)–4 inhibitory inputs from GoC via the GABAergic synapses, one per granule neuron dendrite. The ratio is about $4000:1$ [1]. Granular layer processing is fast and usually output spikes are seen in millisecond time intervals. (b) Detailed granule neuron model adapted from [19]. (c) Extracellular mechanism to study extracellular current flow in compartmental models. This mechanism was used to model LFP (see Section 2.6).

2.6. LFP Reconstruction. The extracellular potential of a single granule neuron (see Figure 1(b)) was estimated using NEURON [32] extracellular mechanism. The mechanism adds two RC compartments (see Figure 1(c)). To understand population code, we reconstructed network evoked LFP response using Laplace equation (see (1)):

$$\nabla^2 \varnothing = 0, \tag{9}$$

where $\varnothing$ is extracellular potential, at boundary condition $(1/\rho)\varnothing = J_m \cdot J_m$ is the transmembrane current density and ρ is the extracellular resistivity. Each cell generated an extracellular response corresponding to the activation pattern elicited by the mossy fibers. With the granular layer network, an electrode was assumed to be placed at the center. Temporal and spatial delays due to distance from electrode were assumed to be 0–3 ms [37]. The electrode could measure cells that generated extracellular currents that came with a delay of 0–3 ms (see [20]). Methodology for modeling the latencies used has been detailed elsewhere [13]. Considering the extracellular activity from each granule cell in the region of interest (number of cells = 700, assuming measurements from a tungsten electrode [18]), we reconstructed evoked LFP response using (10) and (11). Equation (10) adds the delay by padding zeros to linearly

Modeling Spike-Train Processing in the Cerebellum Granular Layer and Changes in Plasticity Reveal Single Neuron Effects in Neural Ensembles

179

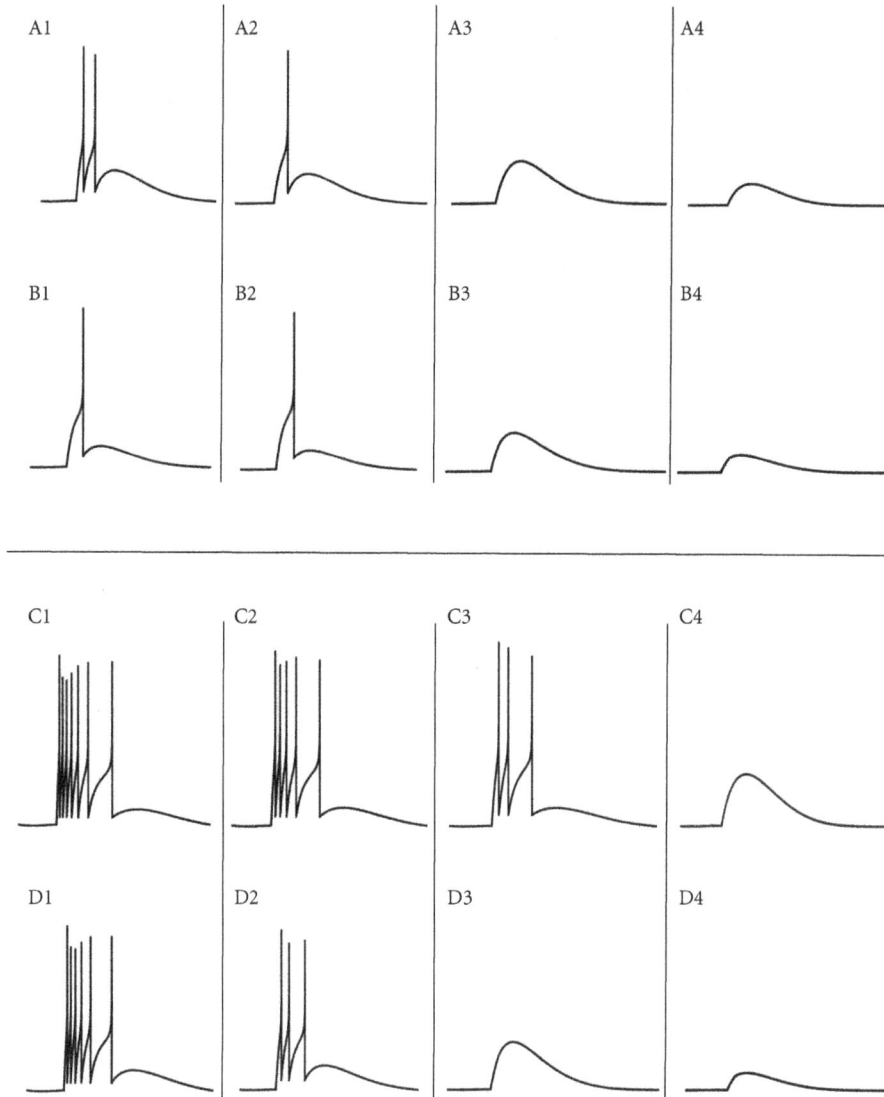

FIGURE 2: Firing patterns observed with AdEx model. (A) and (B) shows the *in vitro* behavior receiving 1 spike through the MF synapses. *In vivo* behavior (burst-burst transmission) is simulated (C) and (D) via bursts through the MF synapses. (A) and (C) show traces with no inhibition while (B) and (D) show traces with inhibition. Responses from left to right indicate input activation patterns from 4 MF excitations to 1 MF excitation. The AdEx model faithfully reproduced granule cell spiking behavior *in vitro* and *in vivo* [9, 19, 28].

time-shift the signal. Equation (11) denotes the process of summing all shifted extracellular signals for all cells linearly. Total signal obtained is the desired evoked LFP:

$$\varnothing_{\text{shifted},i}(t) = \varnothing_i(t - t') \qquad (10)$$

$$\varnothing_{\text{evoked LFP}}(t) = \sum_{i=0}^{n} \varnothing_{\text{shifted},i}(t), \qquad (11)$$

where $\varnothing_i(t)$ is the extracellular potential of ith cell in the neuronal population within the region of interest. $\varnothing_{\text{shifted},i}(t)$ represents the extracellular potential shifted by time delay (0–3 ms) (see [33]). Equations (10) and (11) were calculated separately. The detailed methodology for reconstructing evoked local field potential (LFP) has been described elsewhere [13].

3. Results

We were able to construct two models of granular layer network microcircuit: one using computationally efficient but physiologically limited spiking neurons and other using biophysically detailed multicompartmental neurons and reproduced activation patterns, burst-burst transmission, role of inhibition, and combinatorial coding.

3.1. Time-Windowing Depends on the Feed-Forward Inhibition-Implications from Simple Spiking Network Model. The objective of using a simple spiking model was to understand input-output relationships in terms of firing dynamics in the cerebellar granular layer.

The simulated single granule neuron responses were modeled based on granule neuron electroresponsiveness

[9, 28]. Both *in vitro* (see Figure 2(A)) like behavior with single spike through mossy fibers and *in vivo* (See Figure 2(C)) like response with burst inputs through mossy fibers were simulated. The responses matched experimental data [9, 19, 28]. The role of feed-forward inhibition [31, 38–40] was also modeled. With inhibition, the granule neuron model showed suppression (see Figure 2(B1)) of spike doublet (see Figure 2(A1)). Synaptic inhibition, because of its delayed activation, controlled generation of the second spike in the doublet [18, 41]. In the *in vivo* case, the number of spikes (see Figure 2(C)) was reduced due to inhibition (see Figure 2(D)).

Using the spiking granule neuron model, the 1680 granule cell network was reconstructed. The synaptic input in the mossy fibers were reproduced using either a single pulse to mimic electrical stimulation for *in vitro* simulations or short high-frequency trains mimicking punctuate sensory stimulation for *in vivo* simulations The network model showed 720 spikes (without inhibition) and reproduced the synaptic activation of the granular layer.

The spiking neuron network model was activated with a center-surround activation pattern [8], and the raster of spikes were observed in individual cells. Among 1680 cells, 144 cells with 4 MF active, 432 cells with 3 MF, 144 with 2 MF, and 432 with 1 MF active. The configuration was based on Voltage-Sensitive Dye (VSD) imaging [8] and results matched our previous findings [42]. With LTP and LTD, the numbers of spikes in the network change significantly (see Table 3). The spiking neuron simulations supported a burst-burst transmission modality (see Figure 3) in which high-frequency spike trains are more reliably transmitted.

The time-windowing [1] depended on the feed-forward inhibitory loop, regulated by the Golgi synapses impinging on the granule neurons. In this model, inhibition was modeled with a delay of 4 ms to account the MF-GoC-GrC circuit. As expected, the feed-forward inhibition reduced the number of spikes from 720 to 576 (see Table 3). The spike raster in the simulations (see Figure 3) showed selective inhibition of granule firing due to blocking of the second input as reported in [1, 2, 18]. The increase in number of spikes *in vivo* in the network supports the frequency-modulated transition from LTP to LTD [43]. The model also was computationally efficient in comparison to the biophysical model (see Section 3.2) and took 3 s for a 100 ms simulation.

3.2. Spike-Burst Generation and Bidirectional Plasticity.
Although simple spiking models allow reconstruction of frequency and amplitude information in terms of firing of constituent cells, role of plasticity and selective pharmacological effects in population code could not be studied. As reported in Section 2.2, we used detailed multicompartmental models [15, 16, 19] to generate a 1680 granule cell network. A detailed model allows to focus on understanding how specific temporal dynamics and the geometry of connections will eventually determine the circuit output, as indicated by the evident anomalies in network functioning and behavior caused by single-gene mutations altering the physiology

TABLE 3: Total number of spikes in the granular layer network with biophysically detailed neuron models.

Condition	In vitro		In vivo	
	Without inhibition	With inhibition	Without inhibition	With inhibition
LTP	2736	864	4032	2304
Control	720	576	3600	2160
LTD	0	0	2016	864

Total number of spikes observed in the network under different conditions like LTP (high intrinsic excitability and higher release probability), control, and LTD (lower intrinsic excitability and lower release probability). Observe that there are no spikes *in vitro* during LTD. Golgi inhibition operates a time-window causing a significant reduction in spikes.

of single molecules or neurons [2]. The simulations also attempt to understand and indicate certain functional theories of feed-forward inhibition, sparse recoding via, spikes and long-term plasticity.

LTP in granule cells [21] comprises of variation in release probability and intrinsic excitability. The network model was *modified* with higher intrinsic excitability observed by changes to sodium channel properties and release probabilities of MF synapses, thereby simulating granule cell LTP. LTD [43] was also simulated by combining lower intrinsic excitability and low release probability. With varying amounts of excitation, cells generated different number of spikes. In the simulations, granule cells generated repetitive nonadapting spike discharge in response to a continuous stimulus [27, 28, 45]. Simulations on single granule neurons *in vitro* suggested that granule neurons allowed spike burst generation and resonance in a low-frequency band (between 4 and 10 Hz) [9, 46]. High-frequency bursting [44, 47–49] was simulated to characterize network properties of granule cells *in vivo* [26, 33].

3.3. Network Excitability Changes with Varying Excitatory Release Probabilities.
During events such as in epileptic seizures, heterogeneous spiking activity is noticed [50]. To understand the nature of spiking, we simulated the role of excitation via mossy fiber. We simulated single spikes (low frequency) and bursts (high frequency) so as to understand spiking behavior *in vitro* and *in vivo*.

In the case of simulating *in vitro* behavior (see Section 2.3) in granular layer network, with release probability 0.416 (control) the cells with normal intrinsic excitability receiving 4 excitatory inputs produced spike doublet and cells receiving 3 excitatory inputs produced single spikes (see granule neuron electroresponsiveness in [19]). Cells receiving 2 excitatory and 1 excitatory inputs did not produce any spikes [19].

Decreasing MF synapse release probability from 0.3 to 0.1, many granule cells in the network did not generate spikes. With increased release probability of MF synapses from 0.42 to 0.6, there was an increase in number of spikes (the number of spiking cells increased from 192 to 432) with no significant change in first spike latency and spike amplitude [36].

Modeling Spike-Train Processing in the Cerebellum Granular Layer and Changes in Plasticity Reveal Single
Neuron Effects in Neural Ensembles

181

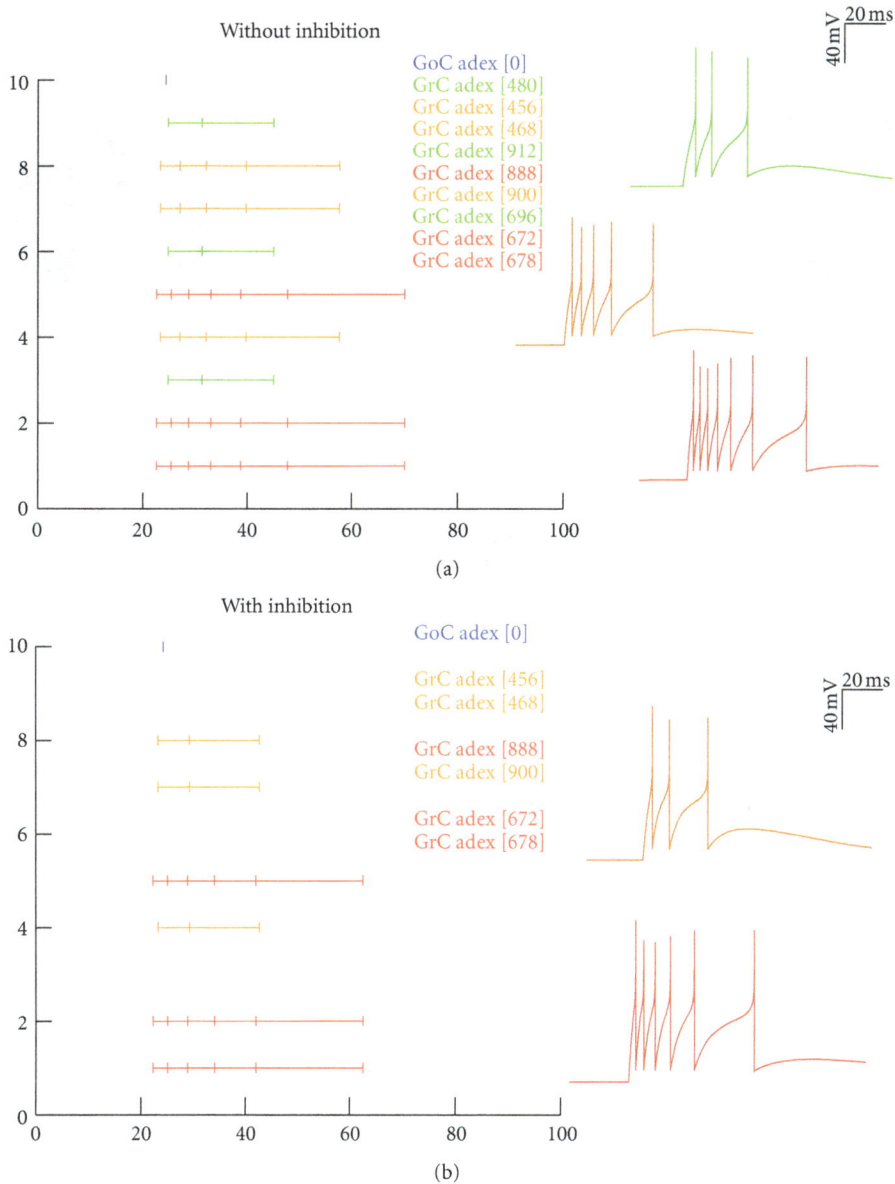

FIGURE 3: Spike raster plots for the network *in vivo* with AdEx models. Network model using AdEx [14] neurons reproduce the spike raster for *in vivo* firing dynamics. A short burst of 5 spikes at 500 Hz was given as inputs through the MF. Feed-forward inhibition affected the network by reducing number of spikes (b). Network without inhibition (a) shows 1–7 spikes and simulates the role of Gabazine that blocks GABAergic synapse.

With the higher release probabilities like 0.7, 0.8 of MF synapses, the number of spikes saturated and the number of spiking cells remained the same (as seen in 0.6 release probability of MF synapses). Varying synaptic release probabilities, it was possible to generate selective responses.

Increasing intrinsic excitability from normal to higher excitability by modifying sodium gating properties (see Section 2.5) showed a significant increase in the spike amplitude (~6%) for all spiking cells, and an increased number of spikes was observed only for the cells with higher release probability and number of active MF synapses. This change corresponded to long-term potentiation in granule neurons [21] confirming the mechanisms role in spiking and

bursting. The number of spiking cells varied from 192 to 432 in the network of 1680 granule cells.

With *in vivo* inputs, the number of cells showed a greater sensitivity with LTP (see Table 3) and the number of nonspiking cells decreased. LTD showed decrease in firing-nonfiring [43] cell ratio. The ratio of firing cells did not change with the length of the burst (see Table 4).

3.4. Inhibition and Spike-Count Modulation. Golgi cells can control both the temporal dynamics and the spatial distribution of information transmitted through the cerebellar granular layer network [43]. The strength of the inhibition depends on the number of inhibitory connections and

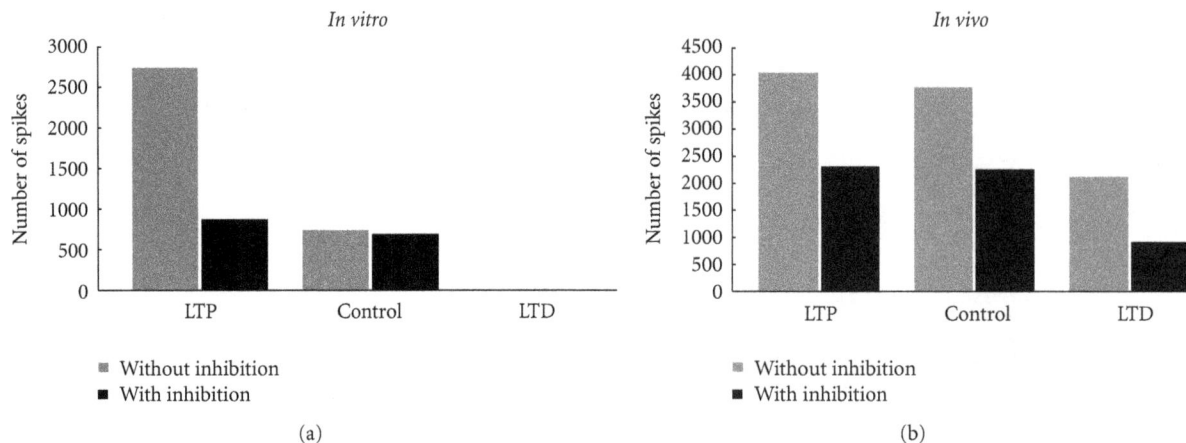

FIGURE 4: Histogram showing the effect of LTP and LTD on spiking in the network built with detailed biophysical models. Under *in vitro* (a) like spike input via MF, the number of spikes changed from 720 to 2736 (gray bars). Control refers to the excitatory release probability, U, which was set to 0.416. The presence of inhibition showed a sharp modulation, and the number of spikes seen in the network was 576 and during LTP it increased to 874 (black bars). LTD showed no spikes. Under *in vivo* (b), the change in number of spikes from control to LTP was 3600 to 4032 and during LTD was 2016 (gray bars). With inhibition (black bar), the number of spikes in the 1680 cell network decreased to 2304 (LTP), 2160 (control), and 864 (LTD).

TABLE 4: Modulation of spiking cells *in vivo* with varying release probability in the detailed network model.

MF release probability	MF input, 5 spikes/burst		MF input, 9 spikes/burst	
	Number of spiking cells	Number of nonspiking cells	Number of spiking cells	Number of nonspiking cells
0.1	1416	264	1416	264
0.2	1416–840	840–264	1416–840	840–264
Control, 0.4	1416–840	840–264	1416–840	840–264
0.5–0.8	1416–840	840–264	1416–840	840–264

synaptic release probability. The dynamics of the granule cells-Golgi cell circuit were explained by the simultaneous activation of both neurons through the mossy fibers, followed by activation of the feed-forward and feed-back inhibitory loops [18, 51]. The granule and Golgi cell received excitatory inputs from mossy fiber (MF) at the same time. There are two basic patterns of mossy fiber activity that can activate the Golgi cells, namely, protracted frequency-modulated discharges and short high-frequency bursts [48, 52]. The inhibitory input from Golgi cell reaches the granule cell with a loop delay of approximately 4 ms [53] compared to the mossy fiber input through GABAergic synapses [2]. The inhibition-based time-windowing in granule cells allow one or more spikes and is seemingly regulated by varying inhibitory inputs.

Golgi cells converging through lateral connections onto some granule cell subsets could generate combined inhibition [2, 8]. The impact of the inhibition on granular layer circuitry differs with respect to two different properties: amount of inhibitory connections and the GABAergic release probability. The variation in the number of spikes with and without inhibition was significant in both cases *in vitro* (see Figure 4(a)) and *in vivo* (see Figure 4(b)). As expected, LTP showed increased number of spikes compared to control,

while LTD showed reduced number of spikes. *In vitro* LTD suppressed spikes (see Figure 4(a)).

During simulations as seen *in vitro* (see Figure 4(a)), increased inhibition regulated the spike count rather than affecting the number of spiking cells. Short burst through MF produced ~7 spikes in single neurons, but inhibition showed a sharp modulation by regulating the time-window. A long burst produced slower modulation of spikes in the network.

The increase in inhibitory connections (see Table 5 and Figure 5) to granule cells in the underlying network model decreased number of spikes (see spike count in Figure 5, control refers to release probability being set at normal condition, $U = 0.416$), spike amplitude (if the spike rises after the 4 ms time-window when inhibitory inputs reaches the granule cell) and decreased spike latency.

Changing inhibitory (GABAergic) synapse release probability (U_{inh}), spike amplitude, and first spike latency were affected [42]. Spike amplitude decreases whereas spike latency remains unchanged, when U_{inh} varied [42].

The increase of inhibitory input increases the number of silent cells (visible by the blue plateau in Figure 4), therefore reducing the number of active cells. The simulations indicate that the response of those granule cells that are intensely activated will favor with the generation of a burst, regulated

Modeling Spike-Train Processing in the Cerebellum Granular Layer and Changes in Plasticity Reveal Single
Neuron Effects in Neural Ensembles

183

(a)

(b)

(c)

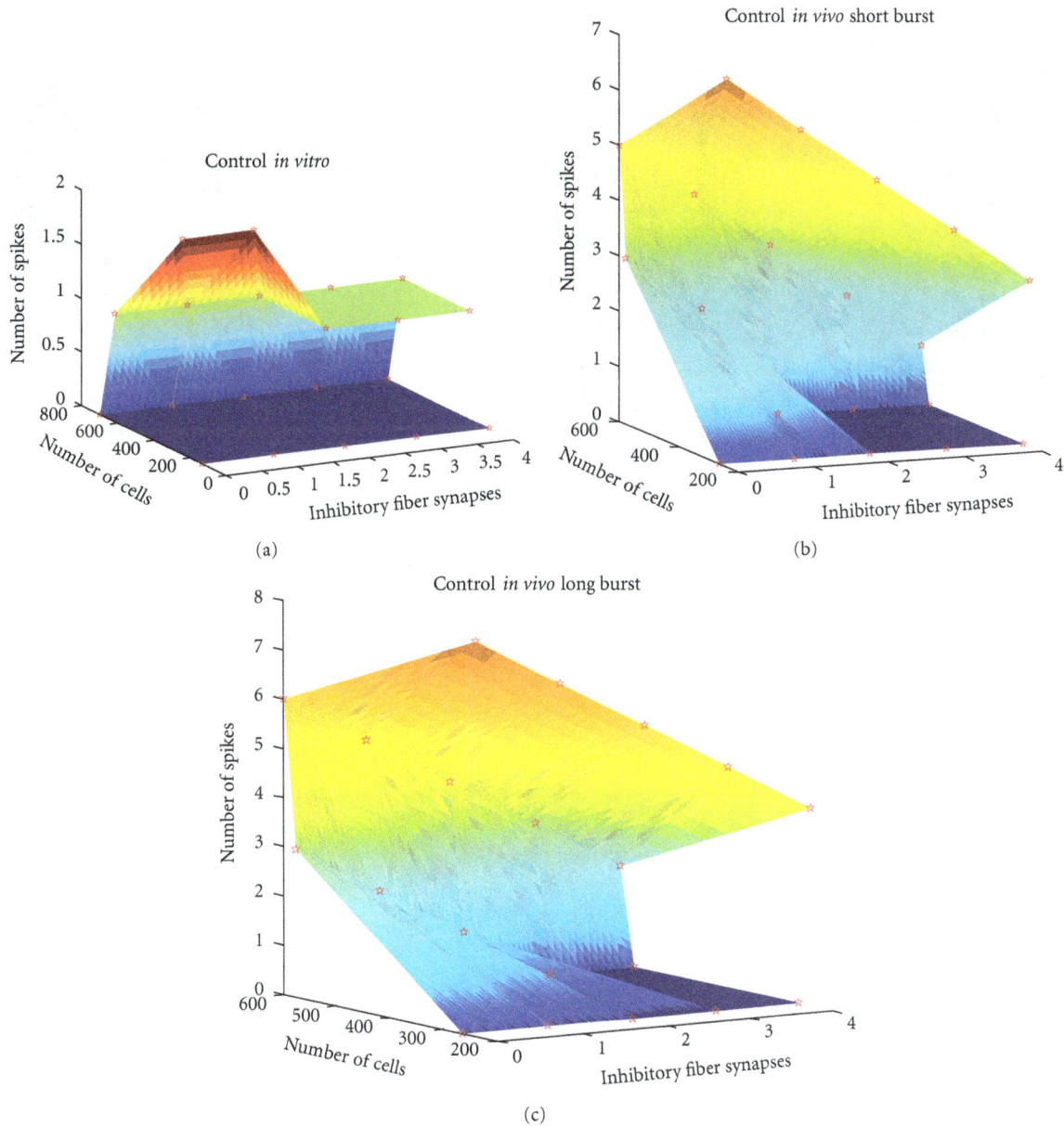

FIGURE 5: Effect of inhibition on variation in number of spikes and spiking cells. With varying inhibition, the number of spiking cells and total number of spikes varied. With 1 spike via MF as input ((a), *in vitro* behaviour), the total number of spiking cells varied from 200–600 cells and 1-2 spikes modulated by the inhibitory inputs (*x*-axis). Tactile stimulation induced two types of bursts *in vivo* [33]. (b) shows the number of spiking cells *in vivo* (as short burst of 5 spikes at 500 Hz via MF) with respect to changes in number of spikes as inhibitory inputs (*x*-axis) were changed. Variation in the number of spiking cells and number of spikes is shown. With a longer burst (9 spikes at 500 Hz via MF) *in vivo* (c), there were more spikes and inhibition did not cause a sharp change in number of spikes or spiking cells in the network (see also Table 5). The number of active cells can be observed also by looking at the increase of the number of "silent" cells. The granule cells favour a better role as signal-to-noise enhancers in the network [44] and facilitate burst-burst transmission.

mainly by feed-forward Golgi cell inhibition (see Figure 5 and [48]).

3.5. Center-Surround Excitation in Populations of Granule Cells.
To understand combinatorial effects in the granular network layer and impacts of double mossy fiber bundle stimulation, combined excitation-inhibition in the network was simulated. The "spots" are maps of excitatory activity as

seen in the cerebellar granular layer [8] when MF rosettes were stimulated [8]. In the model configuration (see Table 6), the center of the spot receives stronger excitatory inputs and the consecutive peripheral neurons receive weaker excitatory input, thereby expressing a center-surround configuration (see Figure 6). Both network models could reproduce the firing dynamics [8] as well as the center-surround structure (see movie, Supplementary Material available online at

(a)

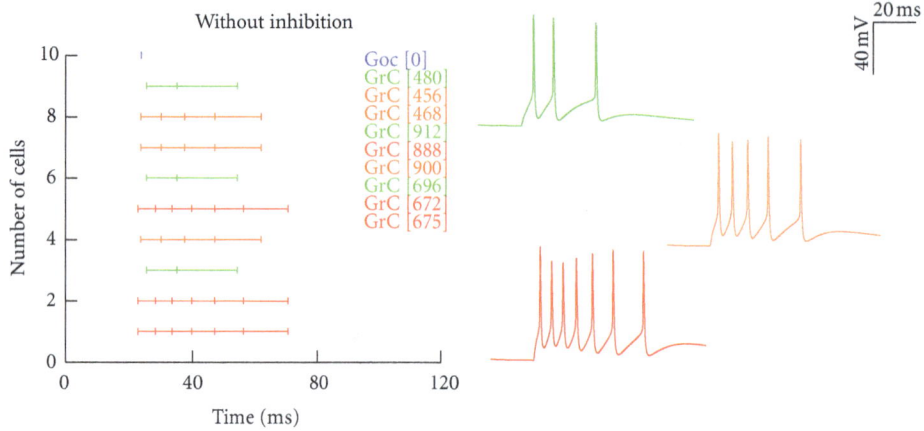

(b)

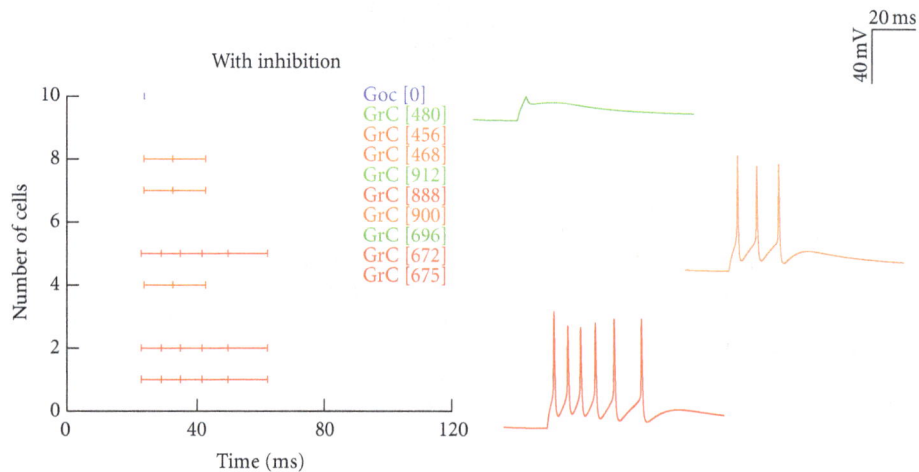

(c)

FIGURE 6: Center-surround "spot" activation. Varying levels of synaptic excitation in the spot (a) as mossy fiber inputs to granule cells in the network were reconstructed. Three spots of which each spot had 384 granule cells (see Section 2.4), and the excitation potential was indicated by the colormap. In the model configuration (a), the center of the spot receives stronger excitatory inputs and the consecutive peripheral neurons receive weaker excitatory input, thereby expressing a center-surround configuration. Compared to the surround, the center detects burst on a broader band and emits bursts with shorter lag, higher frequency, and longer duration [1]. Network model using detailed granule neuron models reproduces the spike raster for *in vivo* firing dynamics and was similar to Figure 2. A short burst of 5 spikes at 500 Hz is passed through the MF as inputs. Inhibition (b) blocked the spikes. Network without inhibition (a) shows 1–7 spikes via granule cells (Grcs).

Modeling Spike-Train Processing in the Cerebellum Granular Layer and Changes in Plasticity Reveal Single
Neuron Effects in Neural Ensembles

185

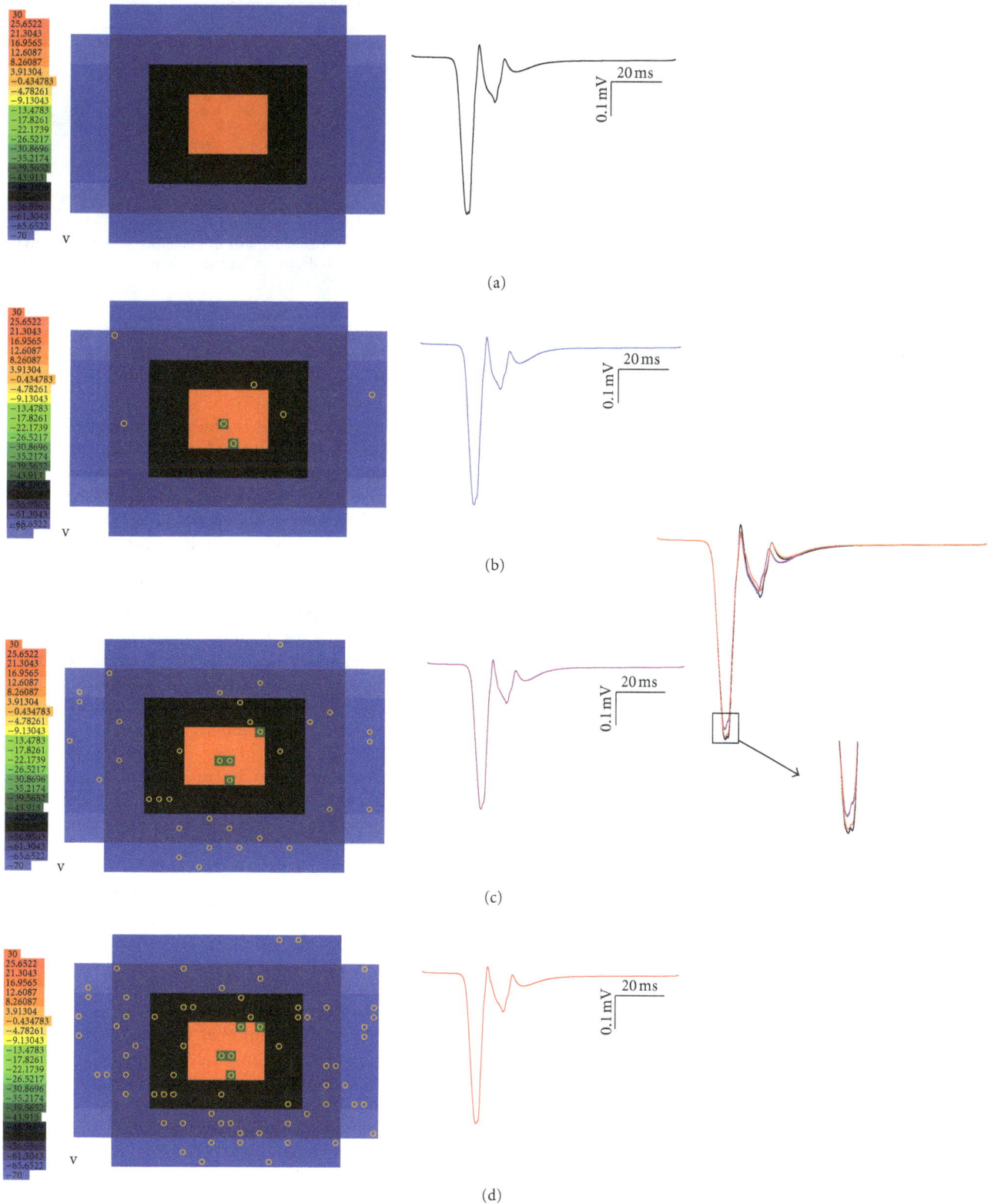

FIGURE 7: "Spot" activation and evoked LFP with selective NMDA dysfunction. Population code via evoked LFP *in vitro* response was reconstructed. Control (a) shows the clear reproducibility of N_{2a} and N_{2b} waves [18]. NR2A/NR2B knockouts show selective dysfunction of NMDA receptors. 1% (b), 5% (c), and 10% (d) cells with NMDA receptor blocked do not show much difference in population code although intracellular spiking remains altered (not shown). A small decrease in peak amplitudes was noticed. The robustness of the population code during spiking in granule cells adds to the sparse recoding theory [4] and clearly granule neurons favour their role as signal-to-noise enhancers for sensory and tactile information received via the mossy fibers (MFs).

TABLE 5: Effect of inhibition on number of spikes in the detailed network model.

Number of inhibitory fiber synapses	*In vitro*				*In vivo* (short burst)				*In vivo* (long burst)			
	4 MF	3 MF	2 MF	1 MF	4 MF	3 MF	2 MF	1 MF	4 MF	3 MF	2 MF	1 MF
0	2	1	0	0	7	5	3	0	8	6	3	0
1	2	1	0	0	6	4	2	0	7	5	2	0
2	1	1	0	0	5	3	0	0	6	4	1	0
3	1	1	0	0	4	2	0	0	5	3	0	0
4	1	1	0	0	3	1	0	0	4	2	0	0

The table shows the effect of inhibitory synapses on the spikes under different conditions (*in vitro, in vivo* (short burst); *in vivo* (long burst)). Under each condition, first row denotes increasing number (1–4) of excitatory synapses from right to left while increasing number of inhibitory synapses (0–4 in first column). With increase in number of inhibitory synapses (0–4), the number of spikes observed in the cells decreases. A gradual decrease in number of spikes with increased inhibition can be observed.

TABLE 6: Center-surround pattern and spiking cells of detailed granular layer network.

Number of cells	Number of active MF synapses	Number of spikes	
		Network without inhibition	Network with inhibition
144	4	7 spikes/burst	6 spikes/burst
432	3	5 spikes/burst	3 spikes/burst
144	2	2 spikes/burst	EPSP
432	1	EPSP	EPSP

[a]Cells with 4 excitatory inputs produced 7 spikes/burst when inhibitory synapse was switched off and produced 5 spikes/burst when it was switched on.

doi: 10.115/2012/359529). Spiking activity was reconstructed with the morphology (see Section 2.4). A single spike through the mossy fiber activates the center followed by the periphery and the Golgi-granule circuit.

Simulation of LTP and LTD induction *in vitro* and *in vivo* on the center-surround spots was modeled by varying release probabilities and intrinsic excitability. The cells in the granular layer network receive GABAergic synaptic inputs equal to the number of excitatory inputs given to the cells in the granular layer network (see Table 5). The high reproducibility indicates that the center-surround organization was a consequence of alternating transitions between burst and silent states at granule cells was not due to the temporal jitter of MFs [54].

The center-surround structures have complex transmission properties: compared to the surround, the center detects burst on a broader band and emits bursts with shorter lag, higher frequency, and longer duration [1]. Purkinje cells overlaying above these structures may be activated, at the same time, enhance inhibition around them, explaining the spot-like organization of molecular layer responses *in vivo* [8].

3.6. Local Field Potential and Selective Blocking of NMDA. Understanding population code through reconstructions was essentially done to suggest how encoding of spike information may happen in cerebellar cortex. Currently, encoding of population activity is explored in microcircuits via comparisons of spatiotemporal properties of simulated neural activity and with experimental measurements using multielectrode recordings [55, 56] or two-photon imaging of activity in blocks of tissue [57, 58]. Evoked responses from granular layer *in vitro* [19] have been reconstructed computationally [13, 26]. We used the "spot" to generate and test nature of local field potentials. The postsynaptic evoked LFP response varied as per input pattern and for a combination of 3 MF and 4 MF synaptic activation, spikes were generated. With 4 MF synapses active, a doublet was seen [19]. Correspondingly the responses generated N_{2a} wave and the doublet caused the N_{2b} wave. Inhibition at time = 24 ms via GABAergic synapses suppressed the spike doublet and thereby suppressed the N_{2b} wave [8] in the evoked LFP response.

Different segments of the network generated varied evoked LFP signals due to the nature of excitatory-inhibitory balance in the network reflecting a relationship different from the extracellular components of a single neuron (see Figure 9 in [13]). We assumed extracellular space in granular layer to be isopotential [26] due to close packing of granule neurons. The simulations closely followed experimental results [8], suggesting that electrotonic compactness of granule neurons contribute to the seemingly linear relationship from granule cell clusters in the granular layer extracellular space. The variations in the nature of spike with number of spiking cells could suggest that sparse coding could be preserved as suggested by Marr [4] and Albus [3].

Blocking NMDA receptors [59] in granule neurons showed reduced excitation. Selective disabling of NMDA receptors, as noticed in mice with NR2A/NR2B [59] mutations, showed decreased number of spikes which is also seen as a change in N_{2a} amplitude compared to control (Figure 7(a)). In order to predict on the nature of such mutations affecting network computation (in addition to affecting the number of spikes), we randomly disabled (1%, 5%, and 10% of total cells) NMDA receptors in the network (see circles in Figures 7(b), 7(c), and 7(d)) and reconstructed the local field response.

Disabling NMDA receptors in 1–10% of cells showed a 2.5% decrease in number of spikes in a spot (the number of spiking neurons in a 720-cell "spot" changed from 240 to 234). The network model clearly showed a *"seemingly linear"* outlook in propagating the nonlinearities of individual neurons in population code (evoked LFP, see Figures 7(a) to 7(d)). This "sense of linearity" in population code was

Modeling Spike-Train Processing in the Cerebellum Granular Layer and Changes in Plasticity Reveal Single
Neuron Effects in Neural Ensembles

187

observed also when the number of affected neurons was very low (neurons with NMDA disabled were only 1–10% of total cells). NMDA knockout mice show errors in cerebellar motor learning [59]. Plasticity changes were also reflected in the evoked LFP waves. As seen in reconstructions based on single granule neuron simulations [26], the LFP simulations on the network showed that LTP and LTD were accompanied by changes in the proportion of discharging granule cells (data not shown).

4. Discussion

Exploring the geometry of excitation and inhibition in cerebellar granular layer, the simulations highlight the modulatory role of inhibitory inputs on the activities of granule cells. The paper details the effects of combined excitation and combined feed-forward inhibition [8] on spiking in the granular layer. The study did not simulate feed-back inhibition coming from Golgi cell since it did not affect the modulation of 4 ms time-window that happens because of the early mossy inputs.

The simulations varying inhibition suggest that granule neurons can generate selective responses by varying synaptic strengths. The increase in spikes and modulation during plasticity indicates that the circuit is well adapted to generate enhanced responses such as in theta-burst patterns [9].

Both *in vitro* and *in vivo* simulations indicate that inhibitory input cannot completely block excitation in the network. However, it acts as a modulator that regulates the postsynaptic excitability. Both models support that burst-burst transmission modality in granule neuron and the granular layer through which high-frequency spike trains are more reliably transmitted. The consequences of transformation of spike inputs from mossy fibers to corresponding codes suggest the variable impulse response scheme indicated by previous study [10] and suggest the granular layer network also operates as an adaptive filter.

The variations of excitatory inputs (without combination of inhibition) showed differences in number of spikes and spike amplitude and did not show variations in first-spike latency [60]. The most promising outcome in variation of spikes and network spiking behavior was with the induction of LTP/LTD where both intrinsic excitability and excitatory release probabilities [21, 43] change the nature of information flow.

Simple spiking neuron models can be tuned to function as network models for accessing timing information. Spatial information in network models [19] was not seen while using spiking models. Synaptic functions in spiking models are not very reliable. Artificial models have limitations unlike biophysical models for understanding certain population activities like generation of LFP [13].

The detailed model was used as a test bench to explore the parameter space and induced plasticity. Epileptic seizure-like symptoms seen in voltage-gated sodium channel binding-related knockout mice granule neurons [7, 50] suggests that sparse and asynchronous neuronal activity can evolve into a single hypersynchronous cluster with elevated spiking rates at seizure initiation. The detailed network model suggests

that LTP favors burst-burst transmission favoring high-frequency spikes. The presynaptic mechanism coexisted with postsynaptic regulation of ionic channels, which played a major role in determining the granule cell output firing frequency. Intrinsic bursting and modulatory effects of inhibition can be seen by mechanistic control of number of spikes in a granule cell.

With increased excitation, along with an increase in spikes, first-spike latency also decreased. This will also impact the local field potential and could probably explain the observations *in vitro* [18]. Both *in vitro* and *in vivo* simulations indicate that the number of spikes was dependent on the release probability of the synapses, while higher or lower intrinsic excitability caused slight change in spike amplitude.

The key role of local circuit inhibition in determining granular layer combinatorial operations was supported by several model-based predictions. Increasing active inhibitory connections saw lesser number of spikes in the network. *In vivo* bursts along mossy fibers combined with inhibitory input showed a consistent reduction of at least one spike as inhibition increased. The simulations indicated that the response of those granule cells that are intensely activated will favor with the generation of a burst, whose duration is limited by a brisk feed-forward inhibition in the Golgi cell. Inhibition controlled the number of spikes, thereby modulating spike transmission in the granular layer. The simulations suggest that erratic spikes in the mossy fibers will not be efficiently transmitted so that the burst-burst mechanism would indeed play a role in secure transmission along the mossy fiber pathway [48]. The studies also show that excitation and inhibition may consequently allow complex patterns to be processed [11].

The paper also shows population signals and effects of mechanism changes on individual neuron affecting population code generated by the network. Reconstructing extracellular properties indicated that plasticity may have similar mechanisms of burst regulation as granule cell burst initiation and may implement an adaptable delay affecting downstream activation into circuitry. The granular layer model indicates a *"seemingly linear"* tendency to propagate the nonlinearities of individual neurons via the population code even when the variations are little (affected cells 1–10% of total). The simulations suggest that a combined mechanism of NMDA blocking the After-hyperpolarization (AHP) and role of inhibition can help reconstruct transient suppression of spikes *in vitro* reported during seizures.

The studies on intensity of mossy fiber synapses and inhibitory synapses help to understand spatiotemporal operations [8] in the cerebellar granular layer. Combining granule neurons and Golgi cell, this study will help to reveal coincidence detection properties and spatial pattern separation [3]. This work is a preliminary start in modeling to understanding long sought spatiotemporal filtering predicted by the motor learning theory [61].

5. Conclusion

Simulations suggest how cerebellum granular layer processes spike information and how afferent information may reach cerebellar cortex and predict how spikes are processed as indicated in the sparse recoding hypothesis [4]. The role of inhibition and plasticity may help fine tune the "sparseness" of the code as indicated in Marr's theory [3, 4]. To evaluate the exact role of firing, a closer view of cells in the region of interest may be needed. The experimental testing of these predictions will require further electrophysiological and imaging investigations of granular layer activity and computational modelling of the cerebellum [1] and of the cerebro-cerebellar control loops [62, 63].

Acknowledgments

This work derives direction and ideas from the chancellor of Amrita University, Sri Mata Amritanandamayi Devi. The authors would like to thank Priyanka James, Nimshitha Abdulmanaph, Harilal Parasuram, and Sergio Solinas for their work and support in making this paper. This work is supported by Grants SR/CSI/49/2010 and SR/CSI/60/2011 from the Department of Science and Technology, Government of India.

References

[1] S. Solinas, T. Nieus, and E. D. Angelo, "A realistic large-scale model of the cerebellum granular layer predicts circuit spatio-temporal filtering properties," *Frontiers in Cellular Neuroscience*, vol. 4, article 12, 2010.

[2] E. D'Angelo and C. I. De Zeeuw, "Timing and plasticity in the cerebellum: focus on the granular layer," *Trends in Neurosciences*, vol. 32, no. 1, pp. 30–40, 2009.

[3] J. S. Albus, "A theory of cerebellar function," *Mathematical Biosciences*, vol. 10, no. 1-2, pp. 25–61, 1971.

[4] D. Marr, "A theory of cerebellar cortex," *Journal of Physiology*, vol. 202, no. 2, pp. 437–470, 1969.

[5] F. Prestori, P. Rossi, B. Bearzatto et al., "Altered neuron excitability and synaptic plasticity in the cerebellar granular layer of juvenile prion protein knock-out mice with impaired motor control," *Journal of Neuroscience*, vol. 28, no. 28, pp. 7091–7103, 2008.

[6] S. P. Perrett, B. P. Ruiz, and M. D. Mauk, "Cerebellar cortex lesions disrupt learning-dependent timing of conditioned eyelid responses," *Journal of Neuroscience*, vol. 13, no. 4, pp. 1708–1718, 1993.

[7] M. Goldfarb, J. Schoorlemmer, A. Williams et al., "Fibroblast growth factor homologous factors control neuronal excitability through modulation of voltage-gated sodium channels," *Neuron*, vol. 55, no. 3, pp. 449–463, 2007.

[8] J. Mapelli, D. Gandolfi, and E. D'Angelo, "Combinatorial responses controlled by synaptic inhibition in the cerebellum granular layer," *Journal of Neurophysiology*, vol. 103, no. 1, pp. 250–261, 2010.

[9] E. D'Angelo, T. Nieus, A. Maffei et al., "Theta-frequency bursting and resonance in cerebellar granule cells: experimental evidence and modeling of a slow K^+-dependent mechanism," *Journal of Neuroscience*, vol. 21, no. 3, pp. 759–770, 2001.

[10] P. Dean, J. Porrill, C. F. Ekerot, and H. Jörntell, "The cerebellar microcircuit as an adaptive filter: experimental and computational evidence," *Nature Reviews Neuroscience*, vol. 11, no. 1, pp. 30–43, 2010.

[11] J. F. Medina and M. D. Mauk, "Computer simulation of cerebellar information processing," *Nature Neuroscience*, vol. 3, pp. 1205–1211, 2000.

[12] R. Maex and E. De Schutter, "Synchronization of Golgi and granule cell firing in a detailed network model of the cerebellar granule cell layer," *Journal of Neurophysiology*, vol. 80, no. 5, pp. 2521–2537, 1998.

[13] H. Parasuram, B. Nair, G. Naldi, E. D'Angelo, and S. Diwakar, "A modeling based study on the origin and nature of evoked post-synaptic local field potentials in granular layer," *Journal of Physiology Paris*, vol. 105, no. 1–3, pp. 71–82, 2011.

[14] R. Naud, N. Marcille, C. Clopath, and W. Gerstner, "Firing patterns in the adaptive exponential integrate-and-fire model," *Biological Cybernetics*, vol. 99, no. 4-5, pp. 335–347, 2008.

[15] S. Solinas, L. Forti, E. Cesana, J. Mapelli, E. De Schutter, and E. D' Angelo, "Fast-reset of pacemaking and theta-frequency resonance patterns in cerebellar Golgi cells: simulations of their impact in vivo," *Frontiers in Cellular Neuroscience*, vol. 1, article 4, 2007.

[16] S. Solinas, L. Forti, E. Cesana, J. Mapelli, E. De Schutter, and E. D' Angelo, "Computational reconstruction of pacemaking and intrinsic electroresponsiveness in cerebellar Golgi cells," *Frontiers in Cellular Neuroscience*, vol. 1, article 2, 2007.

[17] D. A. McCormick, Z. Wang, and J. Huguenard, "Neurotransmitter control of neocortical neuronal activity and excitability," *Cerebral Cortex*, vol. 3, no. 5, pp. 387–398, 1993.

[18] J. Mapelli and E. D'Angelo, "The spatial organization of long-term synaptic plasticity at the input stage of cerebellum," *Journal of Neuroscience*, vol. 27, no. 6, pp. 1285–1296, 2007.

[19] S. Diwakar, J. Magistretti, M. Goldfarb, G. Naldi, and E. D'Angelo, "Axonal Na^+ channels ensure fast spike activation and back-propagation in cerebellar granule cells," *Journal of Neurophysiology*, vol. 101, no. 2, pp. 519–532, 2009.

[20] N. Abdulmanaph, H. Parasuram, B. Nair, and S. Diwakar, "Modeling granular layer local field potential using single neuron and network based approaches to predict LTP/LTD in extracellular recordings," in *Proceedings of Neurocomp*, Lyon, France, 2010.

[21] T. Nieus, E. Sola, J. Mapelli, E. Saftenku, P. Rossi, and E. D'Angelo, "LTP regulates burst initiation and frequency at mossy fiber-granule cell synapses of rat cerebellum: experimental observations and theoretical predictions," *Journal of Neurophysiology*, vol. 95, no. 2, pp. 686–699, 2006.

[22] L. Chen and L. Y. M. Huang, "Protein kinase C reduces Mg^{2+} block of NMDA-receptor channels as a mechanism of modulation," *Nature*, vol. 356, no. 6369, pp. 521–523, 1992.

[23] C. J. McBain and R. Dingledine, "Dual-component miniature excitatory synaptic currents in rat hippocampal CA3 pyramidal neurons," *Journal of Neurophysiology*, vol. 68, no. 1, pp. 16–27, 1992.

[24] N. Ropert, R. Miles, and H. Korn, "Characteristics of miniature inhibitory postsynaptic currents in CA1 pyramidal neurones of rat hippocampus," *Journal of Physiology*, vol. 428, pp. 707–722, 1990.

[25] T. S. Otis, Y. De Koninck, and I. Mody, "Characterization of synaptically elicited $GABA_B$ responses using patch-clamp recordings in rat hippocampal slices," *Journal of Physiology*, vol. 463, pp. 391–407, 1993.

[26] S. Diwakar, P. Lombardo, S. Solinas, G. Naldi, and E. D'Angelo, "Local field potential modeling predicts dense activation in cerebellar granule cells clusters under LTP and LTD control," *PLoS ONE*, vol. 6, no. 7, Article ID e21928, 2011.

Modeling Spike-Train Processing in the Cerebellum Granular Layer and Changes in Plasticity Reveal Single
Neuron Effects in Neural Ensembles

189

[27] E. D'Angelo, G. De Filippi, P. Rossi, and V. Taglietti, "Synaptic excitation of individual rat cerebellar granule cells in situ: evidence for the role of NMDA receptors," *Journal of Physiology*, vol. 484, no. 2, pp. 397–413, 1995.

[28] E. D'Angelo, G. De Filippi, P. Rossi, and V. Taglietti, "Ionic mechanism of electroresponsiveness in cerebellar granule cells implicates the action of a persistent sodium current," *Journal of Neurophysiology*, vol. 80, no. 2, pp. 493–503, 1998.

[29] W. Rall, "Branching dendritic trees and motoneuron membrane resistivity," *Experimental Neurology*, vol. 1, no. 5, pp. 491–527, 1959.

[30] P. Rossi, E. Sola, V. Taglietti et al., "NMDA receptor 2 (NR2) C-terminal control of NR open probability regulates synaptic transmission and plasticity at a cerebellar synapse," *Journal of Neuroscience*, vol. 22, no. 22, pp. 9687–9697, 2002.

[31] L. Mapelli, P. Rossi, T. Nieus, and E. D'Angelo, "Tonic activation of GABA$_B$ receptors reduces release probability at inhibitory connections in the cerebellar glomerulus," *Journal of Neurophysiology*, vol. 101, no. 6, pp. 3089–3099, 2009.

[32] M. L. Hines and N. T. Carnevale, "The NEURON simulation environment," *Neural Computation*, vol. 9, no. 6, pp. 1179–1209, 1997.

[33] L. Roggeri, B. Rivieccio, P. Rossi, and E. D'Angelo, "Tactile stimulation evokes long-term synaptic plasticity in the granular layer of cerebellum," *Journal of Neuroscience*, vol. 28, no. 25, pp. 6354–6359, 2008.

[34] E. Sola, F. Prestori, P. Rossi, V. Taglietti, and E. D'Angelo, "Increased neurotransmitter release during long-term potentiation at mossy fibre-granule cell synapses in rat cerebellum," *Journal of Physiology*, vol. 557, no. 3, pp. 843–861, 2004.

[35] S. Armano, P. Rossi, V. Taglietti, and E. D'Angelo, "Long-term potentiation of intrinsic excitability at the mossy fibergranule cell synapse of rat cerebellum," *Journal of Neuroscience*, vol. 20, no. 14, pp. 5208–5216, 2000.

[36] P. James, N. Abdulmanaph, B. Nair, and S. Diwakar, "Exploring input-output characteristics of the cerebellar granule neuron: role of synaptic inhibition, spike timing and plasticity," in *Proceeding of Neurocomp*, Lyon, France, 2010.

[37] R. A. Silver, S. G. Cull-Candy, and T. Takahashi, "Non-NMDA glutamate receptor occupancy and open probability at a rat cerebellar synapse with single and multiple release sites," *Journal of Physiology*, vol. 494, no. 1, pp. 231–250, 1996.

[38] P. Rossi, L. Mapelli, L. Roggeri et al., "Inhibition of constitutive inward rectifier currents in cerebellar granule cells by pharmacological and synaptic activation of GABA$_B$ receptors," *European Journal of Neuroscience*, vol. 24, no. 2, pp. 419–432, 2006.

[39] J. C. Eccles, M. Ito, and J. Szentagothai, *The Cerebellum as a Neuronal Machine*, Springer, Berlin, Germany, 1967.

[40] C. Palay and S. L. Palay, *Cerebellar Cortex: Cytology and Organization*, Springer, Berlin, Germany, 1974.

[41] E. D'Angelo, "The critical role of Golgi cells in regulating spatio-temporal integration and plasticity at the cerebellum input stage," *Frontiers in Neuroscience*, vol. 2, pp. 35–46, 2008.

[42] S. Subramaniyam, C. Medini, B. Nair, and S. Diwakar, "Modeling spatio-temporal processing in cerebellar granular layer and effects of controlled inhbition on plasticity," in *Proceedings of Neurocomp*, Lyon, France, 2010.

[43] A. D'Errico, F. Prestori, and E. D'Angelo, "Differential induction of bidirectional long-term changes in neurotransmitter release by frequency-coded patterns at the cerebellar input," *Journal of Physiology*, vol. 587, no. 24, pp. 5843–5857, 2009.

[44] H. Jörntell and C. F. Ekerot, "Properties of somatosensory synaptic integration in cerebellar granule cells in vivo," *Journal of Neuroscience*, vol. 26, no. 45, pp. 11786–11797, 2006.

[45] S. G. Brickley, S. G. Cull-Candy, and M. Farrant, "Development of a tonic form of synaptic inhibition in rat cerebellar granule cells resulting from persistent activation of GABA$_A$ receptors," *Journal of Physiology*, vol. 497, no. 3, pp. 753–759, 1996.

[46] J. Magistretti, L. Castelli, L. Forti, and E. D'Angelo, "Kinetic and functional analysis of transient, persistent and resurgent sodium currents in rat cerebellar granule cells in situ: an electrophysiological and modelling study," *Journal of Physiology*, vol. 573, no. 1, pp. 83–106, 2006.

[47] P. Chadderton, T. W. Margie, and M. Häusser, "Integration of quanta in cerebellar granule cells during sensory processing," *Nature*, vol. 428, no. 6985, pp. 856–860, 2004.

[48] E. A. Rancz, T. Ishikawa, I. Duguid, P. Chadderton, S. Mahon, and M. Häusser, "High-fidelity transmission of sensory information by single cerebellar mossy fibre boutons," *Nature*, vol. 450, no. 7173, pp. 1245–1248, 2007.

[49] N. H. Barmack and V. Yakhnitsa, "Functions of interneurons in mouse cerebellum," *Journal of Neuroscience*, vol. 28, no. 5, pp. 1140–1152, 2008.

[50] W. Truccolo, J. A. Donoghue, L. R. Hochberg et al., "Single-neuron dynamics in human focal epilepsy," *Nature Neuroscience*, vol. 14, no. 5, pp. 635–641, 2011.

[51] A. Maffei, F. Prestori, P. Rossi, V. Taglietti, and E. D'Angelo, "Presynaptic current changes at the mossy fiber-granule cell synapse of cerebellum during LTP," *Journal of Neurophysiology*, vol. 88, no. 2, pp. 627–638, 2002.

[52] M. Kase, D. C. Miller, and H. Noda, "Discharges of Purkinje cells and mossy fibres in the cerebellar vermis of the monkey during saccadic eye movements and fixation," *Journal of Physiology*, vol. 300, pp. 539–555, 1980.

[53] A. Volny-Luraghi, R. Maex, B. Vos, and E. De Schutter, "Peripheral stimuli excite coronal beams of Golgi cells in rat cerebellar cortex," *Neuroscience*, vol. 113, no. 2, pp. 363–373, 2002.

[54] T. Yamazaki and S. Tanaka, "A spiking network model for passage-of-time representation in the cerebellum," *European Journal of Neuroscience*, vol. 26, no. 8, pp. 2279–2292, 2007.

[55] G. Buzsáki, "Large-scale recording of neuronal ensembles," *Nature Neuroscience*, vol. 7, no. 5, pp. 446–451, 2004.

[56] M. A. L. Nicolelis and S. Ribeiro, "Multielectrode recordings: the next steps," *Current Opinion in Neurobiology*, vol. 12, no. 5, pp. 602–606, 2002.

[57] K. Ohki, S. Chung, Y. H. Ch'ng, P. Kara, and R. C. Reid, "Functional imaging with cellular resolution reveals precise microarchitecture in visual cortex," *Nature*, vol. 433, no. 7026, pp. 597–603, 2005.

[58] C. Stosiek, O. Garaschuk, K. Holthoff, and A. Konnerth, "In vivo two-photon calcium imaging of neuronal networks," *Proceedings of the National Academy of Sciences of the United States of America*, vol. 100, no. 12, pp. 7319–7324, 2003.

[59] C. E. Andreescu, F. Prestori, F. Brandalise et al., "NR2A subunit of the N-methyl d-aspartate receptors are required for potentiation at the mossy fiber to granule cell synapse and vestibulo-cerebellar motor learning," *Neuroscience*, vol. 176, pp. 274–283, 2011.

[60] R. L. Jenison, "Decoding first-spike latency: a likelihood approach," *Neurocomputing*, vol. 38–40, pp. 239–248, 2001.

[61] M. Fujita, "Adaptive filter model of the cerebellum," *Biological Cybernetics*, vol. 45, no. 3, pp. 195–206, 1982.

[62] M. Uusisaari and E. de Schutter, "The mysterious microcir-
     cuitry of the cerebellar nuclei," *Journal of Physiology*, vol. 589,
     no. 14, pp. 3441–3457, 2011.

[63] L. Bao, A. M. Rapin, E. C. Holmstrand, and D. H. Cox, "Elim-
     ination of the BK_{Ca} channel's high-affinity Ca^{2+} sensitivity,"
     Journal of General Physiology, vol. 120, no. 2, pp. 173–189,
     2002.

Multiobjective Optimization of Evacuation Routes in Stadium Using Superposed Potential Field Network Based ACO

Jialiang Kou,[1] **Shengwu Xiong,**[1] **Zhixiang Fang,**[2] **Xinlu Zong,**[3] **and Zhong Chen**[1]

[1] *School of Computer Science and Technology, Wuhan University of Technology, Wuhan 430070, China*
[2] *State Key Laboratory for Information Engineering in Surveying, Mapping and Remote Sensing, Wuhan University,*
  Wuhan 430079, China
[3] *School of Computer Science and Technology, Hubei University of Technology, Wuhan 430068, China*

Correspondence should be addressed to Shengwu Xiong; xiongsw@whut.edu.cn

Academic Editor: Cheng-Jian Lin

Multiobjective evacuation routes optimization problem is defined to find out optimal evacuation routes for a group of evacuees under multiple evacuation objectives. For improving the evacuation efficiency, we abstracted the evacuation zone as a superposed potential field network (SPFN), and we presented SPFN-based ACO algorithm (SPFN-ACO) to solve this problem based on the proposed model. In Wuhan Sports Center case, we compared SPFN-ACO algorithm with HMERP-ACO algorithm and traditional ACO algorithm under three evacuation objectives, namely, total evacuation time, total evacuation route length, and cumulative congestion degree. The experimental results show that SPFN-ACO algorithm has a better performance while comparing with HMERP-ACO algorithm and traditional ACO algorithm for solving multi-objective evacuation routes optimization problem.

1. Introduction

The evacuation planning in large-scale public area usually possesses two difficult points:

(1) large scale: the large-scale public area has a complex flat structure. And it can hold thousands of people.

(2) multisource and multisink: in evacuation process, the evacuees often start at different places in public area and run away from different exits.

In a word, the evacuation planning in large-scale public area is a challenging problem. For solving this problem, researchers have put forward some effective methods. Shi et al. [1] used agent-based model to simulate and analyze evacuation process in large public building under fire conditions. Chen and Miller-Hooks [2] employed Benders decomposition to determine a set of evacuation routes and the assignment of evacuees to these routes for large building. Tayfur and Taaffe [3] utilized linear programming relaxation to model and solve a resource requirements and scheduling problem during hospital evacuations with the objective of

minimizing cost within a prespecified evacuation completion time. Fang et al. [4] modeled evacuation process in a teaching building with multiexits, simulated it by cellular automata, and analyzed the multiexits choice phenomenon to find out the optimal exits choice combination for all evacuees. Usually, multiple macroscopic objectives are required to be considered in actual evacuation planning, and a set of nondominated plans are needed for decision making. Thus, evacuation planning problem could be transformed into multi-objective optimization problem. However, just a few researches, such as the literature [5–7], focused on that. Among these pieces of literature, the literature [7] successfully solved the multiobjective evacuation routes optimization problem in stadium using HMERP-ACO algorithm. Fang et al. [7] abstracted evacuation zone as a hierarchical directed network according to the feature that the evacuees usually move far away from the center of evacuation zone in evacuation process. However, another feature, namely, each evacuee often moves toward and eventually reaches one of exits, was not considered in Fang's paper. Then, how to take these two features into consideration? In physics, the potential of a point in the space generated by multiple point charges can be calculated by the superposition principle of electric potentials [8]. Inspired by this, we abstracted the center point of stadium as positive point charge and each exit as a negative point charge and used superposition principle of electric potentials to get the two features mentioned previous together. On the basis of superposed potential, we abstracted the Wuhan Sports Center stadium as a superposed potential field network (SPFN). And on the basis of SPFN, we proposed the SPFN-ACO algorithm to solve the multi-objective evacuation routes optimization problem. Compared with HMERP-ACO [7] and ACO [9], the SPFN-ACO shows much better optimization performance for solving multi-objective evacuation routes optimization problem.

The remainder of this paper is organized as follows. Section 2 introduces the state of the art evacuation planning using swarm intelligence. Section 3 defines multi-objective evacuation routes optimization problem. Section 4 introduces the SPFN. Section 5 states SPFN-ACO algorithm. Section 6 verifies optimization performance of SPFN-ACO by experiment and contains some analyses. Section 7 concludes this paper and looks into the future direction of this research.

2. Related Works

People in large-scale public areas are in danger because of a lot of manmade or natural accidents, such as fire, hurricane, and bomb [10]. For coping with these emergencies, many scientists and engineers have paid much attention to the researches about evacuation routes planning. In these researches, the application of swarm intelligence technologies to evacuation routes planning is a hot topic because evacuation process itself is a collective behavior. Swarm intelligence technology mainly includes particle swarm optimization (PSO) [11] technology and ant colony optimization (ACO) [9] technology. The swarm intelligence technology is mainly used in two aspects: the simulation of evacuation process and the optimization of evacuation routes [7, 12]. On one hand, swarm intelligence technologies have natural advantages to simulate collective behavior such as evacuation process [13]. On the other hand, the optimization mechanism of swarm intelligence algorithms can effectively optimize evacuation objectives by iterating the configuration of factors that affect evacuation efficiency [7, 14]. The factors that affect evacuation efficiency includes pheromone [7], location of shelters in evacuation zone [15], the direction of lanes [16], the placement of road barriers [17], and the scheduling of evacuation for each evacuee [18].

Besides, evacuation routes optimization problem usually needs to consider multiple objectives, such as total clearance time [19] total number of survivals [20]. A few researches [5, 6] have involved the multi-objective evacuation routing optimization problem. Some of them applied swarm intelligence technologies to solve this kind of problem [7, 14].

3. Problem Formulation

In this paper, the evacuation zone is divided into many subzones. Each evacuation plan is composed of each evacuee's route. So each evacuation plan $\mathbf{EP}_i$ can be represented as

$$\mathbf{EP}_i = [\mathbf{er}_1 \ \mathbf{er}_2 \ \cdots \ \mathbf{er}_j \ \cdots \ \mathbf{er}_{N_E}], \\ i = 1, 2, \ldots, N_P, \tag{1}$$

where N_p is the number of plans, $\mathbf{er}_j$ is the route of the evacuee j, which can be described as

$$\mathbf{er}_j = [s_{j\text{Start}} \ s_{j1} \ \cdots \ s_{jk} \ \cdots \ s_{j\text{End}}], \\ j = 1, 2, \ldots, N_E, \tag{2}$$

where N_E is the number of evacuees, $s_{j\text{Start}}$ and $s_{j\text{End}}$ are respectively, the start and the end subzone on the jth evacuee's route. s_{jk} is the kth interim subzone on the jth evacuee's route. The end subzone is one of the exits in the evacuation zone.

Thus, the multi-objective evacuation routes optimization problem in this paper could be formulated as in Algorithm 1.

The evacuation routes optimization problem involves three objectives that need to be achieved simultaneously, namely, minimization of total evacuation time, minimization of total evacuation route length, and minimization of cumulative congestion degree.

Total evacuation time (TET) is given by

$$\text{TET} = \sum_{i=1}^{N_E} \text{ET}_i, \tag{3}$$

where ET_i $(i = 1, 2, \ldots, N_E)$ is the evacuation time of evacuee i.

Total evacuation route length (TERL) is given by

$$\text{TERL} = \sum_{i=1}^{N_E} \text{ERL}_i, \tag{4}$$

where ERL_i $(i = 1, 2, \ldots, N_E)$ is the evacuation route length of evacuee i.

Cumulative congestion degree (CCD) is given by

$$\text{CCD} = \sum_{t=1}^{N_T} \sum_{i=1}^{N_S} \frac{N_{E_i}(t)}{C_i}, \tag{5}$$

where $N_{E_i}(t)$ is the number of evacuees in subzone i at tth time step, C_i is the evacuees capacity of subzone i, N_T is the number of time steps, and N_S is the number of subzones.

4. Superposed Potential Field Network (SPFN)

The electric potential field of the point charge is shown in Figure 1. If zero of potential at infinity is chosen, the potential u [8] at a distance r from a point charge Q is

$$u = \frac{kQ}{r}, \tag{6}$$

for positive point charge (Figure 1(a)) or

$$u = -\frac{kQ}{r}, \tag{7}$$

for negative point charge (Figure 1(b)).

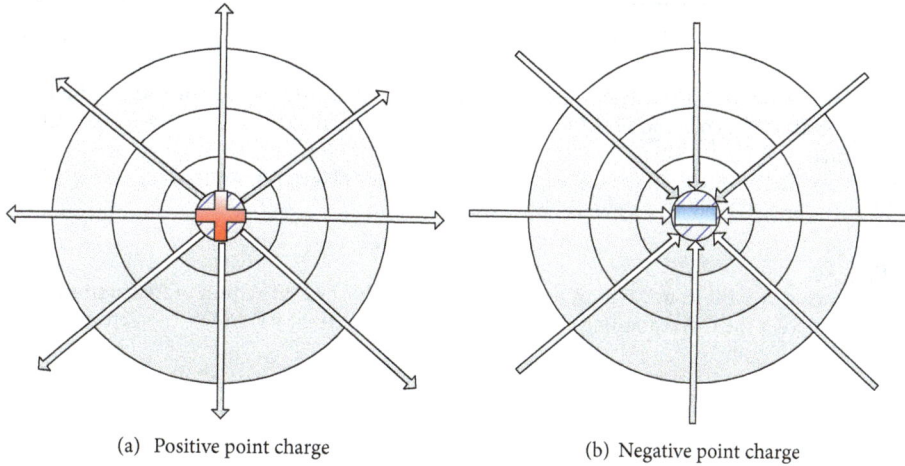

(a) Positive point charge

(b) Negative point charge

FIGURE 1: Potential field of point charge.

Finding the pareto optimal set [21] of evacuation plans, make

$$f_1 = \min(\text{TET})$$

$$f_2 = \min(\text{TERL})$$

$$f_3 = \min(\text{CCD})$$

Subject to

$$N_{E_i}(t) \leq C_i,$$

where, $N_{E_i}(t)$ is the number of evacuees in subzone i at tth time step, C_i is the capacity of subzone i.

ALGORITHM 1: Formulation of multi-objective evacuation routes optimization problem.

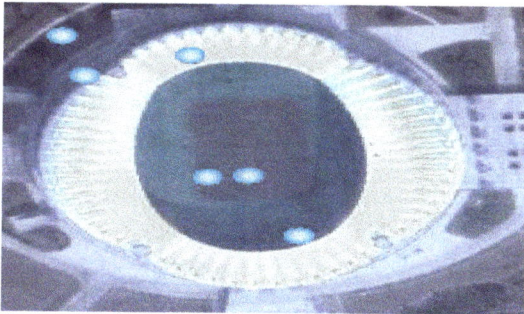

FIGURE 2: Wuhan Sports Center (http://www.wuhansport.com/).

The center point of the stadium could be seen as a positive point charge, and each exit could be seen as a negative point charge. The Wuhan Sports Center (Figure 2) could be seen in a superposed potential field. According to the superposition principle of electric potentials, the superposed potential of a point in stadium u_S could be derived by

$$u_S = u_c + \sum_{j=1}^{N_{\text{Exits}}} u_j = \frac{C_C}{r_C} - \sum_{j=1}^{N_{\text{Exits}}} \frac{C_j}{r_j}, \quad (8)$$

where N_{Exits} is the number of exits, u_j is the potential of exit j, $u_j = -C_j/r_j$. C_j is the capacity of the exit j, r_j is the distance to the exit j, u_C is the potential of center point. r_C is the distance to the center point, and C_C is the capacity of the center point.

Based on the superposed potential, we proposed the superposed potential field network (SPFN) to abstract the stadium. This model is partly based on the point model used in [5]. The SPFN could be formulated as

$$G = (H, Z, U, C), \quad (9)$$

where H is the set of nodes, Z is the set of links, U is the set of potential of each node, and C is the set of capacity of each node.

The stadium is divided into 157 subzones. Each subzone is abstracted as a node in SPFN. Each link between two nodes represents a connection relationship between two subzones. The potential of each node is the potential of the center point of the corresponding subzone. The capacity of each node is the capacity of the corresponding subzone. The coordinate of each node is the coordinate of the center point of the corresponding subzone. If an evacuee or a group of evacuees is seen as a positive test charge, it would always move from high potential node to low potential node. There are 216 links and 157 nodes in the SPFN of the Wuhan Sports Center stadium, including 10 exits nodes and 42 bleachers nodes. Figure 3 shows the potential distribution of SPFN of the Wuhan Sports Center stadium.

5. SPFN-ACO

For solving the multi-objective evacuation routes optimization problem mentioned in Section 3, on the basis of SPFN, we propose SPFN-ACO algorithm.

5.1. The Main Procedure of SPFN-ACO Algorithm. The main procedure of SPFN-ACO algorithm is listed in Algorithm 2.

S1. Initializes initial PVs population P_0 as all-zero vectors. The population size of P_0 is $2 * N_p$. The number of ants is N_A. $m = 1$.

S2. For each PV, do:

    S2.1. Each ants simultaneously finds its evacuation route under current PV by **Simulation of Evacuation Process**;

    S2.2. All ants' routes construct the corresponding evacuation plan under current PV and the objectives' values of this plan are calculated.

S3. Non-dominated sort $2 * N_p$ PVs according to corresponding route plan's objectives. And, select the top N_p PVs.

S4. Update top N_p PVs. The updated top N_p PVs construct PVs population P_m

S5. $m = m + 1$.

S6. For each PV in P_{m-1}, do:

    S6.1. Each ants simultaneously finds its evacuation route under current PV by **Simulation of Evacuation Process**;

    S6.2. All ants' routes construct the corresponding evacuation plan under current PV and the objectives' values of this plan are calculated.

S7. Update N_p pheromone vectors in P_{m-1}. The updated top N_p pheromone vectors construct pheromone vectors population Q_m. The P_{m-1} and Q_m construct R_m, namely $R_m = P_{m-1} \cup Q_m$. The population size of R_m is $2 * N_p$.

S8. Non-dominated sort R_m according to corresponding route plan's objectives. And, select the top N_p pheromone vectors to construct new population P_m.

S9. If $m \leqslant m\_$Max, go to **S5**. Or else, terminate the algorithm and output final Pareto optimal set of evacuation plans.

Note: m is the number of generations; $m\_$Max is the maximum number of generations.

ALGORITHM 2: Procedure of SPFN-ACO.

For each PV, there is a corresponding evacuation plan generated as follows:

S1. $t = 0$;

S2. Set the pheromone amounts on all connections by current PV.

S3. Randomly initialize each ant's position and velocity, and select interim destination node for each ant by **Superposed Potential Field based Roulette Wheel Method**;

S4. $t = t + 1$;

S5. For each ant, do:

    S5.1. move one step towards the center point of its destination subzone;

    S5.2. If this ant reaches its interim destination node, select new interim destination node by **Superposed Potential Field based Roulette Wheel Method**;

    S5.3. If this ant reaches one of exits, this ant stop move;

S6. If all ants have reached exits, quit and output each ant's evacuation route; or else, go to S4.

ALGORITHM 3: Simulation of evacuation process.

We use pheromone vector to represent pheromones configuration on each link in the network. The pheromone vector (PV) is given by

$$PV = \begin{bmatrix} \tau_1 & \tau_2 & \tau_3 & \cdots & \tau_{N_{\text{links}}} \end{bmatrix}, \tag{10}$$

where τ_k is the pheromone on kth link connecting node i and node j and N_{links} is the total number of links between nodes in network.

5.2. Superposed Potential Field Based Roulette Wheel Method for Node Selection.

The main procedure of Superposed Potential Field Based Wheel Method is listed in Algorithm 3. There are N_{S_C} allowed visit neighbor nodes. S_C is the set of allowed visit neighbor nodes. s_k is the kth candidate node in S_C, $k = 1, 2, \ldots, N_{S_C}$. s_j is the node which the ant i is in currently. S_C can be given by

$$S_C = \left\{ k \mid N_{E_k} \leq C_k, u_k < u_j \right\}. \tag{11}$$

The neighbor nodes in S_C must fit two conditions: the capacity constraint and the potential constraint.

The capacity constraint is given by

$$N_{E_k} \leq C_k. \tag{12}$$

N_{E_k} is the number of evacuees in node s_k, which is given by

$$N_{E_k} = N_{A_k} * \mu. \tag{13}$$

N_{A_k} is the number of ants in node s_k. Each ant represents μ evacuees.

C_k is the capacity of node s_k, which is calculated through

$$C_k = \frac{\text{Area}_k}{\text{Area}_E}. \tag{14}$$

Area_k is the area of subzone$k$. Area_E is the average area which an evacuee usually occupies. By the literature [22], each evacuee occupies $0.3 \, \text{m}^2$.

The potential constraint is given by

$$u_k < u_j, \tag{15}$$

S1. Calculate the distance to neighbor nodes.

The distance D_{jk} between the current node s_j and allowed visit neighbor node s_k is given by:

$$D_{jk} = \sqrt{\left(x_j - x_k\right)^2 + \left(y_j - y_k\right)^2}, \quad s_k \in S_C,$$

where, $\left(x_k, y_k\right)$ is the coordinate of the neighbor node s_k; $\left(x_j, y_j\right)$ is the coordinate of the current node s_j; D_{jk} is the length of link jk.

S2. Calculate the congestion degrees of neighbor nodes.

The allowed visit neighbor node s_k's congestion degree $\mathrm{CD}_k(t)$ at the tth time step is given by:

$$\mathrm{CD}_k(t) = \frac{N_{E_k}(t)}{c_k}, \quad s_k \in S_C,$$

where, $N_{E_k}(t)$ is the number of evacuees in neighbor node s_k at the tth time step. c_k is the capacity of node s_k.

S3. Calculate transition probability from node s_j to s_k

The transition probability $P^i_{jk}(m, n, t)$ from node s_j to s_k at the tth time step is given by:

$$P^i_{jk}(m, n, t) = \begin{cases} \dfrac{\tau^\alpha_{jk}(m, n)\, \eta^\beta_{jk}(m, n, t)}{\sum_{s_w \in S_C} \tau^\alpha_{jw}(m, n)\, \eta^\beta_{jw}(m, n, t)}, & s_k \in S_C \\ 0, & \text{otherwise,} \end{cases}$$

where, s_j is the subzone which the ant i is in currently. $\tau^\alpha_{ij}(m, n)$ is the pheromone amount on connection ij at mth generation under nth pheromone vector; $\eta^\beta_{ij}(m, n, t)$ is the heuristic information related with link ij at tth time step, under nth pheromone vector, at mth generation; S_C is the set of candidate nodes; α and β are the parameters to control the relative importance between the pheromone and the heuristic information.

The heuristic information $\eta_{ij}(m, n, t)$ on link ij at tth time step is given by:

$$\eta_{ij}(m, n, t) = \frac{1}{D_{ij} * \left(\left(N_{E_j}(m, n, t) + 1\right) / \left(C_j + 1\right)\right)},$$

where, D_{ij} is the length of link ij; $N_{E_j}(m, n, t)$ is the number of evacuees in node j at the tth time step; C_j is the capacity of node j.

S4. Select one of candidate nodes according to cumulative transition probability

According to roulette wheel selection, the node s_k would be selected only and if only when $PP_{j(k-1)}(m, n, t) < \text{rand} \leq PP_{jk}(m, n, t)$, rand is a random real number between 0 and 1. $PP_{jk}(m, n, t)$ is the cumulative transition probability, which is given by:

$$PP_{jk}(m, n, t) = \sum_{w=0}^{k} P_{jw}(m, n, t), \quad s_k \in S_C$$

Besides, we rule that $PP_{j0}(m, n, t) = 0$.

ALGORITHM 4: Superposed potential field based roulette wheel method.

where u_j is the potential of the current visit node and u_k is the potential of the next visit node. The potential constraint indicates that the ant should move from high-potential node to low-potential node, namely, the potential u_k of next visit node s_k should be less than the potential u_j of current visit node s_j.

The procedure of superposed potential field based roulette wheel method is shown in Algorithm 4. Its principle could be explained by an example in Figure 4. In Figure 4, the digit on each node is the value of potential. The red node is the node which evacuee i is in. By potential, he could just choose the nodes of which the potential value is lower than the node which he is in as the candidates. So, he could choose three neighbor nodes as allowed visit nodes. The potential of allowed visit nodes is, respectively, 4, 4, and 2. And then, he has to choose one of them as the next visit node by calculating the transition probability and cumulative transition probability of each candidate as shown in Algorithm 4.

5.3. Velocity, Position, and Moving Strategy. When the interim destination node is selected, the ant i begins moving along

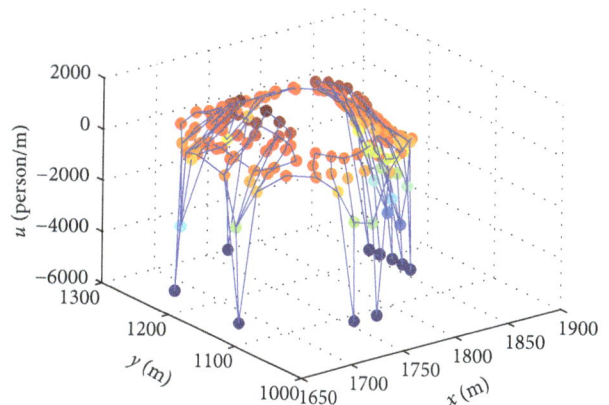

FIGURE 3: Potential distribution of SPFN of Wuhan Sports Center stadium.

link between current node s_j and interim destination node s_k. The moving speed [7] $v^i(t)$ of ant i is given by

$$v^i(t) = v_{\max} * e^{-N_{E_j}(t)/C_j}, \tag{16}$$

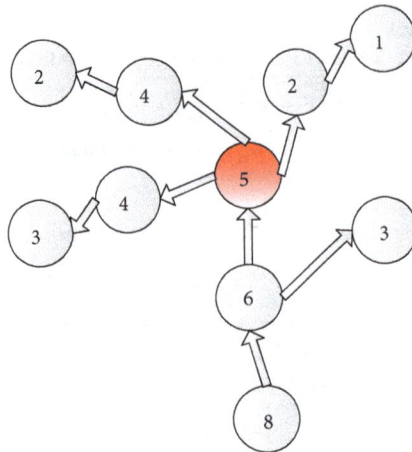

FIGURE 4: An example to show the superposed potential field based roulette wheel method.

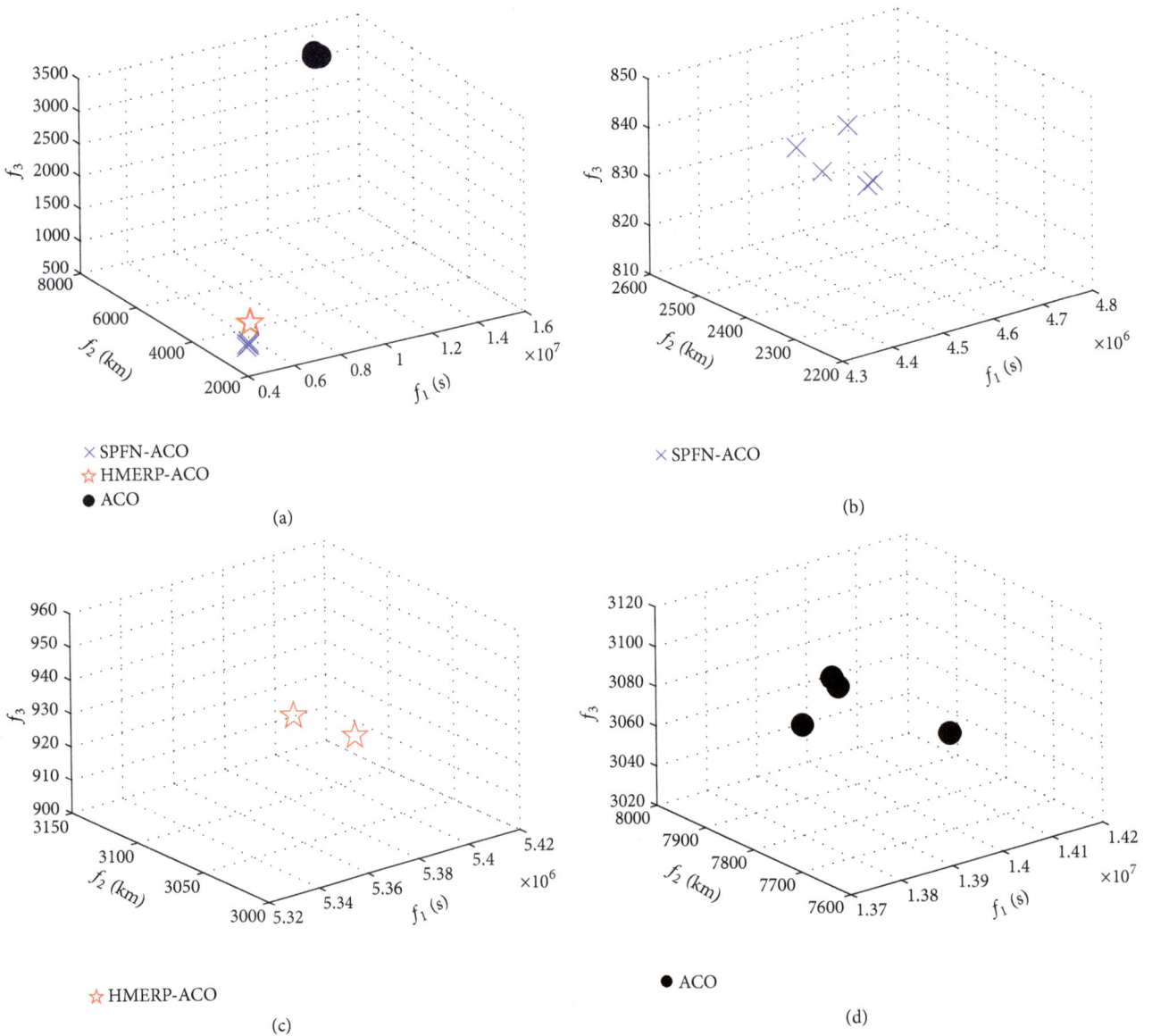

FIGURE 5: f_1, f_2, and f_3 values of nondominated plans derived from three algorithms.

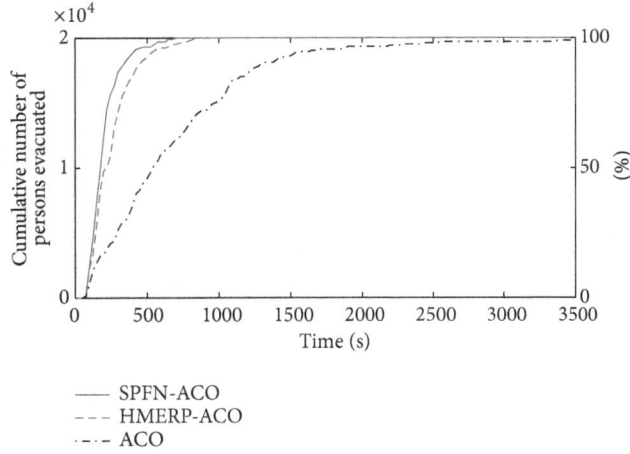

FIGURE 6: Evacuation curves of the three algorithms.

where $N_{E_i}(t)$ is the number of evacuees in node s_j at the tth time ste, C_j is the capacity of node s_j, and v_{max} is the maximum speed of ant i.

We define a concept called remaining distance to interim destination node to measure whether the ant i has already arrived interim destination node. The iterative formula of remaining distance is given by

$$\text{RD}^i(t+1) = \text{RD}^i(t) - v^i(t) * \Delta t, \qquad (17)$$

where $\text{RD}^i(t+1)$ and $\text{RD}^i(t)$ are the remaining distance at $(t+1)$th and tth time step. Δt is the interval of time step, such as ten or twenty seconds. When an ant arrives at interim destination node, the remaining distance is set as the length of link between the interim destination node and the next interim destination node.

5.4. Pheromone Updating. The pheromone on each link between nodes is updated by

$$\tau_{jk}(m+1,n) = (1-\rho)\tau_{jk}(m,n) + \rho\Delta\tau_{jk}(m,n), \qquad (18)$$

where $\tau_{jk}(m+1,n)$ and $\tau_{jk}(m,n)$ are pheromone amount on link jk between nodes s_j and s_k at $(m+1)$th and mth generation under nth pheromone vector. $\Delta\tau_{jk}$ is the variation amount of pheromone on link jk. The variation amount of pheromone $\Delta\tau_{jk}(m,n)$ is given by

$$\Delta\tau_{jk}(m,n) = \frac{1}{D_{jk} * \sum_{t=0}^{N_T(m,n)} \left(N_{E_k}(m,n,t)/C_k \right)}, \qquad (19)$$

where D_{jk} is the length of link jk, $N_{E_k}(m,n,t)$ is the number of evacuees in node s_k at the tth time step, and C_k is the capacity of node s_k.

6. Experiment and Analysis

6.1. The Experiment Design. In this paper, we took a 20000 evacuees' drill in Wuhan Sports Center Stadium as an example to do simulation experiment. This stadium has 42

TABLE 1: Parameter values in SPFN-ACO, HMERP-ACO, and ACO.

m_Max	N_p	v_{max}	Δt	α	β	ρ	N_A	μ
200	10	2 m/s	25 s	1	3	0.5	200	100

bleachers subzones and 10 exits subzones. Ants are randomly allocated to 42 bleachers subzones, and each ant represents 100 evacuees. The maximum speed of each ant is 2 m/s [23] and varies from 0 to 2 m/s along with the congestion degree. The optimization performance of SPFN-ACO was compared with HMERP-ACO and traditional ACO which is used in Fang's paper [7]. By experience, the parameters of the three algorithms are set as Table 1. m_Max is the total number of generations. N_p is the population size of evacuation plans in each generation. v_{max} is the maximum speed of each ant. Δt is the length of each time step. N_A is the total number of ants. Each ant represents μ evacuees. α and β are the parameters to control the relative importance between the pheromone and the heuristic information. ρ is the evaporation rate [24], $\rho \in (0, 1]$.

6.2. The Experimental Result Analysis. Figure 5 shows the $f_1, f_2,$ and f_3 values of non-dominated plans derived from the three algorithms. The "blue cross," "red pentagram," and "black solid circle," respectively, represent the $f_1, f_2,$ and f_3 values of non-dominated plans derived from the SPFN-ACO, the HMERP-ACO, and the ACO algorithm. The $f_1, f_2,$ and f_3 values of non-dominated plans derived from the SPFN-ACO algorithm are smaller than those generated by the other two algorithms. According to Bierlaire's viewpoint [25], the evacuation process could be seen as a series of node selections made by evacuees, and then $f_1, f_2,$ and f_3 values of non-dominated plans would depend on the node selection strategy. For the three algorithms mentioned in this paper, the efficiency of node selection strategy resorts to the transition probabilities. The transition probability is mainly determined by two aspects: the selection of candidate neighbor nodes and the relative importance pheromone versus heuristic information. The latter aspect is determined by the setting of

(a) SPFN-ACO

(b) HMERP-ACO

(c) ACO

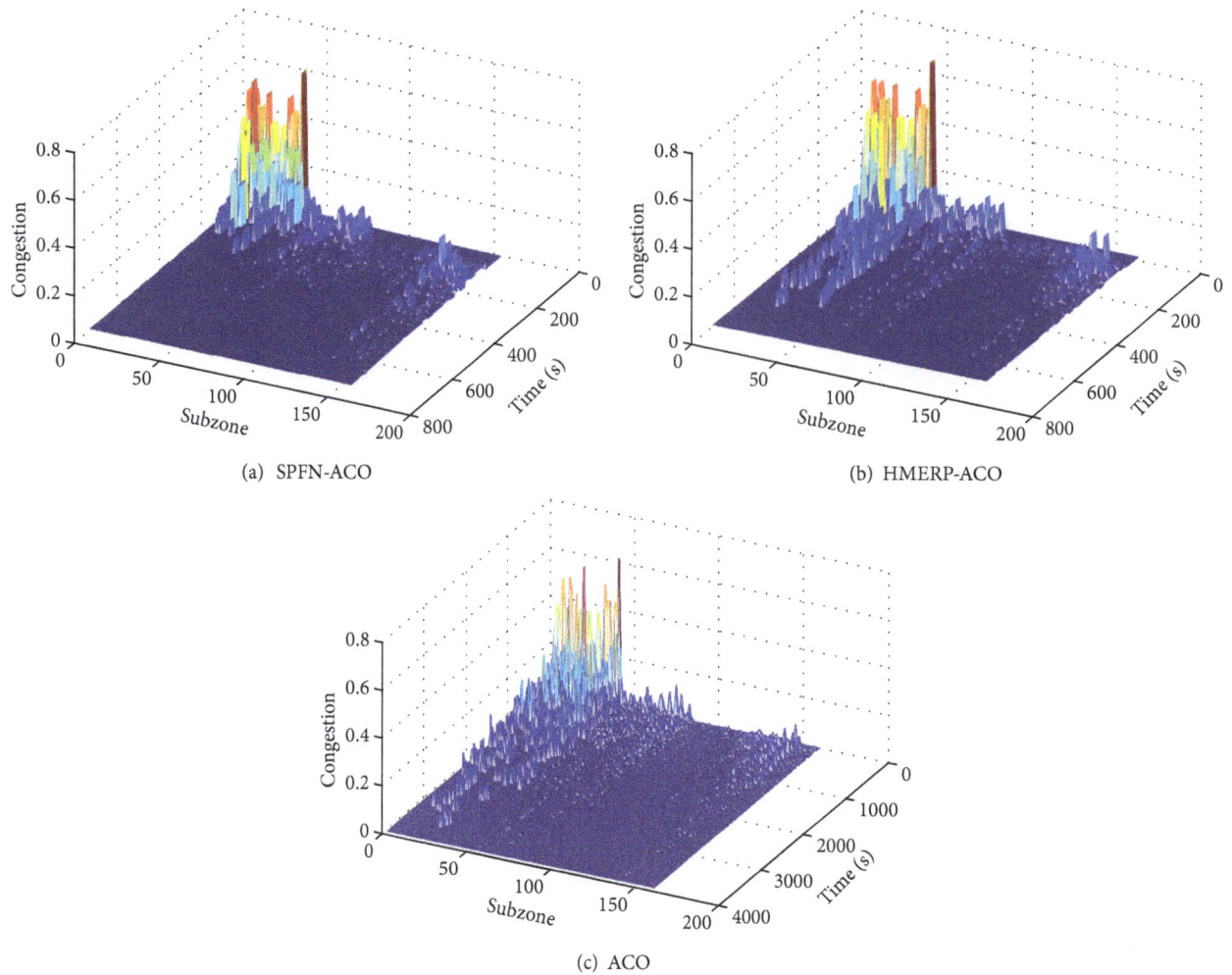

FIGURE 7: Time-varying congestion degrees in three algorithms.

relative importance parameters α and β. The former aspect is determined by the conditions of candidate neighbor nodes selection. Among the conditions, the capacity constraint is same for all the three algorithms. Thus, the difference is another condition: the ACO algorithm adopts the tabu list; the HMERP-ACO adopts the hierarchy defined in Fang's paper [7]; the SPFN-ACO adopts the potential introduced in this paper. The tabu list takes the visited nodes on each ant's route as forbidden visit nodes for this ant. It did not consider any domain knowledge that can raise evacuation efficiency. The hierarchical directed network uses the feature that each evacuee moves far away from the center point of the stadium but without considering another feature that each evacuee moves towards one of the exits. The superposed potential field network takes the two features into account, obviously further raises the evacuation efficiency, and improves optimization objectives. This is the reason why the f_1, f_2, and f_3 values of non-dominated plans derived from the SPFN-ACO are better than those derived from HMERP-ACO and ACO.

Figure 6 shows the evacuation curves [26] of the three algorithms. By SPFN-ACO, 95% of evacuees have left the stadium at 450 seconds, and 100% of evacuees have been

evacuated out of the stadium at 725 seconds. By HMERP-ACO, it, respectively, needs 575 and 875 seconds; by ACO, it even needs 1675 and 3525 seconds. The results indicate that the candidate nodes selection condition using domain knowledge can shorten the evacuation time and raise evacuation efficiency. And if two factors that can facilitate evacuation are taken into account, the evacuation time is less than that just considering one factor. The SPFN-ACO shows a much better evacuation time performance than that of the other two algorithms.

Figure 7 shows the time-varying congestion degrees of the nodes in the three algorithms. At the first X seconds, all the three algorithms show a relatively high congestion in nodes 1 to 100. With the rise of time, the plans generated by the ACO and HMERP-ACO algorithms show a slowly decreased heavy congestion in nodes. But the congestion in nodes decreases sharply for the plan generated by SPFN-ACO. This indicates that, compared with the other two algorithms, the SPFN-ACO can evacuate most of evacuees out of the middle zone of stadium and therefore reduce the congestion degree in the middle zone rapidly. However, in all the three algorithms, it takes a relatively long time

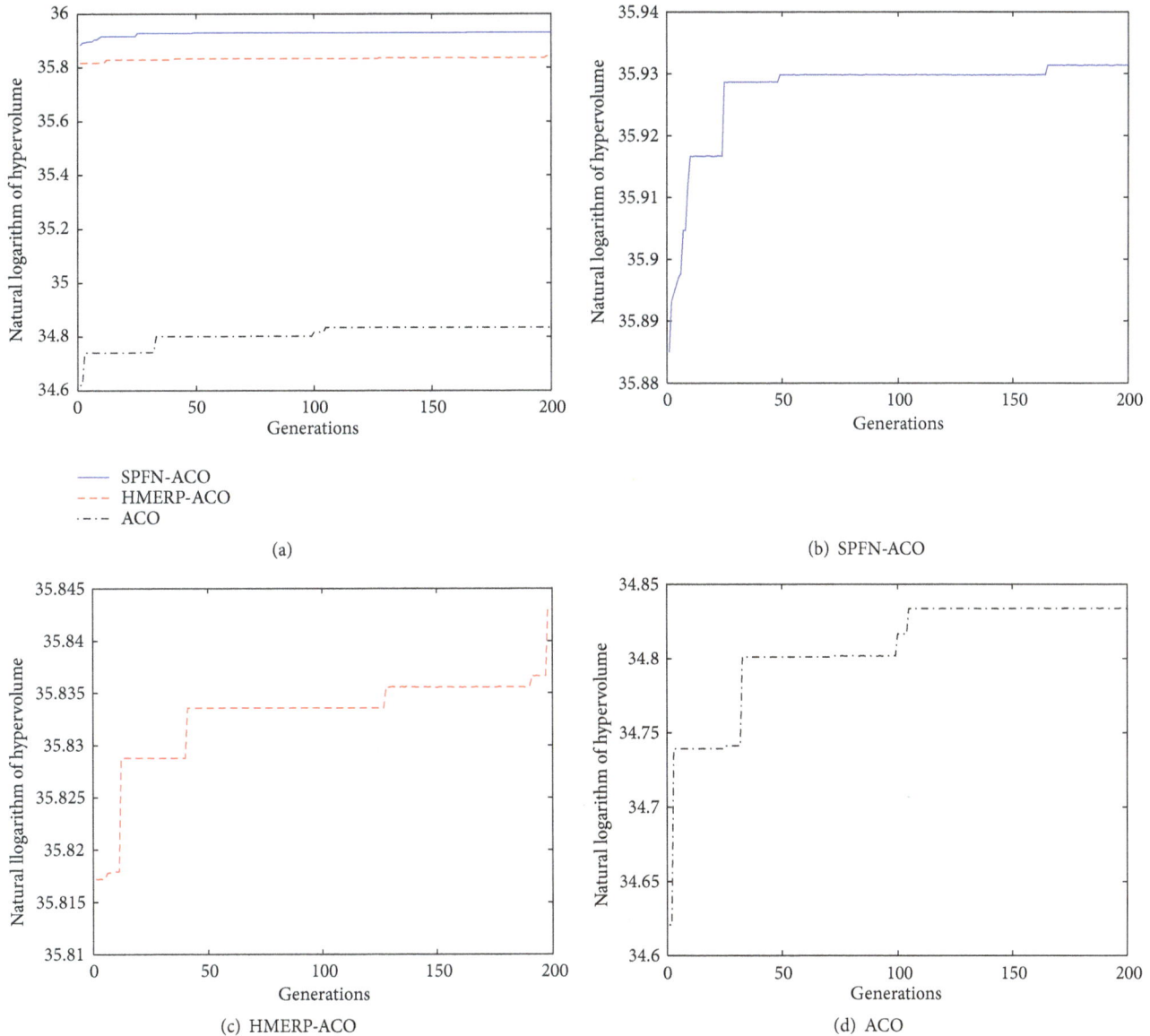

FIGURE 8: The natural logarithm of hypervolume for three algorithms.

to make the congestion degrees in all nodes decrease to zero, although the SPFN-ACO expands the least time. The "long-tail pheromone" indicates that all the three algorithms need a relatively long time (compared with the network clearance time) to take all evacuees out of the stadium. Besides, the SPFN-ACO possesses the smallest cumulative congestion degree in the three algorithms. Therefore, totally speaking, the congestion situation of SPFN-ACO generated plans is better than that of the other two algorithms, but the congestion situation of SPFN-ACO still needs to be improved.

Figure 8 shows the natural logarithm of hypervolume for three algorithms. Horizontal ordinate is the generations of evolution; vertical ordinate is the natural logarithm of the hyper volume (HV). The hyper volume is a metric of convergence [27]. The larger the natural logarithm of hyper volume, the better the convergence of the algorithm. Thus,

from Figure 8, we can conclude that the SPFN-ACO acquires the best convergence performance, the HMERP-ACO comes second, and the ACO has the worst. And, with the rise of generations, the convergence of all three algorithms is improved. It indicates that, with the iteration of pheromones on each link, the evacuation plans generated by all the three algorithms could be gradually slightly improved. However, the relative merits between three algorithms are not changed. This indicates that the relative merits between three algorithms are determined by the selection of candidate nodes and the relative importance pheromone versus heuristic information but not the concrete pheromone value on each link.

Figure 9 shows the proportion of non-dominated plans in all plans derived from three algorithms. As shown in Figure 8, before the 40th generation, for all three algorithms,

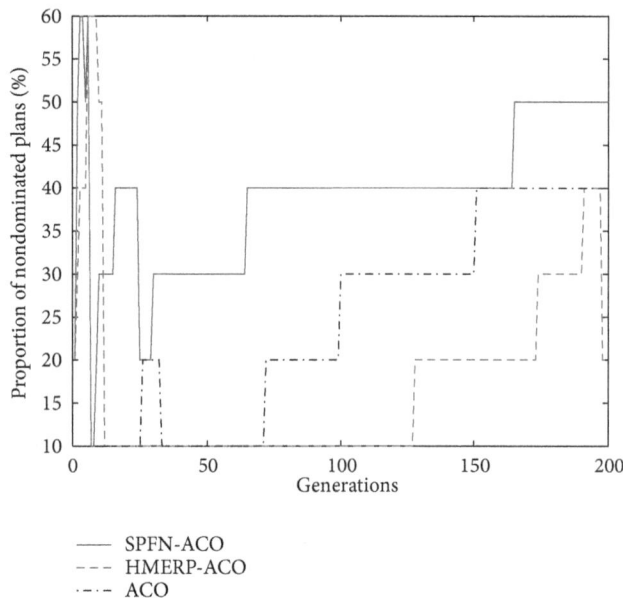

FIGURE 9: The proportion of non-dominated plans derived from three algorithms.

the proportion of non-dominated plans fluctuates; from the 40th to 197th generation, the proportion increases in stage. But at the 198th generation, for HMERP-ACO, the proportion sharply drops down to 20%. Finally, by the evolution of 200 generations, the proportion of non-dominated plans for SPFN-ACO reaches 50%, higher than that for HMERP-ACO (20%) and ACO (40%).

7. Conclusions and Future Works

We proposed a multi-objective optimization algorithm of the evacuation routes SPFN-ACO, which is based on the organization of the evacuees' space-time paths within a superposed potential field network (SPFN). The ACO algorithm organizes evacuees' space-time paths without any domain knowledge that can help improve evacuation efficiency; the HMERP-ACO algorithm merely employs one promotive factor for improving evacuation efficiency; the SPFN efficiently combines two factors together, which can facilitate the raise of evacuation efficiency by reasonably organizing the evacuees' space-time paths. By validation of simulation experiment, compared with HMERP-ACO and ACO algorithms, the SPFN-ACO algorithm is more suitable to solve the multi-objective optimization problem of the evacuation routes.

It is planned to do further researches on the basis of SPFN-ACO, such as defining more realistic evacuation scenarios, studying the effects of grouping size of evacuees and the total number of evacuees on evacuation efficiency, and discussing the influences of the population size of pheromone vectors and the number of evolution generations on algorithm performance.

Acknowledgment

This work was supported in part by the National Science Foundation of China under Grant nos. 61170202, 40971233, and 61202287.

References

[1] J. Shi, A. Ren, and C. Chen, "Agent-based evacuation model of large public buildings under fire conditions," *Automation in Construction*, vol. 18, no. 3, pp. 338–347, 2009.

[2] L. Chen and E. Miller-Hooks, "The building evacuation problem with shared information," *Naval Research Logistics*, vol. 55, no. 4, pp. 363–376, 2008.

[3] E. Tayfur and K. Taaffe, "A model for allocating resources during hospital evacuations," *Computers and Industrial Engineering*, vol. 57, no. 4, pp. 1313–1323, 2009.

[4] Z. Fang, W. Song, J. Zhang, and H. Wu, "Experiment and modeling of exit-selecting behaviors during a building evacuation," *Physica A*, vol. 389, no. 4, pp. 815–824, 2010.

[5] M. Saadatseresht, A. Mansourian, and M. Taleai, "Evacuation planning using multiobjective evolutionary optimization approach," *European Journal of Operational Research*, vol. 198, no. 1, pp. 305–314, 2009.

[6] A. Stepanov and J. M. Smith, "Multi-objective evacuation routing in transportation networks," *European Journal of Operational Research*, vol. 198, no. 2, pp. 435–446, 2009.

[7] Z. Fang, X. Zong, Q. Li, Q. Li, and S. Xiong, "Hierarchical multi-objective evacuation routing in stadium using ant colony optimization approach," *Journal of Transport Geography*, vol. 19, no. 3, pp. 443–451, 2011.

[8] R. S. Elliott, *Electromagnetics: History, Theory, and Applications*, Wiley, Hoboken, NJ, USA, 1999.

[9] M. Dorigo, V. Maniezzo, and A. Colorni, "Ant system: optimization by a colony of cooperating agents," *IEEE Transactions on Systems, Man, and Cybernetics B*, vol. 26, no. 1, pp. 29–41, 1996.

[10] S. Pu and S. Zlatanova, "Evacuation route calculation of inner buildings," in *Geo-Information for Disaster Management*, P. van Oosterom, S. Zlatanova, and E. M. Fendel, Eds., pp. 1143–1161, Springer, Berlin, Germany, 2005.

[11] J. Kennedy and R. Eberhart, "Particle swarm optimization," in *Proceedings of the IEEE International Conference on Neural Networks*, pp. 1942–1948, December 1995.

[12] A. Rahman, A. K. Mahmood, and E. Schneider, "Using agent-based simulation of human behavior to reduce evacuation time," in *Proceedings of the 11th Pacific Rim International Conference on Multi-Agents: Intelligent Agents and Multi-Agent Systems*, pp. 357–369, 2008.

[13] Z. Xue, *A particle swarm optimization based multi-agent stochastic evacuation simulation model [Ph.D. thesis]*, University of New York, 2009.

[14] R. M. Tavares and E. R. Galea, "Numerical optimisation techniques applied to evacuation analysis," in *Pedestrian and Evacuation Dynamics 2008*, W. W. F. Klingsch, C. Rogsch, A. Schadschneider, and M. Schreckenberg, Eds., pp. 555–561, Springer, Berlin, Germany, 2010.

[15] H. D. Sherali, T. B. Carter, and A. G. Hobeika, "A location-allocation model and algorithm for evacuation planning under hurricane/flood conditions," *Transportation Research Part B*, vol. 25, no. 6, pp. 439–452, 1991.

[16] C. Xie, D.-Y. Lin, and S. Travis Waller, "A dynamic evacuation network optimization problem with lane reversal and crossing elimination strategies," *Transportation Research E*, vol. 46, no. 3, pp. 295–316, 2010.

[17] H. Cai and A. Rahman, "A method to develop and optimize the placement of road barriers in emergency evacuation for university campuses," in *Proceedings of the Construction Research Congress: Innovation for Reshaping Construction Practice*, pp. 409–419, May 2010.

[18] H. Sbayti and H. S. Mahmassani, "Optimal scheduling of evacuation operations," *Transportation Research Record*, no. 1964, pp. 238–246, 2006.

[19] S. C. Pursals and F. G. Garzón, "Optimal building evacuation time considering evacuation routes," *European Journal of Operational Research*, vol. 192, no. 2, pp. 692–699, 2009.

[20] P. Lin, S. M. Lo, H. C. Huang, and K. K. Yuen, "On the use of multi-stage time-varying quickest time approach for optimization of evacuation planning," *Fire Safety Journal*, vol. 43, no. 4, pp. 282–290, 2008.

[21] M. Ancău and C. Caizar, "The computation of Pareto-optimal set in multicriterial optimization of rapid prototyping processes," *Computers and Industrial Engineering*, vol. 58, no. 4, pp. 696–708, 2010.

[22] J. Izquierdo, I. Montalvo, R. Pérez, and V. S. Fuertes, "Forecasting pedestrian evacuation times by using swarm intelligence," *Physica A*, vol. 388, no. 7, pp. 1213–1220, 2009.

[23] P.-H. Chen and F. Feng, "A fast flow control algorithm for real-time emergency evacuation in large indoor areas," *Fire Safety Journal*, vol. 44, no. 5, pp. 732–740, 2009.

[24] M. Dorigo and T. Stützle, "Ant colony optimization: overview and recent advances," in *Handbook of Metaheuristics*, M. Gendreau and Y. Potvin, Eds., vol. 146 of *International Series in Operations Research & Management Science*, pp. 227–263, Springer, New York, NY, USA, 2nd edition, 2010.

[25] M. Bierlaire, G. Antonini, and M. Weber, "Behavioral dynamics for pedestrians, in Moving through nets: the physical and social dimensions of travel," in *Proceedings of the 10th International Conference on Travel Behaviour Research*, K. Axhausen, Ed., pp. 1–18, Elsevier, Amsterdam, Netherlands, 2003.

[26] L. D. Han, F. Yuan, and T. Urbanik, "What is an effective evacuation operation?" *Journal of Urban Planning and Development*, vol. 133, no. 1, pp. 3–8, 2007.

[27] A. Auger, J. Bader, D. Brockhoff, and E. Zitzler, "Theory of the hypervolume indicator: optimal μ-distributions and the choice of the reference point," in *Proceedings of the 10th ACM SIGEVO Workshop on Foundations of Genetic Algorithms (FOGA '09)*, pp. 87–102, January 2009.

Permissions

The contributors of this book come from diverse backgrounds, making this book a truly international effort. This book will bring forth new frontiers with its revolutionizing research information and detailed analysis of the nascent developments around the world.

We would like to thank all the contributing authors for lending their expertise to make the book truly unique. They have played a crucial role in the development of this book. Without their invaluable contributions this book wouldn't have been possible. They have made vital efforts to compile up to date information on the varied aspects of this subject to make this book a valuable addition to the collection of many professionals and students.

This book was conceptualized with the vision of imparting up-to-date information and advanced data in this field. To ensure the same, a matchless editorial board was set up. Every individual on the board went through rigorous rounds of assessment to prove their worth. After which they invested a large part of their time researching and compiling the most relevant data for our readers. Conferences and sessions were held from time to time between the editorial board and the contributing authors to present the data in the most comprehensible form. The editorial team has worked tirelessly to provide valuable and valid information to help people across the globe.

Every chapter published in this book has been scrutinized by our experts. Their significance has been extensively debated. The topics covered herein carry significant findings which will fuel the growth of the discipline. They may even be implemented as practical applications or may be referred to as a beginning point for another development. Chapters in this book were first published by Hindawi Publishing Corporation; hereby published with permission under the Creative Commons Attribution License or equivalent.

The editorial board has been involved in producing this book since its inception. They have spent rigorous hours researching and exploring the diverse topics which have resulted in the successful publishing of this book. They have passed on their knowledge of decades through this book. To expedite this challenging task, the publisher supported the team at every step. A small team of assistant editors was also appointed to further simplify the editing procedure and attain best results for the readers.

Our editorial team has been hand-picked from every corner of the world. Their multi-ethnicity adds dynamic inputs to the discussions which result in innovative outcomes. These outcomes are then further discussed with the researchers and contributors who give their valuable feedback and opinion regarding the same. The feedback is then collaborated with the researches and they are edited in a comprehensive manner to aid the understanding of the subject.

Apart from the editorial board, the designing team has also invested a significant amount of their time in understanding the subject and creating the most relevant covers. They scrutinized every image to scout for the most suitable representation of the subject and create an appropriate cover for the book.

The publishing team has been involved in this book since its early stages. They were actively engaged in every process, be it collecting the data, connecting with the contributors or procuring relevant information. The team has been an ardent support to the editorial, designing and production team. Their endless efforts to recruit the best for this project, has resulted in the accomplishment of this book. They are a veteran in the field of academics and their pool of knowledge is as vast as their experience in printing. Their expertise and guidance has proved useful at every step. Their uncompromising quality standards have made this book an exceptional effort. Their encouragement from time to time has been an inspiration for everyone.

The publisher and the editorial board hope that this book will prove to be a valuable piece of knowledge for researchers, students, practitioners and scholars across the globe.

List of Contributors

Penglin Zhang and Jiangping Chen
School of Remote Sensing and Information Engineering, Wuhan University, Wuhan 430079, China

Xubing Zhang
College of Mathematics and Computer Science, Wuhan Textile University, Wuhan 430073, China

Seth A. Herd, Kai A. Krueger, Trenton E. Kriete, Tsung-Ren Huang, Thomas E. Hazy and Randall C. O'Reilly
Department of Psychology, University of Colorado Boulder, Boulder, CO 80309, USA

Subha Fernando
Information Science and Control Engineering, Graduate School of Engineering, Nagaoka University of Technology, 1603-1 Kamitomioka-machi, Nagaoka, Niigata 940-2188, Japan

Koichi Yamada
Management and Information Systems Science, Faculty of Engineering, Nagaoka University of Technology, 1603-1 Kamitomioka-machi, Nagaoka, Niigata 940-2188, Japan

Shinichi Tamura
NBL Technovator Co. Ltd., 631 Shindachimakino, Sennan City, Osaka 590-0522, Japan

Tomomitsu Miyoshi and Hajime Sawai
Department of Integrative Physiology, Graduate School of Medicine, Osaka University, Suita 565-0871, Japan

Yuko Mizuno-Matsumoto
Graduate School of Applied Informatics, University of Hyogo, Kobe 650-0047, Japan

Kyle G. Horn
Program in Neuroscience, Stony Brook Universty, SUNY, Stony Brook, NY 11794-5230, USA
Department of Physiology and Biophysics, Stony Brook Universty, SUNY, Stony Brook, NY 11794-8661, USA

Irene C. Solomon
Department of Physiology and Biophysics, Stony Brook Universty, SUNY, Stony Brook, NY 11794-8661, USA

Heraldo Memelli
Department of Physiology and Biophysics, Stony Brook Universty, SUNY, Stony Brook, NY 11794-8661, USA
Department of Computer Science, Stony Brook Universty, SUNY, Stony Brook, NY 11794-4440, USA

Shinichi Tamura
NBL Technovator Co. Ltd., 631 Shindachimakino, Sennan 590-0522, Japan

Shoji Inabayashi, Waichi Hayakawa and Takahiro Yokouchi
Image Processing Solutions Deptartment, Pacific Systems Corporation, 8-4-19 Tajima, Sakura-Ku, Saitama City 338-0837, Japan

Hiroshi Mitsumoto
Osaka Electro-Communication University, 18-1 Hatsucho, Osaka, Neyagawa 572-8530, Japan

Hisashi Taketani
Tsuyama National College of Technology, 624-1 Numa, Okayama, Ttsuyama 708-8509, Japan

Federico Raimondo and Diego Fernandez Slezak
Departamento de Computacion, Pabellon I, Ciudad Universitaria, C1428EGA Ciudad Autonoma de Buenos Aires, Argentina

Juan E. Kamienkowski and Mariano Sigman
Laboratory of Integrative Neuroscience, Physics Department, University of Buenos Aires, Buenos Aires, Argentina

Aurel A. Lazar and Yevgeniy B. Slutskiy
Department of Electrical Engineering, Columbia University, New York, NY 10027, USA

Zhongbo Hu
School of Mathematics and Statistic, Hubei Engineering University, Xiaogan, Hubei 432000, China
School of Sciences, Wuhan University of Technology, Wuhan, Hubei 430070, China

Qinghua Su
School of Mathematics and Statistic, Hubei Engineering University, Xiaogan, Hubei 432000, China

Cameron D. Wellock and George N. Reeke
Laboratory of Biological Modeling, The Rockefeller University, 1230 York Avenue, New York, NY 10065, USA

Debotosh Bhattacharjee, Ayan Seal, Suranjan Ganguly, Mita Nasipuri and Dipak Kumar Basu
Department of Computer Science and Engineering, Jadavpur University, Kolkata 700032, India

Meng Hu and Hualou Liang
School of Biomedical Engineering, Science & Health Systems, Drexel University, 3141 Chestnut Street, Philadelphia, PA 19104, USA

Patrick Greene
Graduate Program in Applied Mathematics, University of Arizona, Tucson, AZ 8572, USA

Mike Howard and Rajan Bhattacharyya
HRL Laboratories, LLC, Malibu, CA 90265, USA

Jean-Marc Fellous
Graduate Program in Applied Mathematics, University of Arizona, Tucson, AZ 8572, USA
Department of Psychology, University of Arizona, Tucson, AZ 8572, USA

Alexei V. Samsonovich and Giorgio A. Ascoli
Krasnow Institute for Advanced Study, George Mason University, 4400 University Drive MS 2A1, Fairfax, VA 22030-4444, USA

Chaitanya Medini, Bipin Nair and Shyam Diwakar
Amrita School of Biotechnology, Amrita Vishwa Vidyapeetham (Amrita University), Amritapuri, Clappana, Kollam 690525, Kerala, India

Egidio D'Angelo
Department of Physiology, University of Pavia, Via Forlanini 7, 21000 Pavia, Italy
Brain Connectivity Center, IRCCS C. Mondino, Via Mondino 2, 27100 Pavia, Italy

Giovanni Naldi
Department of Mathematics, University of Milan, Via Saldini 50, 27100 Milan, Italy

Jialiang Kou, Shengwu Xiong and Zhong Chen
School of Computer Science and Technology, Wuhan University of Technology, Wuhan 430070, China

Zhixiang Fang
State Key Laboratory for Information Engineering in Surveying, Mapping and Remote Sensing, Wuhan University, Wuhan 430079, China

Xinlu Zong
School of Computer Science and Technology, Hubei University of Technology, Wuhan 430068, China